Mobile Communications Design Fundamentals

WILEY SERIES IN TELECOMMUNICATIONS

Mobile Communications Design Fundamentals

Second Edition

William C. Y. Lee

Vice President and Chief Scientist
Applied Research and Science
PacTel Corporation

A Wiley-Interscience Publication

JOHN WILEY & SONS, INC.

New York · Chichester · Brisbane · Toronto · Singapore

Copyright © 1993 by John Wiley & Sons, Inc.

Library of Congress Cataloging in Publication Data:

Lee, William C. Y.
    Mobile communications design fundamentals / William C.Y. Lee. —
2nd ed.
      p. cm. — (Wiley series in telecommunications)
    "A Wiley-Interscience publication."
    Includes bibliographical references and index.
    ISBN 0-471-57446-5 (alk. paper)
    1. Mobile communication systems—Design. I. Title. II. Series.
TK6570.M6L36 1993
621.3845′6—dc20   92-21130

Printed in the United States of America

10 9 8 7 6 5 4 3 2

To My Parents

CONTENTS

PREFACE

The first edition of this book was published in 1986 by Howard W. Sams Co., then a subsidiary of ITT. When Sams was sold by ITT, it changed its direction of interest to computers and terminated its list of radio communications books. Since that time, many readers have requested that I reissue this book. I am beholden to John Wiley & Sons, Inc. for its willingness to publish this second edition.

Cellular systems have proven to be both high capacity and high quality systems. However, in a realistic situation, due to the frequency re-use scheme for increasing capacity, cellular operators always struggle between quality and capacity, putting all co-channel cells closer together for capacity or putting all co-channel cells farther apart for quality. In August 1985, I was invited by the FCC to give a public presentation on "Spectrum Efficiency Comparison of FM and SSB." My analysis concluded that because of frequency re-use in cellular systems, there was no advantage in splitting the FM channels for capacity. The spectrum efficiencies of FM and SSB are the same. In 1987, I was the first co-chairperson of a newly formed CTIA subcommittee of Advanced Radio Technology (ARTS) and led the cellular industry in setting the requirements for the first North American Digital Cellular System. TIA, then formed a new group TR45.3 to develop cellular digital standards for North America. I personally preferred FDMA because it was a low-risk approach due to the fact that the cellular analog system is also FDMA and the equalizer is not needed. Furthermore, with ARTS' suggestions of having dual-mode subscriber sets and sharing setup channels, the availability of digital FDMA systems by 1990, as purposely stated in the ARTS UPR (User's Performance Requirement), could be easily met. The major vendors such as AT&T and Motorola were voting for FDMA also.

In September 1987, I was invited by the FCC to speak publicly about how to develop digital cellular for capacity from the cellular operator's point of view and introduced a new radio capacity formula (appearing in the new Chapter 9) to measure spectrum efficiency for digital FDMA and TDMA systems. In February 1989, when Qualcomm came to PacTel to present their first version of CDMA, I emphasized the implementation of power control

xv

for reducing near-far interference. Although PacTel supported TDMA, but due to the relative high-risk of developing TDMA and for the sake of safety, PacTel helped Qualcomm develop an alternative digital cellular CDMA system. CDMA has been theoretically proven to have twenty times more capacity over the current analog cellular system. PacTel was unselfish, following MFJ restrictions, in contributing its technical and financial resources to help Qualcomm, a small but technically strong U.S. company, develop world-leading cellular CDMA technology for capacity. A trial CDMA system was built in six months starting from scratch, and a cellular CDMA demonstration was held in San Diego on Two PacTel sites. Cellular CDMA is introduced in the new Chapter 9.

I also felt the need to develop a microcell technology to further increase capacity in analog and digital systems for future PCS (Personal Communication Service). The difference between conventional microcells and the PacTel patented microcell is that the former are dumb cells and the latter are intelligent cells. The new microcell system is introduced in the new Chapter 10.

Since publication of the first edition in 1986, there has been a tremendous increase in the use of mobile communications. In the United States there were 650 thousand cellular units in operation in 1986 and revenues of $46.2 million, and now in 1992 there are 8 million units and revenues close to $4 billion. In the year 2000 the predicted number of cellular units in operation will be 20 million. In the European community, there were 815 thousand cellular units in 1987 and there are 5 million units now. This rapid growth in wireless communication shows the need for technology that will increase capacity and improve system performance. Also, narrowband and wideband radio access technologies are needed. Furthermore in June 1990, the FCC encouraged the wireless communication industry to look into "Personal Communication Service (PCS)" systems. PCS is a generic name for a future personal wireless communication system. In Europe, current systems such as cellular communications (analog and digital), cordless telephone-2 (CT-2) system, and the personal communication networking (PCN) system are all mobile radio communications. In Japan, digital cellular and PCS are already in the development stage. Therefore, this book, with its new added material, such as CDMA and microcell technologies can aid in understanding and developing all mobile radio systems including the future PCS.

Besides adding two new chapters, 9 and 10, I have also expanded the discussion in Chapter 2 on the microcell prediction model, in Chapter 5 on spectrum efficiency and cellular systems, in Chapter 6 on basestation design, and in Chapter 7 on field component diversity antennas. To make the book suitable for graduate course work, I have added problems to the end of each chapter.

I have written three books covering the *why*, *what*, and *how* of mobile radio system design. My other two volumes in this series of books deal with the *why* and *what*. This volume presents the theoretical framework for radio communications and tells the reader how such present and future systems are designed.

Knowing *why*, *how*, and *what* is critical for developing confidence in any system design. I hope that this book will build your strength and knowledge in designing future mobile radio systems.

Walnut Creek, California   William C. Y. Lee
November, 1992

ACKNOWLEDGMENTS

H.W. Sams & Company is changing its business strategy by moving into computer science and was kind enough to release the copyrights to me for this book. This gave me the opportunity to have John Wiley & Sons, Inc. publish the Second Edition of this book.

I deeply appreciate the inspiration I have received from my professors, C. H. Walter and Leon Peters, who introduced me to the field of communications and wave propagation. I also wish to thank C. C. Cutler and Frank Blecher who gave me valuable advice and encouragement in writing the first edition of this book.

During the revision stage of this second edition, I was encouraged by the engineers who took my courses at George Washington University and by my colleagues at PacTel. I am obliged to Mr. George Telecki and Ms. Cynthia Shaffer of John Wiley & Sons, Inc. who constantly watched my progress during revisions and to Ms. Susan Shaffer who was always patient in typing my untidy manuscript.

Last but certainly not least, I thank my lovely wife, Margaret, and my daughters, Betty and Lily, for their support and patience. I have promised them that I would make up for all the time they have spent alone while I was writing. It has not happened yet, which I have to blame on the rapid advances in mobile communication technologies which keep me on the hook. I hope they will kindly accept my good intentions.

This book is dedicated to the memory of my parents. The many books my father wrote himself were an inspiration to me, and I am proud to carry on in his spirit. Even though he passed away at the beginning of my writing of the first edition of this book, his spirit has been with me—and always will be.

William C. Y. Lee

Mobile Communications Design
Fundamentals

1

THE MOBILE RADIO
ENVIRONMENT

1.1 REPRESENTATION OF A MOBILE RADIO SIGNAL

The mobile radio signals we describe in this book are mainly ground mobile signals. The ground mobile radio medium is unique and complicated. Much research is still being done in this field. But before we can consider the theoretical aspects of a mobile radio signal, we must try to understand the mobile radio environment.

1.1.1 Description of a Mobile Radio Environment

A wave propagation mechanism is closely affected by the wavelengths of the propagation frequencies. In the human-made environment houses and other buildings ranging from 18 to 30 m wide and from 12 to 30 m high (60 to 100 ft and 40 to 100 ft) are found in suburban areas, and there are even larger buildings and skyscrapers in urban areas. Whether suburban or urban, buildings are natural wave scatterers. The sizes of buildings are equivalent over many wavelengths of a propagation frequency, creating reflected waves at that frequency. For the mobile radio environment treated in this book, we assume that all buildings are scatterers as long as the antenna height of a mobile unit is much lower than the height of an average house. Given these conditions, the propagation frequency has to be above 30 MHz in order to form a multipath propagation medium. The frequency range for a mobile radio multipath environment would be 30 MHz and higher. The base-to-mobile link length is usually less than 24 km (15 miles), so no radio horizon (no radio-path loss attributable to the curvature of the earth) needs to be considered. When the interfering signal comes from more than 24 km (15 miles) away, the radio horizon usually contributes an additional radio-path loss, and the effective interference becomes even weaker. The earth's natural curvature helps to reduce interference and makes it easier for a system design to deal with long-distance interference.

1

In designing a mobile radio environment with a large cell radius of 6.5 – 13 km (or 4–8 miles), we would consider the height of the base-station antenna which is usually 30 to 50 m (100 to 150 ft) in small suburban towns, and over 50–91 m (150 ft) in large cities. The height of a mobile-unit antenna is about 2–3 m (6–10 ft). The base-station antenna is usually clear of its surroundings, whereas the mobile-unit antenna is embedded in them. The terrain configuration, as well as the human-made environment in which a communication link between a base station and a mobile unit lies, determines the overall propagation path loss.

From this description of the environment, we might imagine that the mobile site will receive many reflected waves and one direct wave. The reflected waves received at the mobile site would come from different angles equally spaced throughout 360°, as shown in fig. 1.1. Often a direct wave is presented and relatively strong as comparing with the reflected waves. The described situation is called Rician statistical model. However, a mobile communication system design cannot be based upon this optimistic situation; it is based on the case of weak or nondirect waves which normally occur at the fringe area. All the reflected waves received at the mobile unit combine to produce a multipath fading signal. This described situation is called Rayleigh statistical model. Both Rician and Rayleigh statistics are appeared in Sections 1.5.2 and 1.5.3.

1.1.2   Field-Strength Representation[1]

The field strength of a signal can be represented as a function of distance in space (the spatial domain) or as a function of time (the time domain). As soon as the height of a base-station transmitting antenna at a site is fixed (Fig. 1.2A), the field strength* (the envelope $r(x)$ of a received signal $s(x)$ along x-axis in the space) is then defined as illustrated in Fig. 1.2B. The field strength at every point along the x-axis is measured by a mobile receiver whose antenna height is given—approximately 3 m (10 ft) above the ground. The received field strengths along the x-axis show severe fluctuation when the mobile unit is away from the base station. Field strengths $r(x)$ can be studied either by associating them with geographical locations (areas) or by averaging a length of field strength data to obtain a so-called local mean (see Section 1.3.1) at each corresponding point. The speed of the mobile unit V must remain constant while the data are measured. Since the speed is kept constant, the time axis ($t = x/V$) can be converted to the spatial axis. The field strengths $r_1(t)$ and $r_2(t)$, with speeds of 48 and 24 km/h (30 mph and 15 mph), can be seen in Figs. 1.2C and 1.2D, respectively. It is clear that $r_1(t)$ in Fig. 1.2C fluctuates much faster than $r_2(t)$ in Fig. 1.2D. However, both speeds can be scaled to the same spatial axis, as shown in the two figures. If

---

*Field strength expressing in a ratio refers to usually one microvolt/meter as dBμV.

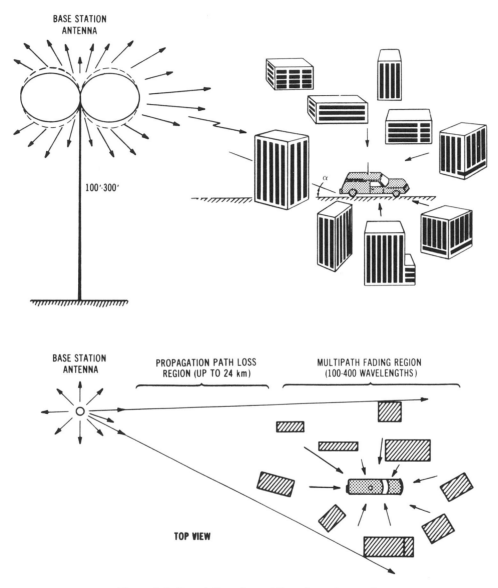

Figure 1.1. Description of a mobile radio environment.

the mobile unit does not maintain a constant speed while receiving the signal, information of changing speed vs time has to be recorded. The field strength with various speeds is shown in Figure 1.2E. The signal field strength $r(t)$ of Fig. 1.2E has to be converted to Figure 1.2B before processing the data. This process is called the *velocity-weighted conversion*. The technique is shown in

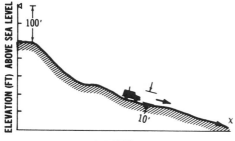
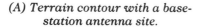
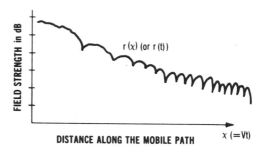

(A) Terrain contour with a base-station antenna site.   *(B) r(x) along x-axis in the space.*

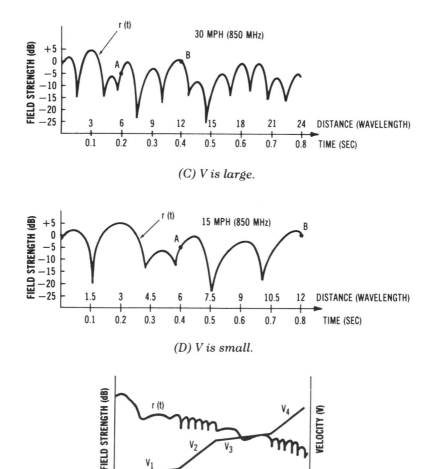

(C) V is large.

(D) V is small.

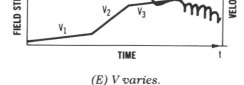

(E) V varies.

Figure 1.2. The characteristics of field strength.

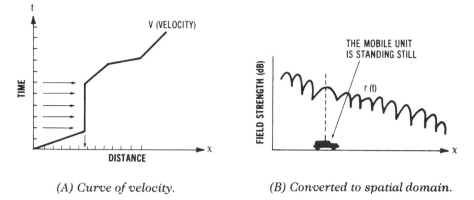

(A) Curve of velocity.   *(B) Converted to spatial domain.*

Figure 1.3. Velocity weight conversion.

Figure 1.3. The data are digitized in the time domain with equal intervals. The curve of velocity shown in Fig. 1.3*A* is then used to convert all the data points from the time domain to the spatial domain (Fig. 1.3*B*).

Another method of converting field strengths from time domain to spatial is to synchronize the turning speed of the vehicle wheels with the speed of the field-strength recording device. This method does not need a velocity-weighted conversion process. Both field-strength representations are useful. The representation $r(t)$ in the time domain is used to study the signal-fading phenomenon. The representation $r(x)$ in the spatial domain is used to generate the propagation path loss curves.

1.1.3 Mobile Radio Signal Representation

The mobile radio signal is received while the mobile unit is in motion. In this situation the field strength (also called the *fading signal*) of a received signal with respect to time t, or space x, is observed, as shown in Fig. 1.2. When the operating frequency becomes higher, the fading signal becomes more severe. The average signal level of the fading signal $\bar{r}(x)$ or $\bar{r}(t)$ decreases as the mobile unit moves away from the base-station transmitter. The average signal level of a fading signal (field strength) will be defined later. This average signal level dropping is called *propagation path loss*.

1.2 CAUSES OF PROPAGATION PATH LOSS

In free space the causes of propagation path loss are merely frequency f and d, as shown in Eq. (1.2.1):

$$\frac{P_{or}}{P_t} = \frac{1}{(4\pi df/c)^2} = \frac{1}{[4\pi(d/\lambda)]^2} \qquad (1.2.1)$$

where c is the speed of light, λ is the wavelength, P_t is the transmitting power, and P_{or} is the received power in free space.

As seen in Eq. (1.2.1), the difference between two received signal powers in free space, Δ_p, received from two different distances becomes

$$\Delta_p = 10 \log_{10}\left(\frac{P_{or2}}{P_{or1}}\right) = 20 \log_{10}\left(\frac{d_1}{d_2}\right) \text{ (dB)} \qquad (1.2.2)$$

If the distance d_2 is twice distance d_1, then the difference in the two received powers is

$$\Delta_p = 20 \log_{10}(0.5) = -6 \text{ dB}$$

Therefore the free-space propagation path loss is 6 dB/oct (octave), or 20 dB/dec (decade). An octave means doubling in distance, and a decade means a period of 10. Twenty dB/dec means a propagation path loss of 20 dB will be observed from a distance of 3 to 30 km (2 to 20 miles).

Example 1.1. What will y dB/oct be when converted to x dB/dec?

$$y = x \cdot \log_{10} 2 \qquad (1.2.3)$$

If $y = 6$ dB/oct, then $x = 20$ dB/dec.

As described previously, in a mobile radio environment the propagation path loss not only involves frequency and distance but also the antenna heights at the base station and the mobile unit, the terrain configuration, and the human-made environment. These additional factors make the prediction of propagation path loss of mobile radio signals more difficult. The prediction of propagation loss will be presented in Chapter 2.

1.3 CAUSES OF FADING

The signal strength $r(t)$ or $r(x)$, shown in Fig. 1.2B, is the actual received signal level in dB. Based on what we know about the cause of signal fading in past studies, the received $r(t)$ can be artificially separated into two parts by cause: long-term fading $m(t)$, and short-term fading $r_0(t)$ as

$$r(t) = m(t) \cdot r_0(t) \qquad (1.3.1)$$

or

$$r(x) = m(x) \cdot r_0(x) \qquad (1.3.2)$$

1.3.1 Long-Term Fading, $m(t)$ or $m(x)^2$

Long-term fading is the average or envelope of the fading signal,* as the dotted curve shows in Fig. 1.4A. It is also called *a local mean* since along the long-term fading each value corresponds to the mean average of the field strength at each local point. The estimated local mean $\hat{m}(x_1)$ at point x_1 along x-axis can be expressed mathematically as

$$\hat{m}(x) = \frac{1}{2L} \int_{x_1-L}^{x_1+L} r(x)\, dx = \frac{1}{2L} \int_{x_1-L}^{x_1+L} m(x) r_0(x)\, dx \qquad (1.3.3)$$

Assume that $m(x_1)$ is the true local mean, then at point x_1 in Fig. 1.4A,

$$m(x = x_1) = \hat{m}(x = x_1) \qquad x_1 - L < x < x_1 + L \qquad (1.3.4)$$

When the length L is properly chosen, Eq. (1.3.3) becomes

$$\hat{m}(x_1) = m(x_1) \cdot \frac{1}{2L} \int_{x_1-L}^{x_1+L} r_0(x)\, dx \qquad (1.3.5)$$

To let $\hat{m}(x_1)$ approach $m(x_1)$ in Eq. (1.3.5), the following relation holds:

$$\frac{1}{2L} \int_{x_1-L}^{x_1+L} r_0(x)\, dx \to 1 \qquad (1.3.6)$$

The length L will be determined after fully understanding the statistical characteristics of short-term fading $r_0(x)$.

The long-term signal fading $m(x)$ is mainly caused by terrain configuration and the built environment between the base station and the mobile unit. Terrain configurations can be classified as

Open area
Flat terrain
Hilly terrain
Mountain area

and the human-made environment as

Rural area
Quasi-suburban
Suburban area
Urban area

---

*A fading signal is an envelope of a received rf signal. Therefore the long-term fading signal is the envelope of a received fading signal.

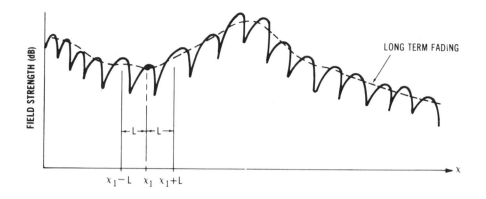

(A) The local mean, integrated from a proper 2L window.

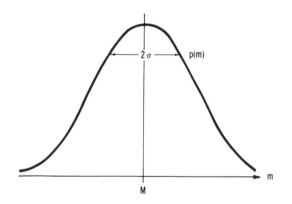

(B) The log-normal distribution.

Figure 1.4. The local mean and terrain contour.

The terrain configuration causes local-mean (long-term fading) attenuation and fluctuation, whereas the human-made environment only causes local-mean attenuation. The human-made environment also causes short-term fluctuation (fading) in signal reception. Short-term fading will be described later. Under certain circumstances the fluctuation of a long-term fade caused by the terrain configuration can form a log-normal distribution because of the statistical nature of the fluctuation shown in Fig. 1.4B. Here we must differentiate between the terms "radio path" and "mobile path": The former is the path that the radio wave travels, and the other is the path that the mobile unit travels. Two cases are shown in Figs. 1.4C and 1.4D. One is when the

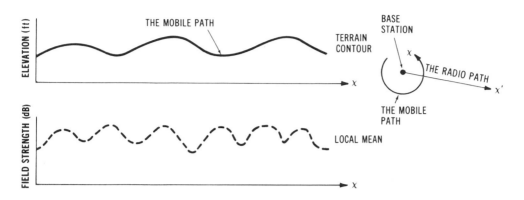

*(C) The mobile-path terrain contour and the local mean
are uncorrelated.*

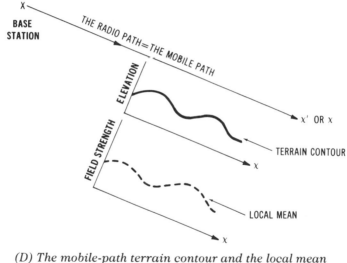

*(D) The mobile-path terrain contour and the local mean
are correlated.*

Figure 1.4 *(Continued)*

mobile unit is circling around the base station; another is when the mobile
unit is moving away from the base station. In the first case the fluctuation of
the long-term fading received at the mobile unit is affected by the circled
terrain contour around the base station. In this case the radio path does not
correspond to the mobile path. In the second case the fluctuation of the long-
term fading received is affected by the radial terrain contour where the mobile
travels in a certain direction. The radio path corresponds to the mobile path.
In the latter case the terrain contour where the mobile unit travels has a
strong correlation with the received signal as shown in Figure 1.4D. The

received signal is strong when the mobile unit is at the top of the hill and weak when the mobile unit is in the valley. The configuration of the terrain affects the standard deviation σ_l (spread) of the log-normal curve representing the local mean signal in that area. The σ_l of the local mean will vary in dB values dependent on the configuration of the terrain.

1.3.2 Short-Term Fading, $r_0(t)$ or $r_0(x)$

Short-term fading is mainly caused by multipath reflections of a transmitted wave by local scatterers such as houses, buildings, and other human-built structures, or by natural obstacles such as forests surrounding a mobile unit. It is not caused by a natural obstruction such as a mountain or hill located between the transmitting site and the receiving site. In explaining the causes of short-term fading, a base-station transmitter and a mobile receiver are assumed. Four cases illustrate the phenomenon.

CASE 1. The mobile receiver is standing still, surrounded by moving objects such as tractors. (See Fig. 1.5A.) The received signal will show fading. The number of fades will depend on the traffic flow of the tractors and the distance between the tractors and the mobile receiver.

CASE 2. The mobile unit is moving with a velocity V, yet there is no single scatterer around the mobile unit, as shown in Figure 1.5B. In this case the received signal can be represented by assuming an incoming signal arriving at an angle θ with respect to the motion of the mobile unit as

$$s_r = A \, \exp[j(2\pi f_t t - \beta x \cos \theta)] \tag{1.3.7}$$

where β is called the wave number, $\beta = 2\pi/\lambda$ and λ is the wavelength. Expression $(j2\pi f_t t)$ is the complex exponential representing a transmitting frequency f_t that propagates in time domain. x is the displacement; $x = Vt$, where V is the speed of the mobile receiver. A is a constant amplitude, and f_t is the transmitting frequency or so-called propagation frequency. Equation (1.3.7) can be rewritten

$$s_r = A \, \exp\left[j2\pi\left(f_t - \frac{V}{\lambda} \cos \theta \right)t \right] \tag{1.3.8}$$

where the Doppler frequency f_D is

$$f_D = \frac{V}{\lambda} \cos \theta \tag{1.3.9}$$

The amplitude of the signal is $|s_r| = A$, where A is a constant at the

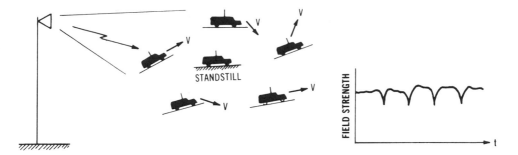

(A) Case 1: the mobile receiver is standing still.

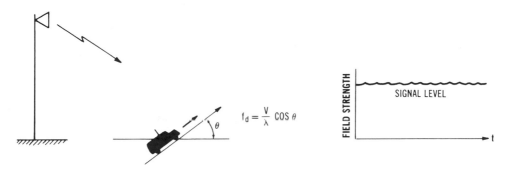

$$f_d = \frac{V}{\lambda} \cos \theta$$

(B) Case 2: the mobile receiver is moving (no scatterers).

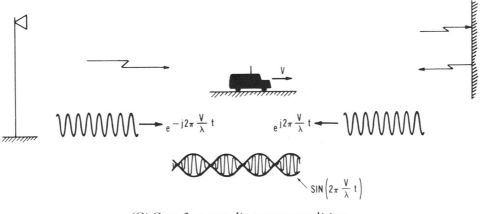

$$e^{-j2\pi \frac{V}{\lambda} t}$$

$$e^{j2\pi \frac{V}{\lambda} t}$$

$$\sin\left(2\pi \frac{V}{\lambda} t\right)$$

(C) Case 3: a standing wave condition.

Figure 1.5. Short-term fading phenomena.

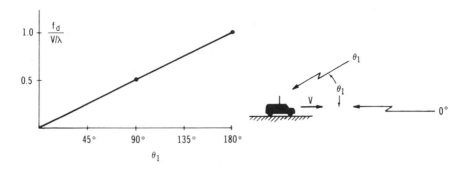

(D) Case 4: a general standing wave condition.

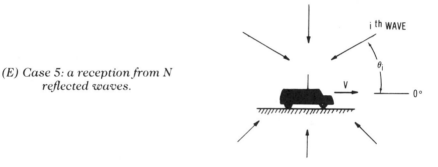

(E) Case 5: a reception from N reflected waves.

Figure 1.5 *(Continued)*

baseband. The received frequency f_r is offset by the transmitting frequency f_t by a Doppler frequency f_D as

$$f_r = f_t - \frac{V}{\lambda} \cos \theta \qquad (1.3.10)$$

When the mobile unit is moving away from the source $\theta = 0°$, the received frequency is $f_r = f_t - V/\lambda$. When the mobile unit is circling around the source $\theta = 90°$, then $f_r = f_t$. When the mobile unit is moving toward the source $\theta = 180°$, and $f_r = f_t + V/\lambda$.

CASE 3. The mobile unit is moving with a velocity V along a road between the transmitter and one scatterer. (See Fig. 1.5C.) The incoming signal from the transmitter, represented by Eq. (1.3.8); with $\theta = 0°$, is

$$A \exp\left[j2\pi\left(f_t - \frac{V}{\lambda} \right)t \right]$$

Assume that the scatterer is so perfect that the wave, reflected back from the opposite direction ($\theta = 180°$), is

$$-A \exp\left[j2\pi\left(f_t + \frac{V}{\lambda}\right)t\right]$$

The resultant signal is the sum of two waves

$$s_r = (Ae^{-j2\pi(V/\lambda)t} - Ae^{j2\pi(V/\lambda)t})e^{j2\pi f_t t} \tag{1.3.11}$$

The envelope of s_r, $|s_r|$ is a standing wave pattern expressed as

$$|s_r| = 2A \sin\left(2\pi \frac{V}{\lambda} t\right) \tag{1.3.12}$$

Example 1.2. Find the difference in plotting standing wave patterns in linear scale and dB scale.

Equation (1.3.12) is graphed so a standing wave pattern, and it can be drawn on both a linear scale and a dB scale, as shown in Figs. E1.1*A* and E1.1*B*, respectively. Since in this simple situation the mobile receiver observes a standing wave pattern instead of a constant amplitude at the baseband, the fading phenomenon can be explained from the standing wave concept. The fading pattern with N reflected waves where N is much greater than 2 is illustrated in Fig. E1.1*C* and discussed in case 5.

Example 1.3. Two waves arrive at the mobile unit from two different directions θ_1 and θ_2. The amplitude of the received signal is

$$|s_r| = 2A \cos\left[2\pi \frac{V}{2\lambda} (\cos \theta_1 - \cos \theta_2)\right]$$

CASE 4. Assume that two incoming waves are not exactly opposite each other. Let the two incoming angles be $\theta = 0°$ and $\theta = \theta_1°$, as shown in Fig. 1.5*D*. Assume that these two waves are reflected waves of equal amplitude. Using Eq. (1.3.8) to represent each of the two waves and summing up the two waves, we calculate the fading frequency at the mobile receiver as follows:

$$s_r = Ae^{j2\pi f_t t}(e^{-j\beta x} + e^{-j\beta x \cos\theta_1})$$

$$= Ae^{j2\pi f_t t} \cdot 2e^{j\beta x(1+\cos\theta_1)/2} \cdot \cos\left(\frac{\beta x}{2} - \frac{\beta x}{2} \cos \theta_1\right) \tag{1.3.13}$$

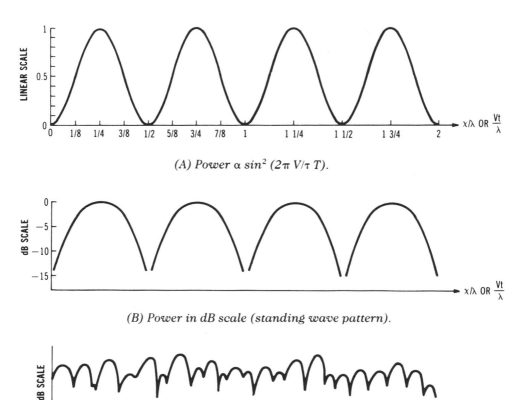

(A) Power α sin² (2π V/τ T).

(B) Power in dB scale (standing wave pattern).

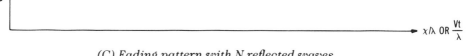

(C) Fading pattern with N reflected waves.

Figure E1.1. Fading illustrations with different scales.

Let $x = Vt$, the standing wave frequency f_d or the angular standing frequency ω_d can be obtained from Eq. (1.3.13) as

$$\omega_d = 2\pi f_d = \frac{\beta x}{2} - \frac{\beta x}{2} \cos \theta_1 = 2\pi \frac{V}{2\lambda} \cdot (1 - \cos \theta_1)$$

or

$$f_d = \frac{V}{2\lambda} (1 - \cos \theta_1) \tag{1.3.14}$$

When $\theta_1 = 0°$, $f_d = 0$. This means that the fading frequency is zero when two waves come from the same direction. The fading frequency is V/λ when

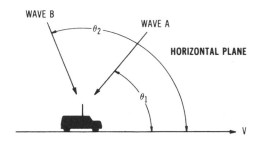

Figure E1.2. Two incoming waves at different angles.

two waves come to the mobile receiver from opposite directions. Equation (1.3.14) is plotted in Fig. 1.5D.

Example 1.4. The general formula of f_d for two incoming waves at different angles θ_1 and θ_2 with respect to the motion of the vehicle, as shown in Fig. E1.2, can be obtained by changing the term $e^{-j\beta x}$ to $e^{-j\beta x\cos\theta_2}$ in Eq. (1.3.13). The result is

$$f_d = \frac{V}{2\lambda}(\cos\theta_2 - \cos\theta_1) \qquad (1.3.14a)$$

In general, the Doppler frequency shown in Eq. (1.3.9) is different from the fading frequency shown in Eq. (1.3.14). The Doppler frequency is the same as the fading frequency only in a case of $\theta = 180°$. Therefore one should be cautious about using these two terms. Many people misuse them when discussing mobile radio problems.

CASE 5. Assume that N reflected waves, but no direct waves, come from N arbitrary directions with equal probability. The signal received at the mobile receiver becomes

$$s_r = \sum_{i=1}^{N} A_i\, e^{j2\pi f_t t} \cdot e^{j\beta V t\cos\theta_i} \qquad (1.3.15)$$

where f_t is the transmitting frequency, V is the velocity of the mobile unit, θ_i is the direction of ith wave arrival, as shown in Fig. 1.5E.

In presenting a fading signal, A_i is a complex random variable with zero mean and variance 1, and θ_i is also a random variable from 0° to 360°. Equation (1.3.15) represents a fading signal. Amplifying the picture shown in Fig. 1.5E, and adding many pairs of standing waves (each one is the same as shown in case 4), the maximum fading frequency can be obtained from Eq. (1.3.14) as

$$f_{max} = \frac{V}{\lambda} \tag{1.3.16}$$

which is the same as the maximum Doppler frequency f_D shown in Eq. (1.3.9).

1.3.3   Classification of Channels[3]

In a dispersive medium, there are two kinds of spread: Doppler spread F and multipath spread δ. Doppler spread F is spreading in frequency, and multipath spread δ is spreading in time. In a strict sense all media are dispersive. However, we can classify a medium's characteristics based upon the signal duration T and the signal bandwidth W of a transmitted waveform in operation. We can also treat all media as channels, since the definition of a channel is a link connecting a transmitter and a receiver.

Nondispersive Channels
A nondispersive but fading channel is created if these two kinds of spread, F and δ, meet the following conditions:

$$F \ll \frac{1}{T} \quad \text{and} \quad \delta \ll \frac{1}{W}$$

A nondispersive fading channel is also called a *flat-flat fading channel*. In many practical systems we choose the values of W and T so that the previous conditions are met and the system is in a nondispersive channel.

Time-Dispersive Channels
Some channels are dispersive only in time, but not in frequency. For a time-dispersive channel to exist, the following conditions must hold:

$$\delta \gg \frac{1}{W} \quad \text{and} \quad \delta \gg T \qquad \text{(dispersive in time)}$$

but

$$F \ll \frac{1}{T} \qquad \text{(not dispersive in frequency)}$$

The illustrations of these time-dispersive channels are shown in Fig. 1.6. They are also called *frequency-selective fading*. This is because at the same time a signal could be faded at one frequency but not necessarily at another. Sometimes they are called *time-flat fading channels*.

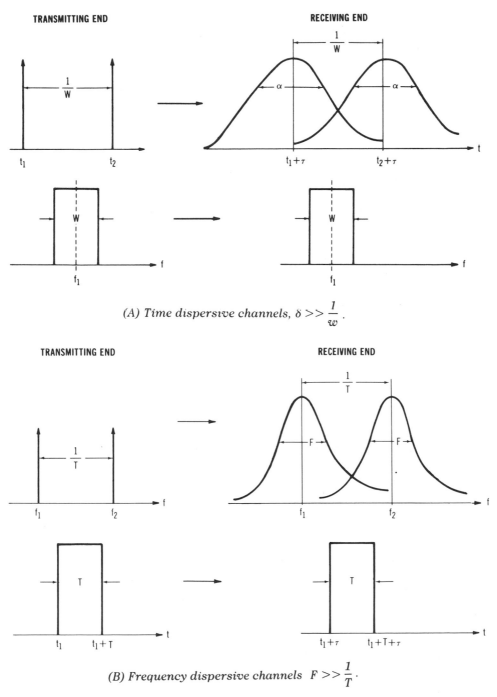

(A) *Time dispersive channels,* $\delta >> \dfrac{1}{w}$.

(B) *Frequency dispersive channels* $F >> \dfrac{1}{T}$.

Figure 1.6. Classification of channels.

Frequency-Dispersive Channels

Some channels are dispersive in frequency, but not in time. The following conditions are expressed for frequency-dispersive channels:

$$F \gg \frac{1}{T} \quad \text{and} \quad F \gg W \qquad \text{(dispersive in frequency)}$$

but

$$\delta \ll \frac{1}{W} \qquad \text{(not dispersive in time)}$$

The frequency-dispersive channel is also called *time-selective fading,* since the channel selectivity alters certain time segments of the transmitted waveform. It is also called *frequency-flat fading,* since all the constituent frequencies of the transmitted waveform are modulated by the same function. This frequency dispersive channel is illustrated in Fig. 1.6.

Doubly-Dispersive Channels

Finally, some channels exhibit both time-selective and frequency-selective fading. The fading is neither time flat nor frequency flat.

1.3.4 Effects of Weather

Heavy ground fog or extremely cold air over warm earth can cause atmospheric refractions to vary with elevation. However, when the base-station antenna is less than 91 m (300 ft) above the ground and the transmission path is less than 23 km (15 miles), atmospheric refraction does not effectively bend the radio path from a straight-line path.

Since radio waves travel in a straight line over a transmission path, the frequency-selective type of fading caused by interference between two or more radio waves in the atmosphere cannot exist. Sometimes in a mobile communications system the interfering signal can come from more than 32 km (20 miles) away. If the radio path of the interfering signal is bent and faded due to variations in the atmosphere, there will be an additional path loss. Knowing this general tendency of bent radio paths helps to ease the interference problems found when designing a system.

Reflection of Microwaves on Snow-Covered Terrain[4]

The degree of reflected energy can be measured by the reflection coefficient, which is the ratio of the incident wave to the reflected wave. A reflection coefficient of 1 means a totally reflected wave. A reflection coefficient of 0 means no reflected wave. When a grazing (incident) angle α (shown in Fig. 1.1) is small (a situation in the mobile radio environment), then the amplitude-

TABLE 1.1   Amplitude-of-Reflection Coefficient on Snow-Covered Terrain

Incident Angle	Height of Settled Snow, 98 cm (39 in.)		Length of Grainy Snow, 64 cm (25 in.)	
	Horizontal Polarization	Vertical Polarization	Horizontal Polarization	Vertical Polarization
$\alpha = 2.5°$	0.764	0.7	0.84	0.743
$\alpha < 1°$	1	1	1	1

TABLE 1.2   Attenuation Loss When Ice Is On Antenna

Ice/Snow Accretion on	Attenuation Loss (dB)
Paraboloid half-surface, 1–3 cm (0.4–1 in.)	16 ~ 4 (frozen snow)
Feedhorn except mica window, 2 cm (0.3 in.)	7 (wet snow) 2.5 (frozen snow)
Feedhorn lower half surface, 2 cm (0.8 in.)	5 (frozen snow)
Feedhorn entire surface, 0.3 cm (0.118 in.)	6 (wet snow) 4 (frozen snow)

of-reflection coefficient of a wave reflected from the ground is always close to 1 regardless of the characteristics of the ground—wet or dry, soil or sand, snow or ice. In Table 1.1 we show empirical data on microwaves reflected from different types of snow forms at 4000 MHz.

The reflection coefficient is close to 1 regardless of the type of snow or polarization as long as the grazing angle is less than 1°. Some general observations on the reflection coefficient are given in Chapter 3. Since the wave will have a 180° phase change after it is reflected from the ground, the resultant reflection coefficient is minus one.

Electromagnetic Effect of the Accretion of Snow on an Antenna

In general, the attenuation of waves propagated through ice or snow is roughly less than 0.95 dB/m. If a parabola antenna is used, the attenuation due to the thickness of snow on the feedhorn or the parabola dish, or both, can result in extensive loss. Since a parabola dish is usually used for a frequency above the C band, the loss can be listed at 7000 MHz. The antenna has a diameter of 1.2 m (3.9 ft), which provides a gain of 38 dB at 7000 MHz. The loss is listed in terms of the thickness of ice in Table 1.2.

In the table it can be seen that attenuation loss due to the accretion of snow on the antenna is proportional to both the frequency and the thickness of the snow on the apparatus. It is also affected by the condition of the ice or snow, and the location of ice or snow on the antenna.

1.4 RECIPROCITY PRINCIPLE

The reciprocity principle states that a signal field strength is received the same at a base-station antenna from a mobile transmitter as it is received at the mobile-unit antenna from the base-station transmitter. The reciprocity principle holds in a mobile radio environment for certain situations. Sometimes one test setup is more easily accomplished than another test setup. We could use reciprocity to predict a result even if the actual setup was done otherwise.

However the base-station antenna is always higher than most of its surroundings, and the mobile antenna is usually only 3 m (10 ft) above the ground. Even though the medium is homogeneous, the signal-to-noise ratio *S/N* received at the mobile unit is different from that of the base station, for the following reason:

Since the produced noise is dominated by automotive ignition noise (see Sections 64. and 7.6), the noise source is closer to the mobile antenna than to the base-station antenna. The noise received at the mobile antenna is thus higher than that at the base-station antenna. Though the signal strengths received at both sites (base and mobile) are the same, following the principle of reciprocity, the carrier-to-noise ratios at the two sites are different.

Therefore the reciprocity principle does not hold for carrier-to-noise ratio *C/N* in the mobile radio environment. Often in a mobile radio environment the characteristic values obtained at the mobile unit will not be the same as the ones obtained at the base station. So when we say that the reciprocity principle holds, we refer only to the "field strengths" with the understanding that transmitters, receivers, and antennas at both sites (terminals) remained unchanged.

Other characteristics such as *C/N* (Sections 6.4 and 7.6), correlation separation between antennas (Section 6.2), directional antennas (Section 7.4), and antenna pattern ripple effects (Section 6.3) are not reciprocal. We will describe these topics in later chapters.

1.5 DEFINITIONS OF NECESSARY TERMS AND THEIR APPLICATIONS

1.5.1 Averages

In real data processing there are sample averages $\bar{x}(t)$ and biased time averages $\hat{x}(t)$. In statistics there are ensemble averages $E[x]$ and unbiased time averages $\langle x(t) \rangle$.

Sample Average ($\bar{x}$)
The samples are averaged as a conventional arithmetical average

$$\bar{x} = \frac{\sum_{i=1}^{N} x_i}{N} \tag{1.5.1}$$

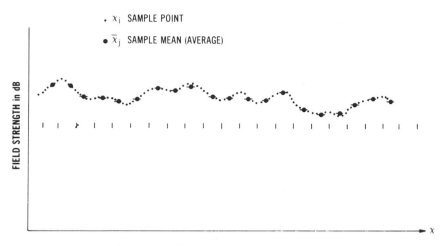

Figure 1.7. Illustration of sample averages.

where x_i is a random variable. N should be large in order to claim $\bar{x}$ as a meaningful average.

Using Eq. (1.5.1) we learn the distribution of $\bar{x}$. Assume that there is a set of M variables, x_i, where $M \gg N$. Then after a sample average, there are M/N number of $\bar{x}$. Each variable $\bar{x}$, $\bar{x}_j$

$$\left(\bar{x}_j = \sum_{i=(j-1)N+1}^{jN} \frac{x_i}{N} \right)$$

is a new variable. Regardless of the distribution of random variables x_i, as long as N is large (more than 10) the new random variable $\bar{x}_j$ becomes a Gaussian variable. If x_i is in dB scale, then $\bar{x}_j$ is a log-normal variable shown in Fig. 1.7. A log-normal distribution will be described in Section 1.5.2.

Statistical Average

The statistical average is also called the ensemble average when N in Eq. (1.5.1) approaches infinity:

$$E[x] = \lim_{N \to \infty} \frac{\sum_{i=1}^{N} x_i}{N} \tag{1.5.2}$$

To determine the number of N such that $\bar{x}$ is approaching $E[x]$, we write

$$(E[x] - \bar{x})^2 < \delta \tag{1.5.3}$$

where δ is a tolerated error. In the ensemble domain we assume that x_i and

x_{i+1} are independent variables, that is, x_{i+1} cannot be predicted from x_i. Then the number of N can be determined by

$$\left(\frac{\sum_{i=1}^{N+1} x_i}{N+1} - \frac{\sum_{i=1}^{N} x_i}{N}\right)^2 < \delta \tag{1.5.4}$$

Biased Time Average
When a random process $x(t)$ is recorded in a time scale, the average can be obtained as

$$\bar{x}(t) = \frac{1}{T}\int_0^T x(t)\, dt \tag{1.5.5}$$

This is the time average we get when using an integrator.

Unbiased Time Average
The unbiased time average is when the time interval T of Eq. (1.5.5) becomes infinite:

$$\langle x(t) \rangle = \lim_{T \to x} \frac{1}{T}\int_0^T x(t)\, dt \tag{1.5.6}$$

Since T cannot be infinite in real measurements, we have to determine the interval T such that

$$\bar{x}(t) \xrightarrow[T \to \text{large}]{} \langle x(t) \rangle$$

or

$$[\langle x(t) \rangle - \bar{x}(t)]^2 < \delta$$

where δ is an infinite decimal unit which is approaching to zero. Actually, we use the following relation to determine the interval T:

$$\left[\frac{1}{T+t}\int_0^{T+t} x(t)\, dt - \frac{1}{T}\int_0^T x(t)\, dt\right]^2 < \delta \tag{1.5.7}$$

Ergodic Process
If a statistical average value obtained in the ensemble domain is the same as in the time domain, we call this kind of random process an ergodic process:

$$E[x(t)] = \langle x(t) \rangle$$
$$E[x^2(t)] = \langle x^2(t) \rangle \tag{1.5.8}$$
$$E[x^n(t)] = \langle x^n(t) \rangle$$

If a communication system setup remains unchanged in an ergodic process, the data values (signal strengths of a mobile radio signal) received at each location will also remain unchanged in different time intervals. Fortunately the mobile radio fading signal can be assumed to be an ergodic process.[5] Since it is easier and more effective for us to process an average in a time domain than in an ensemble domain, we will find the statistical average in a time domain. Also for simplifying notations in this book we use $\bar{x}$ to mean $\langle x \rangle$ unless a different use is mentioned.

1.5.2 Probability Density Function (pdf)

A typical multipath signal fading is shown in Fig. 1.8A with N sample points. The vertical scales are in dB. First, we equally divided a dB scale into dB bins, one dB per a bin. Then we counted the sample points in each bin and plotted the counts vs the levels shown in Fig. 1.8B.

Usually the probability density function (pdf) generated from the experimental data in dB values needs a proper normalization factor to turn into a linear value before it can be compared with the theoretical one. Assume that the pdf of x in the decibel scale is $p(x)$ and that the pdf of y in the linear scale is $p(y)$; then

$$p(y) = \left(\frac{20 \log_{10} e}{y} \right) p(x)$$

Normally we always try to obtain the experimental cumulative probability distribution (CPD) directly (Fig. 1.8C). Then the experimental pdf can be found by taking a derivative of CPD, which will be described in Section 1.5.3.

The theoretical pdf defined here are three pdf used to describe the mobile radio environment.

Log-Normal pdf
The log-normal pdf represents long-term fading or local means.

$$p(y) = \frac{1}{\sqrt{2\pi}\sigma_y} \exp\left[-\frac{(y - m)^2}{2\sigma_y^2} \right] \qquad (1.5.9)$$

where the parameters (as in Eq. 1.5.9)—the log-normal variable y, its mean m, and its standard deviation σ_y—are in dB scales. The log-normal pdf are always symmetrical with respect to the mean level (see Fig. 1.9A).

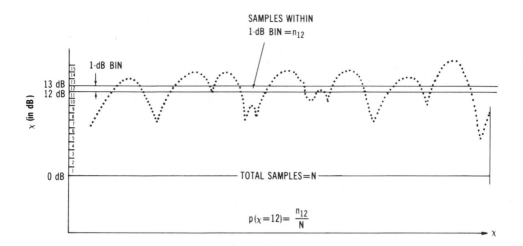

(A) Count the pdf at each level.

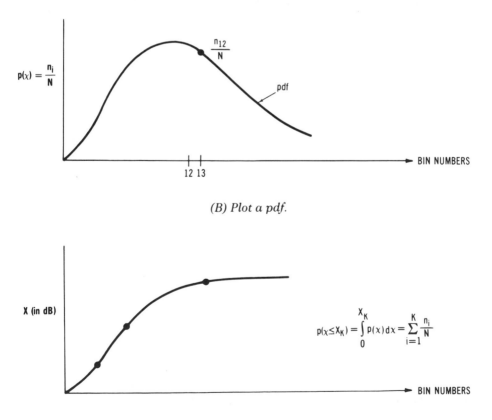

(B) Plot a pdf.

(C) Plot CPD.

Figure 1.8. The typical steps for obtaining pdf and CPD.

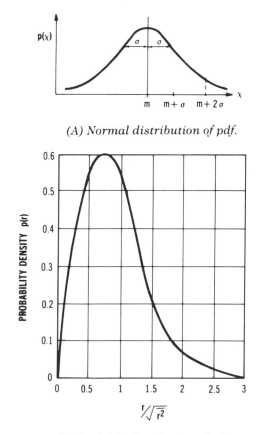

(A) *Normal distribution of pdf.*

(B) *Rayleigh distribution of pdf.*

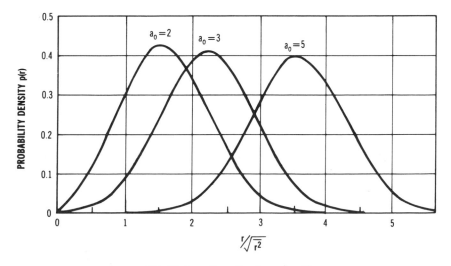

(C) *Rician distribution of pdf.*

Figure 1.9. Different distributions of pdf.

Rayleigh pdf

The Rayleigh pdf represents short-term fading or multipath fading:

$$p(r) = \frac{r}{\dfrac{\sqrt{\overline{r^2}}}{2}} \exp\left(-\frac{r^2}{\overline{r^2}}\right) \tag{1.5.10}$$

$$p(R) = 2R \exp(-R^2)$$

or

$$p(\gamma) = \frac{1}{\Gamma} \exp\left(\frac{-\gamma}{\Gamma}\right)$$

where $\overline{r^2}$ is the average power of the short-term fading, or $\sqrt{\overline{r^2}}$ is an rms value of r. The value of r cannot be below zero. The Rayleigh pdf is not a symmetrical function (see Fig. 1.9B). Rayleigh $R = \sqrt{r^2/\overline{r^2}}$, *where R is* the amplitude variation with respect to its rms value $\sqrt{\overline{r^2}}$. In the above equation γ is the signal-to-noise ratio and Γ is the average S/N ratio. The relationship between r, γ, and R is

$$\frac{\gamma}{\Gamma} = \frac{r^2/N}{\overline{r^2}/N} = R^2$$

The standard deviation σ_r is

$$\sigma_r = \frac{\sqrt{4 - \pi}}{2}\left(\sqrt{\overline{r^2}}\right) \tag{1.5.11}$$

The mean of r is

$$m = \frac{\sqrt{\pi}}{2}\left(\sqrt{\overline{r^2}}\right) \tag{1.5.12}$$

The average power is at the 63% level, which means 63% of the signal is below the average power level.

Rician pdf[6]

The Rician pdf represents a direct wave plus reflected waves:

$$p(r) = 2\frac{r}{\overline{r^2}} \exp\left(-\frac{r^2 + a^2}{\overline{r^2}}\right) I_0\left(\frac{r}{\sqrt{\overline{r^2}/2}} \cdot \frac{a}{\sqrt{\overline{r^2}/2}}\right) \tag{1.5.13}$$

where r is the envelope of the fading signal, $\overline{r^2}$ is the average of the fading signal, a is the amplitude of a direct wave, and $I_0(\,.\,)$ is the modified Bessel function of zero order, which can be expressed mathematically as

$$I_0(z) = \sum_{n=0}^{\infty} \frac{z^{2n}}{2^{2n}n!n!} \qquad (1.5.14)$$

For $z \gg 1$, Eq. (1.5.14) can be expressed by

$$I_0(z) = \frac{e^z}{(2\pi z)^{1/2}}\left(1 + \frac{1}{8z} + \frac{9}{128z^3} + \cdots + \right) \qquad (1.5.15)$$

Figure 1.9C is the Rician pdf of a fading envelope r with different values of $a_0 = a/\sqrt{\overline{r^2}/2}$. When $ra \gg \sqrt{\overline{r^2}}$, Eq. (1.5.13) becomes

$$p(r) = \frac{\sqrt{2}}{\sqrt{\overline{r^2}}}\left(\frac{r}{2\pi a}\right)^{1/2} \cdot \exp\left(-\frac{(r-a)^2}{\overline{r^2}}\right) \qquad (1.5.16)$$

When a is large, r is near a, then Eq. (1.5.16) is approximately Gaussian. When the direct wave does not exist, a becomes zero, and Eq. (1.5.13) becomes Rayleigh pdf.

1.5.3 Cumulative Probability Distribution (CPD)

From the same signal fading shown in Fig. 1.8A, we count N_1 sample points below a certain level L_1. The percentage N_1/N is obtained for level L_1 where N is the total number of samples. The percentages of sample points below other levels, say, $L_2, L_3, \ldots, L_n$ can be also easily obtained by counting the number of samples below those levels. The plot of the percentages vs the levels is called *cumulative probability distribution* (CPD). The CPD is shown in Fig. 1.8C. It can be also plotted on a Rayleigh paper, as shown in Fig. 1.10. The Rayleigh paper gets its name from the fact that the Rayleigh curve drawn on this paper is a straight line. Any outcome can be easily evaluated by visually comparing it with the Rayleigh line. The theoretical expressions for three CPDs are log-normal CPD, Rayleigh CPD, and Rician CPD.

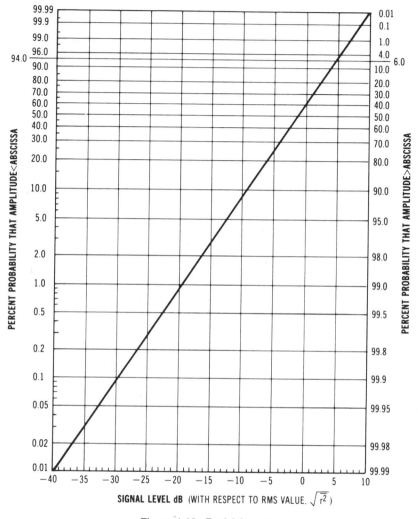

Figure 1.10. Rayleigh paper.

Log-Normal CPD

There is no analytical solution for log-normal CPD. We can write

$$P(y \le L) = \int_{-\infty}^{L} p(y) \, dy = \int_{-\infty}^{L} \frac{1}{\sqrt{2\pi}\sigma_y} \exp\left(-\frac{(y-m)^2}{2\sigma_y^2}\right) dy \qquad (1.5.17)$$

and insert a normalized $z = (y - m)/\sigma_y$ in Eq. (1.5.17),

$$P\left(z \le Z = \frac{L-m}{\sigma_y}\right) = \frac{1}{2\pi} \int_{-\infty}^{Z} \exp\left(\frac{-z^2}{2}\right) dz \qquad (1.5.18)$$

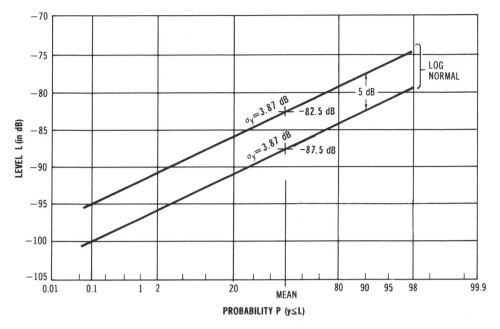

Figure 1.11. CPD of log-normal.

For a large value of Z, Eq. (1.5.18) becomes

$$P(z \le Z) = 1 - \frac{e^{-z^2/2}}{\sqrt{2\pi}Z}\left(1 - \frac{1}{Z^2} + \frac{1 \cdot 3}{Z^4} - \frac{1 \cdot 3 \cdot 5}{Z^6} + \dots\right)$$

$$\simeq 1 - \frac{e^{-z^2/2}}{\sqrt{2\pi}Z} \qquad (1.5.19)$$

Equation 1.5.19 is graphed in Fig. 1.11 with two different means, -82.5 dB and -87.5 dB, and the same standard deviation $\sigma_y = 3.87$ dB. The value of σ_y can be obtained from Problem 1.9.

Rayleigh CPD
The Rayleigh CPD is obtained by integrating Eq. (1.5.10):

$$P(r \le R) = \int_0^R p(x)\, dr$$

$$= \int_0^R \frac{r}{\overline{r^2}/2} \exp\left(-\frac{r^2}{\overline{r^2}}\right) dr$$

$$= 1 - \exp\left(\frac{R^2}{\overline{r^2}}\right) \qquad (1.5.20)$$

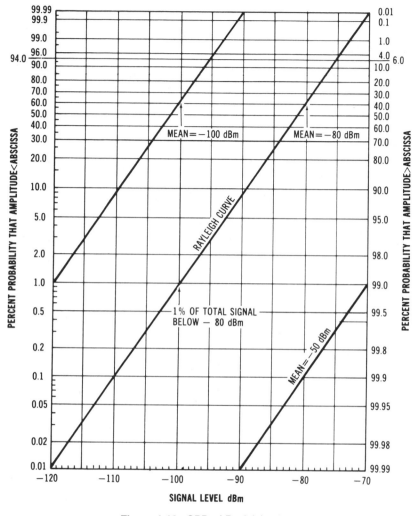

Figure 1.12. CPD of Rayleigh curve.

Equation (1.5.20) is graphed in Fig. 1.12. Changing the mean power levels shifts the curves, but the slopes remain the same. In other words, the mean power level of each curve can be found by checking the 63% level on the graph.

Rician CPD
The Rician CPD is obtained by integrating Eq. (1.5.13):

$$P(r \le R) = \int_0^R p(r)\, dr$$

$$= \int_0^{R_0} r_0 \exp\left(-\frac{r_0^2 + a_0^2}{2}\right) I_0(a_0 r_0)\, dr_0 \qquad (1.5.21)$$

where r_0, a_0, and R_0 are normalized parameters of r, a, and R, respectively, with

$$r_0 = \frac{r}{\sqrt{\overline{r^2}/2}}$$

$$a_0 = \frac{a}{\sqrt{\overline{r^2}/2}} \tag{1.5.22}$$

$$R_0 = \frac{R}{\sqrt{\overline{r^2}/2}}$$

When $ar \gg \sqrt{\overline{r^2}}$, we substitute the expansion series of Eq. (1.5.15) into Eq. (1.5.21) and neglect the terms after z^{-3} to obtain the approximate solution for a large a:

$$
\begin{aligned}
P(r \leq R) \simeq\ &\frac{1}{2} + \frac{1}{2}\operatorname{erf}\left(\frac{R_0 - a_0}{\sqrt{2}}\right) \\
&- \frac{1}{\sqrt{8\pi}a_0}\left[1 - \frac{R_0 - a_0}{4a_0} + \frac{1 + (R_0 - a_0)^2}{8a_0^2}\right] \\
&\times \exp\left[-\frac{(R_0 - a_0)^2}{2}\right]
\end{aligned}
\tag{1.5.23}
$$

The Rician CPD with different values of a_0 is shown in Fig. 1.13.

1.5.4 Level-Crossing Rate (lcr) and Average Duration of Fades (adf)[7]

We count the crossings at the positive slopes at a level A. The total number of crossings N over a T-second length of data divided by T seconds becomes the level-crossing rate (lcr):

$$n(r - A) = \frac{N}{T} \tag{1.5.24}$$

The level-crossing rate of a typical fading signal can be calculated and is shown in Fig. 1.14. The theoretical equation of lcr will be expressed in Chapter 3.

The average duration of fades (adf) is defined as the sum of N fades at level A dividing by N:

$$\bar{t}(r - A) = \frac{\displaystyle\sum_{i=1}^{N} t_i}{N} \tag{1.5.25}$$

where t_i is the individual fade shown in Fig. 1.14.

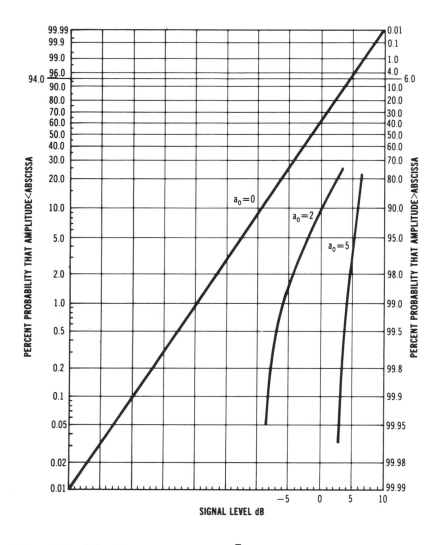

Figure 1.13. CPD of Rician distribution ($a_0/\sqrt{2}$ = amplitude of direct path/rms value).

Now the product of Eqs. (1.5.24) and (1.5.25) becomes a CPD as shown here:

$$n(A) \cdot \bar{t}(A) = \frac{N}{T} \cdot \frac{\sum\limits_{i=1}^{N} t_i}{N} = \frac{\sum\limits_{i=1}^{N} t_i}{T}$$

$$= P(r \leq A) \qquad (1.5.26)$$

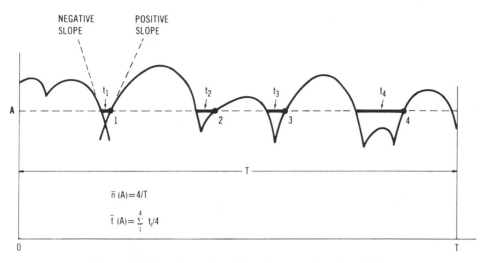

Figure 1.14. Level-crossing rate and average duration of fades.

Equation (1.5.26) provides a relationship among three parameters: lcr, adf, and CPD.

$$(\text{lcr})(\text{adf}) = \text{CPD} \qquad (1.5.27)$$

Because they are functions of time related to vehicle speeds, lcr, and adf are the second-order statistics. CPD is a first-order statistic; it is not a function of time. Equation (1.5.27) shows that the product of two second-order statistical functions become a first-order statistical function.

1.5.5 Correlation and Power Spectrum

Correlations

There are two kinds of correlations, autocorrelation and crosscorrelation. Their general expressions are as follows:

Autocorrelation Functions. Let two random variables x_1 and x_2 be two random processes, $x(t_1)$ and $x(t_1 + \tau)$, respectively.

$$x_1 = x(t_1)$$

$$x_2 = x(t_2) = x(t_1 + \tau)$$

An autocorrelation function is obtained from an ensemble average of the product of x_1 and x_2. The difference between a random variable x_1 and a random process $x(t_1)$ is that the former does not need any sequential order of randomness, but the latter does need a sequential order of randomness in a time domain. The autocorrelation function obtained from an ensemble average of two random variables x_1 and x_2 is expressed

$$R_x(t_1, t_1 + \tau) = E[x_1 x_2] = \int dx_1 \int x_1 x_2 P(x_1, x_2)\, dx_2 \qquad (1.5.28)$$

For a stationary process it becomes

$$R_x(\tau) = R_x(t_1, t_1 + \tau) \qquad (1.5.29)$$

The autocorrelation function obtained from a time average of the product of two random processes $x(t_1$ and $x(t_1 + \tau)$ can be expressed as

$$\tilde{R}_x(t_1, t_1 + \tau) = \lim_{T \to \infty} \frac{1}{2T} \int_{-T}^{T} x(t_1) x(t_1 + \tau)\, dt_1 = \langle x(t_1) x(t_1 + \tau) \rangle \qquad (1.5.30)$$

For a stationary process it becomes

$$\tilde{R}_x(\tau) = \tilde{R}_x(t_1, t_1 + \tau) \qquad (1.5.31)$$

For an ergodic process (applied to the mobile radio medium),

$$R_x(\tau) = \tilde{R}_x(\tau) \qquad (1.5.32)$$

Since the mobile radio signal is an ergodic process, the autocorrelation functions obtained from an ensemble average and from a time average have the same results. We can choose to process either one of two kinds of averages. From now on, the notation $R(\tau)$ will be used for both the time and the ensemble average of the product $x_1 x_2$. The average power of x_1 can be obtained from $R(\tau)$ by setting $\tau = 0$:

$$E[x^2] = \langle x^2 \rangle = R(0) \qquad (1.5.33)$$

We use the notation $\overline{x^2}$ to mean $\langle x^2 \rangle$ in most places in this book to simplify the notations unless a different use of $\overline{x^2}$ is mentioned and $R(0)$ is the maximum value of $R(\tau)$:

$$R(0) \geq R(\tau) \qquad (1.5.34)$$

Autocorrelation Coefficients. The autocorrelation coefficients are obtained

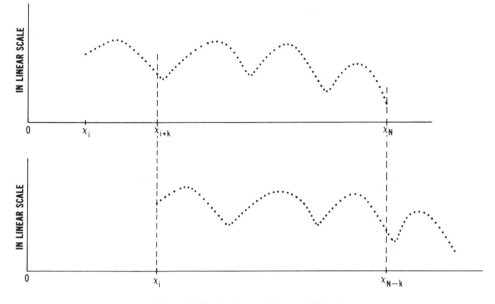

Figure 1.15. Autocorrelation coefficient.

from the autocorrelation functions. The following expressions are used based on the formats of the random process $x_1 = x(t_1)$, either in analog or in digital:

$$\rho_x(\tau) = \frac{R(\tau) - \langle x_1 \rangle^2}{R(0) - \langle x_1 \rangle^2} \qquad \text{(for an analog format)} \qquad (1.5.35)$$

$$\rho_x(k) = \frac{\left[\sum_{i=1}^{N-k} x_i x_{i+k} \bigg/ (N - k) \right] - \left[\left(\sum_{i=1}^{N} x_i \right)^2 \bigg/ N^2 \right]}{\sum_{i=1}^{N} (x_i^2/N) - \left[\left(\sum_{i=1}^{N} x_i \right)^2 \bigg/ N^2 \right]} \qquad \text{(for a digital format)}$$

$$(1.5.36)$$

An outline of how we obtained Eq. 1.5.36 is shown in Fig. 1.15. The range of $|\rho_x(\tau)|$ or $|\rho_x(k)|$ is equal to or less than 1. Usually $|\rho_x(\tau)| \rightarrow 0$ as $\tau \rightarrow \infty$ if a signal is a nonperiodic function, as in the case of a mobile fading signal being received. The time separation in processing $\rho_x(\tau)$ is equivalent to the space separation Δd in processing $\rho_x(\Delta d)$, since $\tau = \Delta d/V$, where V is the velocity of the vehicle. One application of Δd is for the antenna separation on the vehicle, and Δd can be determined based on a required value of $\rho_x(\Delta d)$.

The correlation coefficients shown in many places in this book are mostly processed among the fading signal envelopes. When $\rho_x(\Delta d)$ approaches zero, it means that two fading signal envelopes received by two antennas separated by a spacing Δd are uncorrelated. Under this condition the two fading signals combined could result in a considerably reduced fading signal.

Crosscorrelation Functions and Coefficients. Let two random variables x and y be

$$x_1 = x(t_1)$$

$$y_2 = y(t_2) = y(t_1 + \tau)$$

Then the crosscorrelation functions can be expressed as

$$R_{xy}(\tau) = E[x_1 y_2]$$

$$= \int_{-\infty}^{\infty} dx_1 \int_{-\infty}^{\infty} x_1 y_2 p(x_1, y_2)\, dy_2 \qquad \text{(ensemble average)} \qquad (1.5.37)$$

$$\tilde{R}_{xy}(\tau) = \lim_{T \to \infty} \frac{1}{2T} \int_{-T}^{T} x(t_1) y(t_1 + \tau)\, dt_1 \qquad \text{(time average)} \qquad (1.5.38)$$

Since the random signal fading received in a mobile radio environment is an ergodic process.

$$R_{xy}(\tau) = \tilde{R}_{xy}(\tau) \qquad (1.5.39)$$

and the following relationship always holds:

$$|R_{xy}(\tau)| \leq |R_x(0) R_y(0)|^{1/2} \qquad (1.5.40)$$

The crosscorrelation coefficient can be expressed as follows:

$$\rho_{xy}(\tau) = \frac{R_{xy}(\tau) - \langle x \rangle \langle y \rangle}{\sqrt{\langle x^2 \rangle - \langle x \rangle^2}\,\sqrt{\langle y^2 \rangle - \langle y \rangle^2}}$$

$$\text{(for an analog format)} \qquad (1.5.41)$$

$$\rho_{xy}(k) = \frac{\sum_{i=1}^{N-k} [x_i y_{i+k}/(N-k)] - \left(\sum_{i=1}^{N} (x_i/N)\right)\left(\sum_{i=1}^{N} (y_i/N)\right)}{\sqrt{\sum_{i=1}^{N} (x_i^2/N) - \left(\sum_{i=1}^{N} (x_i/N)\right)^2}\,\sqrt{\sum_{i=1}^{N} (y_i^2/N) - \left(\sum_{i=1}^{N} (y_i/N)\right)^2}}$$

$$\text{(for a digital format)} \qquad (1.5.42)$$

Equations (1.5.41) and (1.5.42) have expressions similar to Eqs. (1.5.35) and (1.5.36), respectively. The range of $|\rho_{xy}(\tau)|$ or $|\rho_{xy}(k)|$ is also equal to or less than unity. Two measured signals or data received from two different antennas or sources are always processed by using the crosscorrelation coefficient, $\rho_{xy}(0)$. If $\rho_{xy}(0)$ approaches zero, it means that two signals are dissimilar at $\tau = 0$.

Autocorrelation Estimates Using FFT Computations.[8] The double use of fast Fourier transformation (FFT) procedures can compute the correlation function more efficiently than direct procedures as described previously. The speed ratio between the direct procedure and FFT procedures can be obtained from the following calculation.

Using a straightforward calculation of Eq. (1.5.36) of a set of N samples for m lags requires approximately Nm real multiply-add operations. However, the FFT computation requires approximately $8\,Np$ real multiply-add operations where p is from $N = 2^p$. The speed ratio is

$$\text{Speed ratio} = \frac{\text{number of operations (direct)}}{\text{number of operations (FFT)}} = \frac{Nm}{8Np} = \frac{m}{8p} \qquad (1.5.43)$$

When N is large, m is always larger than 8 p. For example, $n = 2^{10}$ and $m = (0.2) \cdot N = 204$; the speed ratio is 2.5 ($= 204/80$). Thus the FFT computation is faster than the direct calculation by 2.5 times. When N and m become larger, the speed ratio increases.

Power Spectrum
The direct computation of power spectrum is obtained by using the Fourier transformation method:

$$S(f) = \int_{-\infty}^{\infty} R(\tau)e^{-j\omega\tau}\, d\tau \qquad (1.5.44)$$

From Fourier integral theory, the following is true:

$$R(\tau) = \int_{-\infty}^{\infty} S(f)e^{+j\omega\tau}\, d\tau \qquad (1.5.45)$$

and the dc power can be obtained from Eq. (1.5.44) as

$$S(0) = \int_{-\infty}^{\infty} R(\tau)\, d\tau \qquad (1.5.46)$$

The FFT computations are more efficient than direct calculation. For an estimate of the power spectrum of a single record $x(t)$ over a time period T, choose the sampling interval $h = \Delta t$, such that

$$h = \frac{1}{2f_s} \le \frac{1}{2f_d} \qquad (1.5.47)$$

where f_s is the sampling frequency and where f_d is the highest frequency anticipated in the data. The Fourier components from the sampled data x_n can be obtained from

$$x_k = \sum_{n=0}^{N-1} x_n \exp\left(-j\frac{2\pi kn}{N}\right) \qquad (1.5.48)$$

where

$$N = \frac{T}{h} \qquad (1.5.49)$$

The power spectrum estimate becomes

$$S(f_k) = \frac{2h}{N}|x_k|^2 \qquad (1.5.50)$$

The speed ratio between the direct computation and the FFT computation for the power spectrum is

$$\text{Speed ratio} = \frac{\text{number of computations (direct)}}{\text{number of computations (FFT)}} = \frac{Nm}{4Np} = \frac{m}{4p} \qquad (1.5.51)$$

Comparing Eqs. (1.5.43) and (1.5.51), it is apparent that the power spectrum is computed twice as fast when computer correlated with FFT.

1.5.6 Delay Spread, Coherence Bandwidth, Intersymbol Interference

The three parameters—delay spread, coherence bandwidth, and intersymbol interference—are interrelated as follows:

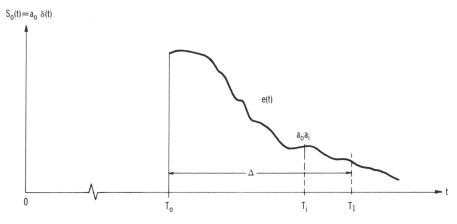

Figure 1.16. Delay spread.

Delay Spread

Due to the time-dispersive medium, the typical delay envelope $e(t)$ of an impulse response at the reception occurs as shown in Fig. 1.16. The mean delay time T_d and the delay spread Δ can be calculated as

$$T_d = \int_0^\infty t \cdot e(t) \, dt \tag{1.5.52}$$

and

$$\delta^2 = \int_0^\infty t^2 \cdot e(t) \, dt - T_d^2 \tag{1.5.53}$$

respectively, where $e(t)$ is the resultant impulse signal received from an impulse signal $s_0(t) = a_0 \cdot \delta(t)$.

$$e(t) = \left[a_0 \sum_{i=1}^{N} a_i \cdot \delta(t - T_i) \right] e^{-j\omega t} = E(t) e^{j\omega T} \tag{1.5.54}$$

where T is the time delay, a_i is the reflection coefficient of the ith paths, and $\delta(t)$ is the impulse function, as shown in Fig. 1.16. The following mean delay spread data are listed as an aid to the reader.

Type of Environment[9,10]	Delay Spread (Δ)
In-building	$<0.1 \ \mu s$
Open area	$<0.2 \ \mu s$
Suburban area	$0.5 \ \mu s$
Urban area	$3 \ \mu s$

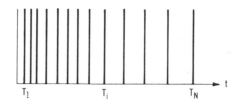

(A) Most arrive closely near T_1.

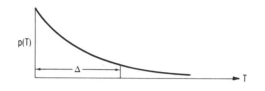

(B) Delay spread $p(Ti)$.

Figure 1.17. The distribution of delay spread.

The previous values are unchanged for any operating frequency above 30 MHz, because above 30 MHz the wavelengths are always much less than the sizes of human-made structures. All architectural structures can be treated as reflectors for any operating frequency above 30 MHz. When the number of reflectors is the same for those operating frequency above 30 MHz, then the same reflected-wave paths occur among those frequencies, the delay spread is the same. A delay-spread model can be expressed as

$$p(T_i) = \frac{1}{\Delta} \exp\left(-\frac{T_i}{\Delta}\right) \qquad (1.5.55)$$

where T_i is the time delay. This model is assumed by N equal amplitude reflected waves, most of which arrive closely earlier on. Very few arrive later, as shown in Fig. 1.17A. The distribution of delay spread $p(T_i)$ is shown in Fig. 1.17B. Another model uses the exponential amplitudes at equal time intervals. The two models are equivalent, but the model of Eq. (1.5.55) is easier to use for mathematical manipulation.

Coherence Bandwidth[11]
In a time-dispersed medium the fades of two received envelopes will coincide in time if the frequency separation ΔF is small enough. This means that the ΔF is within the coherence bandwidth. If a coherence bandwidth can be found, by choosing two frequencies that are far from the coherence bandwidth, we can cause the two received signals to fade independently. The coherence

bandwidth is derived from a correlation function $R(\Delta f)$ of two fading signal envelopes at two frequencies f_1 and f_2, respectively,

$$R(\Delta f) = \langle s(f_1) \cdot s(f_2) \rangle \qquad \Delta f = |f_1 - f_2| \qquad (1.5.56)$$

and $\rho(\Delta f)$ is a correlation coefficient after normalizing $R(\Delta f)$, using Eq. (1.5.35), and replacing τ with Δf. Let $\rho(\Delta f_1) \le 0.5$ as a criterion corresponding to a value of Δf_1, which we call the *coherence bandwidth* (B_c):

$$B_c = \Delta f_1 \qquad \text{for } \rho(\Delta f_1) = 0.5 \qquad (1.5.57)$$

We also can find a correlation coefficient from two random phases of two fading signals following the same steps as shown in Eq. (1.5.56). For two fading amplitudes to vary uncorrelatedly, the frequency separation should be greater than the coherence bandwidth B_c:[12]

$$\Delta f > B_c = \frac{1}{2\pi\Delta} \qquad (1.5.58)$$

The coherence bandwidth is different in suburban and urban areas. Because their time delay spreads (Δ) are different due to different human-made structures in the mobile radio environment. For two random phases to vary uncorrelatedly, the frequency separation should be greater than the different coherence bandwidth B_c':[13]

$$\Delta f > B_c' = \frac{1}{4\pi\Delta} \qquad (1.5.59)$$

Both Eqs. (1.5.58) and (1.5.59) involve delay spreads.

Intersymbol Interference

In a time-dispersive medium the transmission rate R_b in a digital transmission is limited by the delay-spread phenomenon. Since Δ is the mean delay spread, the transmission rate should be based on the maximum delay spread, which can be about 2Δ, if a low bit-error rate performance is required:

$$R_b < \frac{1}{2\Delta} \qquad (1.5.60)$$

In a real situation R_b would be determined from the required bit-error rate, which is based on the delay spread. (See Section 3.6.)

1.5.7 Confidence Interval

The range of an interval within which a degree of certainty (in percent) can be determined between a true value and an estimated value, such as a true

TABLE 1.3 Values of z_1 with Different Confidence Intervals

$z_1 = \dfrac{x_1 - m}{\sigma}$	$P(z_1) = P(-z_1 \leq z \leq z_1)$	Equivalent to the Number of σ Intervals
2.58	99%	2.58
2	95.46%	2
1.65	90%	1.65
1.28	80%	1.28
1	68%	1
0.5	38%	0.5

mean and a sample mean, is called a *confidence interval*. If the sample mean is in linear value on a normal distribution, the confidence interval is obtained; if the sample mean is in dB value, the confidence interval is obtained based on a log-normal distribution.

If x is log-normal distributed, then according to Eq. (1.5.18) its CPD can be expressed as

$$P(x \leq x_1) = \int_{-\infty}^{x_1} \frac{1}{\sqrt{2\pi}\sigma} \exp\left(-\frac{(x-m)^2}{2\sigma}\right) dx \qquad (1.5.61)$$

Applying a limit (z_1, z_2) to Eq. (1.5.18) yields

$$P(-z_1 \leq z \leq z_1) = \int_{-z_1}^{z_1} \frac{1}{\sqrt{2\pi}} \exp\left(-\frac{z^2}{2}\right) dz \qquad (1.5.62)$$

where

$$z_1 = \frac{x_1 - m}{\sigma} \qquad (1.5.63)$$

Table 1.3 lists the values of z_1 with different confidence intervals. Another way of expressing confidence intervals is as follows:

$$P(m - 2\sigma_1 \leq x \leq m + 2\sigma_1) = P(z_1) = 95.46\%$$

$$P(m - \sigma_1 \leq x \leq m + \sigma_1) = P(z_1) = 68\%$$

For example, if $m = 5$ dB and $\sigma = 1$ dB, then x lies between (4, 6) based upon 1σ interval with a 68% confidence; x lies between (3, 7) based upon 2σ interval with a 95.46% confidence.

1.5.8 False-Alarm Rate and Word-Error Rate

Signaling is used as a communication link between two parties. The signal format affects both the false-alarm rate and the word-error rate. The false-alarm rate is the occurrence rate of a false recognizable word that may cause a malfunction in a system. The reduction of the false-alarm rate is based on the signaling format. If two code words have a length of L bits and are different from each other by d bits, we say that the Hamming distance is d bits for a total length of L bits.

With a given L and d, the false-alarm rate can be expressed as

$$P_f = \text{false-alarm rate} = P_e^d(1 - P_e)^{L-d} \tag{1.5.64}$$

where P_e is the bit-error rate in a mobile radio environment. This is described in Section 3.6. In general, as d becomes larger, the false-alarm rate becomes lower. The signaling and address formats have to be designed based on both the required false-alarm rate and the fading medium, especially in a multipath fading medium in a mobile radio environment. The detailed false-alarm rate is described in Section 8.2.

The word-error rate is obtained by considering a word that may not be detected due to an error introduced by the medium. In a Gaussian noise environment, the word-error rate is directly related to the bit-error rate. The word-error rate P_w of an L bit-length word can be expressed as

$$P_w = 1 - (1 - P_e)^L \tag{1.5.65}$$

where L is the number of information bits, and P_e is the bit-error rate. Equation (1.5.65) is only valid if all the bits in a word are uncorrelated in fading errors. However, in a Rayleigh fading environment, Rayleigh fading causes additional burst errors. Since the average duration of fading is related to the vehicle speed, the word-error rate and the bit-error rate have no direct relation. When the vehicle speed is slow, the adjacent bits are correlated in fading errors. We describe the cases in Section 8.3. If a word of L inserts g bits in order to correct t bits in error, then the word-error rate of a new N bits $(L + g)$-length word can be expressed as

$$P_{cw} = 1 - \sum_{k=0}^{t} C_k^N P_e^k(1 - P_e)^{N-k} \tag{1.5.66}$$

where

$$C_k^N = \frac{N!}{(N - k)!k!} \tag{1.5.67}$$

The throughput of the coded word is

$$\text{Throughput} = \frac{L}{L + g} \tag{1.5.68}$$

The coded word-error rate P_{cw} is always lower than the word-error rate P_w. The disadvantage to having a coded scheme is slow throughput. In later chapters we will introduce techniques other than coding schemes for fading reduction.

References

1. Lee, W. C. Y., *Mobile Communications Engineering* (McGraw-Hill, 1982), ch. 6.
2. Lee, W. C. Y., and Y. S. Yeh, "On the Estimation of the Second-Order Statistics of Log-Normal Fading in Mobile Radio Environment," *IEEE Trans. Commun.* Com-22 (June 1974): 869–873.
3. Kennedy, R. S., *Fading Dispersive Communication Channels* (Wiley-Interscience, 1969), ch. 3.
4. Asami, Y., *Microwave Propagation in Snowy Districts* (Sapporo, Japan: The Research Institute of Applied Electricity, Hokkaido University, 1958): 73–107.
5. Davenport, W. B., and W. L. Root, *Random Signals and Noise* (McGraw Hill, 1958), 68.
6. Rice, S. O., "Properties of Sine Wave Plus Random Noise," *Bell Sys. Tech. J.* 27 (Jan. 1948): 109–157.
7. Lee, W. C. Y., "Statistical Analysis of the Level Crossings and Duration of Fades of the Signal from an Energy Density Mobile Radio Antenna," *Bell System Technical Journal* 46 (Feb. 1967): 417–448. This article is the first time that the level-crossing rates and average duration of fading of a mobile radio signal were introduced.
8. Bendat, J. S., and A. G. Piersol, *Random Data—Analysis and Measurement Procedures* (Wiley-Interscience, 1971), 312.
9. Cox, D. O., "Delay-Doppler Characteristics of Multipath Propagation at 910 MHz in a Suburban Mobile Radio Environment," *IEEE Trans. Antenna Propagation* 20 (Sept. 1972): 625–635.
10. Cox, D. O., and R. P. Leck, "Distribution of Multipath Delay Spread and Average Excess Delay for 910 MHz Urban Mobile Radio Path," *IEEE Trans. Antenna Propagation* 23 (March 1975): 206–213.
11. Lee, W. C. Y., *Mobile Communications Engineering*, 144.
12. Ibid., 198.
13. Ibid., 219.

PROBLEMS

1.1 The fading characteristics depicted in Fig. 1.2 show that as the vehicle speed increases, the fading changes more rapidly. What is the relationship between the fading frequency and the vehicle speed?

1.2 When a vehicle is standing still in a multipath environment, is the mobile receiver observing the signal fading?

1.3   When the vehicle speed is 96 km/hr (or 60 mph) and the operation frequency is 850 MHz, what is the fading frequency when the vehicle is traveling in a multipath environment?

1.4   What is the difference between radio path and mobile path?

1.5   Why can't the path-loss curve be generated directly from the measured data over the radio paths?

1.6   If one incoming wave is perpendicular to the vehicle motion, what is the fading frequency?

1.7   What are the fading frequencies when two incoming waves with two angles θ_1 and θ_2 shown in Fig. E1.2 are

$$\text{Case 1: } \theta_1 = \theta_2$$
$$\text{Case 2: } \theta_1 = -\theta_2$$
$$\text{Case 3: } \theta_1 = 0^\circ, \theta_2 = 90^\circ$$

1.8   Why does the reciprocity principle only apply to signal strength but not to the carrier-to-noise ratio (C/N)?

1.9   The standard deviation σ_y of a log-normal CPD can be obtained from Eq. (1.5.18) with assistance from the normal distribution table: $P(z \leq z = 1.29) = 90\%$. Verify that $\sigma_y = 3.87$ in Fig. 1.11.

1.10   Two frequencies separating more than a coherent bandwidth B_c possess different uncorrelated signal fading in suburban and urban areas. Which environment has provided a larger B_c?

2

PREDICTION OF PROPAGATION
LOSS

2.1 THE PHILOSOPHY BEHIND THE PREDICTION OF PROPAGATION LOSS

In a mobile radio environment the irregular configuration of the natural terrain, the various shapes of architectural structures, changes in weather, and changes in foliage conditions make the predicting of propagation loss very difficult. In addition the signal is received while the mobile unit is in motion. There is no easy analytic solution to this problem. Combining both statistics and electromagnetic theory helps to predict the propagation loss with greater accuracy.

2.2 OBTAINING MEANINGFUL PROPAGATION-LOSS DATA FROM MEASUREMENTS

As shown in Section 1.3, the local mean can be obtained by averaging a suitable spatial length L over a piece of raw data as shown in Fig. 2.1. The length L can be treated as an average window over a long piece of raw data. If the length L is too short, the short-term variation cannot be smoothed out and will affect the local mean. If the length L is too long, the averaged output cannot represent the local mean since it washes out the detailed signal changes due to terrain variation. Therefore it is essential that the suitable length L be determined.

2.2.1 Determining the Length L^1

Let the short-term fading r_0 be a Rayleigh fading shown in Eq. (1.5.10). Inserting it in Eq. (1.3.5) we obtain

$$\langle \hat{m}(x) \rangle = \sqrt{\frac{\pi}{2}} \sqrt{\frac{\overline{r^2}}{2}} \tag{2.2.1}$$

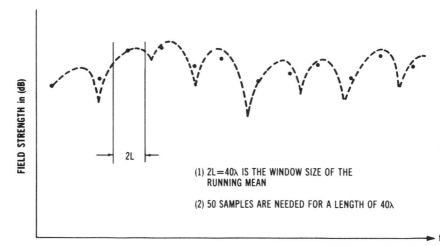

Figure 2.1. Obtaining the local mean.

TABLE 2.1 σ_m **Versus 2L**

2L	σ_m	$1\sigma_m$ Spread (dB)
5	0.165	3
10	0.122	2.1
20	0.09	1.56
40	0.06	1

Equations (2.2.1) and (1.5.12) are the same. They show that the true mean equals the mean of the sample mean; $m(x) = \langle \hat{m}(x) \rangle$.

$$\sigma_{\hat{m}}^2 = \langle \hat{m}^2(x) \rangle - \langle \hat{m}(x) \rangle^2 = \frac{1}{4L} \int_0^{2L} \left(1 - \frac{y}{2L} \right) J_0^2(\beta y) \, dy \qquad (2.2.2)$$

The 1 $\sigma \hat{m}$ spread is defined as

$$1\,\sigma_{\hat{m}} \text{ spread} = 10 \log \frac{1 + \sigma_{\hat{m}}}{1 - \sigma_{\hat{m}}} \text{ (dB)} \qquad (2.2.3)$$

The computed results of Eqs. (2.2.2) and (2.2.3) are given in Table 2.1.

As can be seen from Table 2.1, the length $2L = 40\lambda$ is desirable because its 1 $\sigma_{\hat{m}}$ spread for 40λ approaches 1 dB. The 40λ is considered to be the proper length to use in smoothing out the Raleigh fading. If the length $2L$ is shorter than 40λ, the average output would retain only a weaker portion of Rayleigh fading. If the length $2L$ is greater than 40λ, the excessive length of averaging would smooth out the local-mean information, which it is not supposed to do. Therefore $2L = 40\lambda$ is considered to be the appropriate length. However, in practice, L in the 20λ to 40λ range is acceptable.

2.2.2 Determining the Number of Sample Points Required[2] over 40λ

Since most data processing is done digitally, what is the proper number of samples required for a piece of analog data? Experimental autocorrelation has shown that a separation of 0.8λ is required for a correlation coefficient below 0.2 between two adjacent samples.[2] Then 50 weak-correlated samples would be needed to represent a length of 40λ in digital form. It must be determined whether 50 samples are enough to obtain an average value over a length of 40λ with great confidence. From the ensemble average $\bar{r}_j$ of a set of N variables, r_i along a piece of M-sample data is shown in Section 1.5.1 as

$$\bar{r}_j = \frac{\sum_{i=(j-1)N+1}^{jN} r_i}{N} \qquad 1 \leq j \leq \frac{M}{N} \qquad (2.2.4)$$

We define $\hat{m}$ and $\hat{\sigma}$ as the mean and standard deviation of $\bar{r}_j$, respectively. From Section 1.5.1, $\bar{r}_j$ is always a Gaussian variable if all N variables r_i are added in linear scales. Since r_i itself is a Rayleigh with both mean m and standard deviation σ_r expressed in linear values, it can be shown that

$$\hat{m} = \langle \bar{r}_j \rangle = m \qquad (2.2.5)$$

$$\hat{\sigma} = (\langle \bar{r}_j^2 \rangle - \langle \bar{r}_j \rangle^2)^{1/2} = \frac{\sigma_r}{\sqrt{N}} \qquad (2.2.6)$$

Applying the confidence interval of 90% in Eq. (1.5.62) yields

$$P\left(-1.65 \leq \frac{\bar{r}_j - \hat{m}}{\hat{\sigma}} \leq 1.65\right) = 90\% \qquad (2.2.7)$$

Equation (2.2.7) can be restated

$$P(\hat{m} - 1.65\hat{\sigma} \leq \bar{r}_j \leq \hat{m} + 1.65\hat{\sigma}) = 90\% \qquad (2.2.8)$$

Equation (2.2.7) shows the 90% confidence interval (CI) of $\bar{r}_j$ is within $\hat{m} \pm 1.65\hat{\sigma}$—$\bar{r}_j$ approaches $\hat{m}$ if $\hat{\sigma}$ becomes smaller.

Inserting Eqs. (2.2.5) and (2.2.6) into Eq. (2.2.8) yields

$$P\left(m_r - 1.65 \frac{\sigma_r}{\sqrt{N}} \le \bar{r}_j \le m_r + 1.65 \frac{\sigma_r}{\sqrt{N}}\right) = 90\%$$

or

$$P\left[\left(1 - \frac{1.65}{\sqrt{N}} \cdot \frac{\sigma_r}{m}\right)m \le \bar{r}_j \le \left(1 + \frac{1.65}{\sqrt{N}} \cdot \frac{\sigma_r}{m}\right)m\right] = 90\% \qquad (2.2.9)$$

Inserting the values of m and σ_r from Eq. (1.5.11) and Eq. (1.5.12), respectively, into Eq. (2.2.9) yields

$$P\left[\left(1 - \frac{1.65}{\sqrt{N}} \cdot \sqrt{\frac{4 - \pi}{\pi}}\right)m \le \bar{r}_j \le \left(1 + \frac{1.65}{\sqrt{N}} \cdot \sqrt{\frac{4 - \pi}{\pi}}\right)m\right] = 90\%$$

Simplifying above equation, we obtain the following expression:

$$P\left[\left(1 - \frac{0.8625}{\sqrt{N}}\right)m \le \bar{r}_j \le \left(1 + \frac{0.8625}{\sqrt{N}}\right)m\right] = 90\% \qquad (2.2.10)$$

The 90% confidence interval (CI) expressed in dB is

$$\mathrm{CI} = 20 \log\left\{\frac{[1 + (0.8625/\sqrt{N})]m}{m}\right\} = 20 \log\left(1 + \frac{0.8625}{\sqrt{N}}\right) \qquad (2.2.11)$$

Let $N = 50$; Eq. (2.2.11) becomes

$$90\% \ \mathrm{CI} = 1 \ \mathrm{dB} \qquad (2.2.12)$$

The estimated value of $\bar{r}_j$ with $N = 50$ and $2L = 40\lambda$ for a 90% confidence interval is within 1 dB of its true mean value. If N is reduced to 36, the 90% confidence interval increases to 1.17 dB of its mean.

Example 2.1. Find an estimated value for $\bar{r}_j$, averaging 50 samples in a 99% confidence interval. From the table listed in Eq. (1.5.63) or any mathematical table

$$\frac{\bar{r}_j - \hat{m}}{\hat{\sigma}} \le 2.58$$

Then the expression of Eq. (2.2.7) changes as

$$P\left(-2.58 \le \frac{\bar{r}_j - \hat{m}}{\hat{\sigma}} \le 2.58\right) = 99\% \qquad (\text{E2.1.1})$$

Following the same steps as shown in Eq. (2.2.8) through Eq. (2.2.11), we obtain the result as

$$99\% \text{ CI} = 1.5 \text{ dB} \qquad (E2.1.2)$$

Comparing Eqs. (2.2.12) and (E2.1.2), we find that the percentage of confidence interval increases as the interval at which $\bar{r}_j$ lies increases.

Perhaps using 36 or up to 50 samples in an interval of 40 wavelengths is an adequate averaging process for obtaining the local means. A simpler way of obtaining local means is to use a running mean with a 40λ window. For a low-frequency operation, we may have to take an interval of 20λ for obtaining local means. The reason is that the terrain contour can change at a distance greater than 20λ as the wavelength becomes longer.

2.2.3 Mobile Path and Radio Path

The local means are recorded while the mobile units are traveling along the road (the mobile path) on the x-axis. However, each local mean is based on a radio path x' between the base station and the mobile unit (the radio path) at a corresponding spot, as shown in Fig. 2.2. Since we are calculating a path-loss curve along a radio path x', and not along a mobile path, the local means obtained from averaging a recorded signal along a mobile path have to be converted from the mobile path on x-axis to the radio-path axis on x'-axis shown in Fig. 2.3. We may have to make many runs (on many different mobile paths) and plot them on the x'-axis. A curve fitting can be made on all the experimental data plotted along x'-axis. This is what is called the *path-loss curve* or the *path-loss prediction curve*.

From the experimental data we found that one standard deviation (1 $\sigma_{x'}$) of the data spread on any radio-path length is about 8 dB. This spread is due to the various terrain conditions from which the data are collected at the same radio-path length. The distribution of all the data points at any path length along the radio path follows the log-normal distribution[1] as shown in Fig. 2.3.

This is an area-to-area prediction method. Each curve is generated from many data produced by similar terrain configurations. Since it is a general prediction, 50% of the real measured value will be higher or lower than the value predicted from the curve. The spread in 1 $\sigma_{x'}$ means that the measured values can be spread 8 dB above or below the predicted value, or that the measured values are in the ± 1 $\sigma_{x'}$ uncertainty range. Therefore, if 68% of the measured values fall within a range of ± 8 dB, the predicted path-loss curve can be considered a good one.

Although both the local mean and the path-loss slope follow the log-normal distributions, it should be clearly understood that the spread of the local mean σ_x is dependent on the terrain configuration from which the data are taken. Nevertheless, the spread $\sigma_{x'}$ of the radio-path data on any path length is always 8 dB. It is the difference between σ_x and $\sigma_{x'}$.

The 8 dB spread of radio-path data remains constant. At the close distance

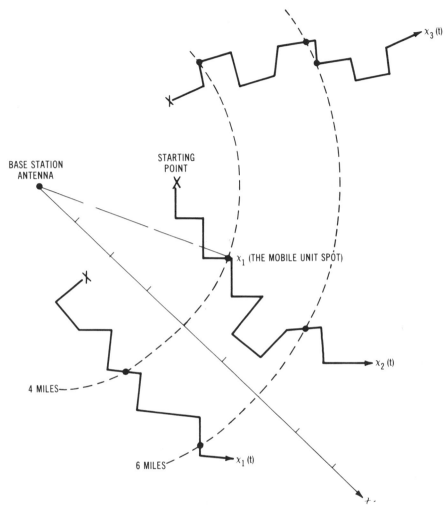

Figure 2.2. The mobile-unit running map.

it is caused by the variety of unique surroundings at individual base-station sites. The surrounding effect diminishes as the distance increases, but the effect of the terrain variation takes over.

2.3 PREDICTION OVER FLAT TERRAIN

2.3.1 Finding the Reflection Point on a Terrain

The wave reflection always conforms to Snell's law

$$N_1 \cos \theta_1 = N_2 \cos \theta_2 = \text{constant} \qquad (2.3.1)$$

where N_1 and N_2 are the refraction indices of two different media.

Since N_1 and N_2 are the same for a direct wave and a reflected wave in the mobile radio environment, the incident angle θ_1 and the reflected angle

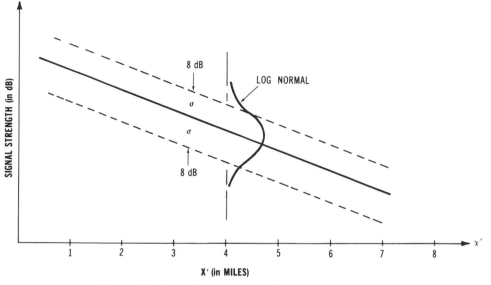

Figure 2.3. Generating a path-loss slope from local means.

θ_2 are identical, as shown in Eq. (2.3.1). The reflection point of a reflected wave on a flat terrain can be obtained two ways. Each approach requires the following steps:

1. *Take an image point from the base-station antenna.*
   a. Find an image point that is equal to the height of the base-station antenna on the image side (below the ground level).
   b. Connect the image point to the mobile-unit antenna.
   c. Obtain the reflection point where the connecting line intercepts the ground level.
2. *Take an image point from the mobile-unit antenna.* The same steps described in the first approach can be used in the second by finding an image point of the mobile-unit antenna below the ground level, then connecting the image point to the base-station antenna. The reflection point is where the connecting line intercepts the ground level.

Figure 2.4 shows how to obtain the reflection unit on flat terrain.

2.3.2 Classification of Terrain Roughness

The phase difference between the two rays shown in Fig. 2.5 can be expressed as

$$\Delta\psi = \beta \cdot (\Delta d) = \frac{2\pi}{\lambda} \cdot (2H \sin \theta) \qquad (2.3.2)$$

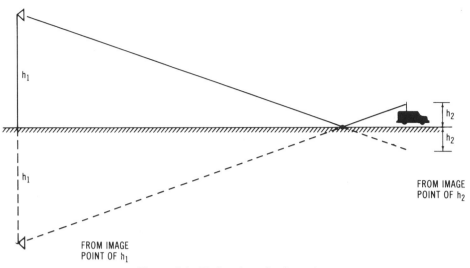

Figure 2.4. Finding the reflection point.

where Δd is the difference in radio-path lengths between two rays from wavefront A to wavefront B, β is the wave number ($= 2\pi/\lambda$), and H is the height of terrain irregularities.

The Rayleigh criterion for roughness is set at

$$\Delta\psi = \frac{\pi}{2} \tag{2.3.3}$$

and the Rayleigh height H_R and the minimum spacing S_R are found from Eq. (2.3.2):

$$H_R = \frac{\lambda}{8 \sin \theta} \, ; \, S_R = \frac{2H_R}{\tan \theta} = \frac{\lambda}{4 \sin \theta \tan \theta} \tag{2.3.4}$$

For a small incident angle θ,

$$H_R = \frac{\lambda}{8\theta} \, ; \, S_R = \frac{\lambda}{4\theta^2} \tag{2.3.5}$$

According to Rayleigh criteria, if the undulation surface height is greater than H_R under the constraint that the spacing between the two noticeable humps is greater than S_R, then it is a rough terrain.

In a mobile radio environment we would assume different criteria for different propagation path lengths. The reason is that the mobile antenna is usually close to the grounds so that the reception of both the direct waves and the specular reflected waves is weak. In this case, even if the phase difference $\Delta\psi$ between direct and specular reflected waves is about $\pi/4$ at a path distance larger than 9.7 km (6 miles), unexpected reflected waves are often received and will further weaken the resultant signal.

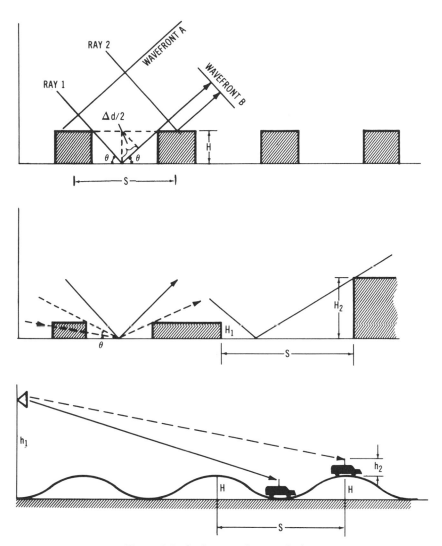

Figure 2.5. Surface roughness criterion.

Therefore the following criteria are suggested for different propagation path lengths:

$$\Delta\psi = \frac{\pi}{2} \text{ (less than 3.2 km or 2 miles)}, H_R = \frac{\lambda}{8\theta}; S_R = \frac{\lambda}{4\theta^2}$$

$$\Delta\psi = \frac{\pi}{3} \text{ (from 3.2 − 9.6 km or 2 to 6 miles)}, H_R = \frac{\lambda}{12\theta}; S_R = \frac{\lambda}{6\theta^2} \quad (2.3.6)$$

$$\Delta\psi = \frac{\pi}{4} \text{ (9.6 km or 6 miles and up)}, H_R = \frac{\lambda}{16\theta}; S_R = \frac{\lambda}{8\theta^2}$$

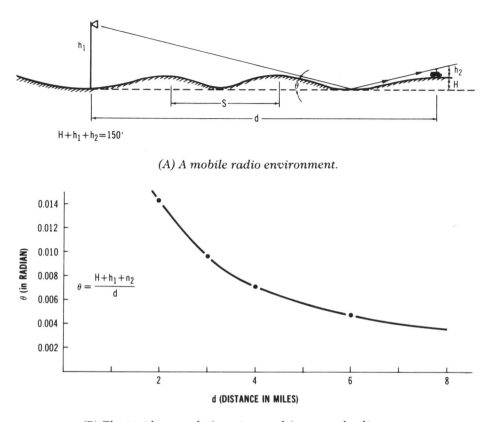

$H+h_1+h_2=150'$

(A) A mobile radio environment.

(B) The incident angle (grazing angle) versus the distance.

Figure 2.6. The terrain irregularities in a mobile radio environment.

The parameters H_R and S_R shown in Eq. (2.3.6) are functions of the incidence (grazing) angle θ. First the incidence angle has to be found from the following expression:

$$\theta = \frac{H + h_1 + h_2}{d}$$

where h_1 and h_2 are antenna heights at the base and the mobile unit, respectively, and H is the difference in elevation between the high and low spots around the mobile unit. All parameters are shown in Fig. 2.6A. The wave incident angle θ, based on a base-station antenna height h_1, a mobile antenna height h_2, and a terrain irregular height H, $[h_1 + h_2 + H = 45.7$ m (150 ft)] becomes a function of distance as shown in Fig. 2.6B. The angle θ decreases as the distance increases.

The roughness is determined by the frequency, incident angle, and the terrain irregularity height and spacing in the terrain expressed in Eq. (2.3.6).

Therefore at one frequency a surface is considered a rough surface but not at another frequency. This also applies to different incident angles.

In the mobile radio environment we use the following criterion to determine terrain roughness. Let the variation of elevation H be the difference of two adjacent extremes in elevations. Then, if

$$H > H_R$$

in the vicinity of $\Delta x = 1/2 \, S_R$ at the mobile site, then the terrain is rough. This may easily be applied as follows: The angle θ is obtained from the base-station antenna height h_1, the distance between the base and the mobile unit d, the terrain irregularity height H, and the mobile antenna height h_2. However, based on the location of the mobile unit, which is on the hump or in the valley, the incident angle θ is within

$$\frac{h_1 + h_2}{d} \leq \theta \leq \frac{H + h_1 + h_2}{d}$$

In calculating the terrain roughness, we always use the highest value of θ in Eq. (2.3.6).

Usually the base-station antenna height is greater than the mobile antenna height; the reflection point is then closer to the mobile unit. The distance $\Delta x = 0.5 S_R$ from the mobile unit toward the base only needs to be searched to see if any variation of elevation is greater than H_R. If one variation of elevation H within Δx is greater than H_R, the terrain is called a rough terrain. (See Fig. 2.6.) The methodologies of path-loss predictions used in a rough terrain and a smooth terrain are different and will be described in Section 2.5.5.

2.3.3 The Reflection Coefficient of the Ground Wave

Concerning ground reflections in a mobile radio environment, the incident angle (or reflected angle) is very small due to the fact that the base-station antenna heights and the mobile-unit antenna heights (in feet or meters) are relatively very short compared with the distance between two antennas (in miles or kilometers). In this case the wave reflection coefficient is always minus one regardless of the type of ground.

This is easy to verify from the following complex reflection coefficients:

$$a_h = R_h e^{-j\psi_h}$$

$$= \frac{\sin \theta_1 - (\varepsilon_c - \cos^2\theta_1)^{1/2}}{\sin \theta_1 + (\varepsilon_c - \cos^2\theta_1)^{1/2}} \qquad \text{(horizontal incidence)} \qquad (2.3.7)$$

$$a_v = R_v e^{-j\psi_v}$$

$$= \frac{\varepsilon_c \sin \theta_1 - (\varepsilon_c - \cos^2\theta_1)^{1/2}}{\varepsilon_c \sin \theta_1 + (\varepsilon_c - \cos^2\theta_1)^{1/2}} \qquad \text{(vertical incidence)} \qquad (2.3.8)$$

where a_h and a_v are complex reflection coefficients with amplitudes R_h and R_v and phases ψ_h and ψ_v. The relative permittivity dielectric constant of the medium is ε_c. Equations (2.3.7) and (2.3.8) are only used for nonferrous media where the permeability μ_r is close to unity so that the refraction index of Eq. (2.3.1) becomes

$$N = \sqrt{\mu_r \varepsilon_c} = \sqrt{\varepsilon_c} \qquad (2.3.9)$$

Since θ_1 is very small, α_h and a_v can be found from Eqs. (2.3.7) and (2.3.8) to be

$$a_v = a_h = -1 \qquad (2.3.10)$$

or

$$R_v = R_h = 1, \ \psi_v = \psi_h = 180° \qquad (2.3.11)$$

regardless of the values of ε_c of the medium. It can be visualized as a rock skipping on the water. When the incident angle is large the rock cannot be skipped. As the incident angle becomes smaller, a rock starts to skip. The smaller the incident angle, the larger the number of skips. It means more energy is reflected from the water surface.

2.3.4 Models for Predicting Propagation Path Loss

There are very few theoretical models, but many empirical models. A theoretical model will be introduced and two empirical models described in this book. A theoretical model is introduced because it is simple to explain and is effective in many ways. Two empirical models are used because they can clearly distinguish predicted path losses, not only in different terrain configurations but also in different built structures. Other models are listed in the reading list at the end of this chapter.

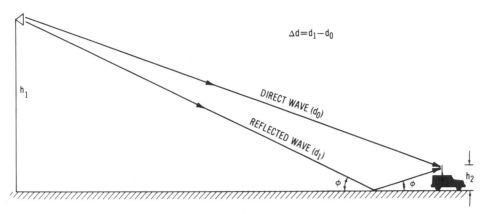

Figure 2.7. A simple theoretical model.

2.3.5 A Theoretical Model for Path Loss[3]

Be aware that the path-loss model is only valid for analyzing path-loss predicting and not for multipath fading. The model for predicting multipath (short-term) fading will be described later.

Assume that the characteristics of a rough earth surface are random in nature and that the radius of the curvature of surface irregularities is large in terms of the wavelength of the incident wave. Then the received signal can be represented by a scattered field E_s, which can be approximated by combining two waves: a direct wave and a reflected wave.[4]

$$E_s = (1 + a_v e^{j\Delta\psi})E \qquad (2.3.12)$$

The reflection coefficient is a_v, and $\Delta\psi$ is the phase difference between the direct wave and reflected wave. This phase difference can be expressed as

$$\Delta\psi = \beta \cdot \Delta d = \frac{2\pi}{\lambda} \cdot \Delta d \qquad (2.3.13)$$

Where β is the wave number ($\beta = 2\pi/\lambda$) and Δd is the difference between two radio paths as shown in Fig. 2.7; E of Eq. (2.3.12) is the direct wave received at the mobile antenna.

According to the free-space propagation path-loss formula, the received power from a direct wave is from Eq. (1.2.1):

$$P_{0r} = \frac{|E|^2}{2\eta_0} = P_t \left(\frac{1}{4\pi d/\lambda}\right)^2 \qquad (2.3.14)$$

where η_0 is the intrinsic impedance of free space.

The received power of the scattered field E_s is

$$P_r = \frac{|E_s|^2}{2\eta_0} \tag{2.3.15}$$

where E_s can be substituted from Eq. (2.3.12). Since in the mobile radio environment a_v of Eq. (2.3.12) is always approximately equal to -1 and $\Delta\psi$ is much less than one radian, Eq. (2.3.15) becomes

$$P_r = P_t\left(\frac{1}{4\pi d/\lambda}\right)^2 |1 - \cos\Delta\psi - j\sin\Delta\psi|^2$$

$$\approx P_t\left(\frac{1}{4\pi d/\lambda}\right)^2 (\Delta\psi)^2 \tag{2.3.16}$$

where

$$\Delta\psi = \beta(\Delta d) = \beta(\sqrt{(h_1 + h_2)^2 + d^2} - \sqrt{(h_1 - h_2)^2 + d^2}) \tag{2.3.17}$$

For $d \gg h_1 + h_2$, Eq. (2.3.17) can be approximated as

$$\Delta\psi \approx \beta\left(1 + \frac{(h_1 + h_2)^2}{2d^2} - 1 - \frac{(h_1 - h_2)^2}{2d^2}\right)d = \frac{2\pi}{\lambda} \cdot \frac{2h_1 h_2}{d}$$

$$= \frac{4\pi h_1 h_2}{\lambda d} \tag{2.3.18}$$

Substituting Eq. (2.3.18) into Eq. (2.3.16) yields

$$P_r = P_t = \left(\frac{h_1 h_2}{d^2}\right)^2 \tag{2.3.19}$$

Equation (2.3.19) is an imperfect formula since it does not involve wavelength. It indicates two correct facts and also shows two weak points.

Two Correct Facts

1. The equation shows a path loss of 40 dB/dec ($\sim d^{-4}$) or 12 dB/oct. This has been verified from the experimental data. Based on the previous rule, we may obtain an additional path loss from a distance d_1 to a distance d_2

$$\text{Path loss} = 40\log_{10}\left(\frac{d_2}{d_1}\right) \tag{2.3.20}$$

2. The equation shows a 60-dB/oct ($\sim h_1^2$) rule for an antenna height gain at the base station. Experiments have proved that (in flat terrain) doubling the antenna height at the base gains 6 dB. If an antenna height h_1 is increased (or decreased) other than twice, then

$$\text{Antenna height gain (loss)} = 20 \log_{10} \frac{h'}{h_1} \qquad (2.3.21)$$

where h' is the new antenna height.

Two Weak Points

1. The frequency or wavelength term is missing in Eq. (2.3.19). However, the measured data show that the empirical path-low formula is a function of frequency as

$$P_r \propto f^{-n} \qquad \text{where } 2 \leq n \leq 3$$

2. The equation shows a 6-dB/oct ($\sim h_2^2$) rule for an antenna height gain at the mobile unit. This is not true. For a mobile antenna height of 3 m (10 ft), cutting its height by one-half only results in a 3-dB loss of power from the experimental result.

2.3.6 An Area-to-Area Path-Loss Prediction Model

The prediction model for area-to-area path loss consists of two parts. First is the area-to-area path-loss prediction described in this section. The second part uses the area-to-area path-loss prediction as a base and develops a point-to-point prediction described in Section 2.3.7.

An area-to-area prediction is usually used to predict a path loss over a general flat terrain without knowing the particular terrain configuration over which the actual path loss is found. If the actual path loss is obtained from a hilly terrain area, then a great difference between the actual value and the value predicted from an area-to-area prediction curve will be expected. The area-to-area path-loss prediction requires two parameters:[5] (1) the power at the 1-mile point of interception P_{r0}, and (2) a path-loss slope γ.

The field strength of the received signal P_r can be expressed as

$$P_r = P_{r0}\left(\frac{r}{r_0}\right)^{-\gamma}\left(\frac{f}{f_0}\right)^{-n}\alpha_0 \qquad \text{(linear expression)}$$

$$= P_{r0} - \gamma \log\left(\frac{r}{r_0}\right) - n \log\left(\frac{f}{f_0}\right) + \alpha_0 \qquad \text{(dB expression)} \quad (2.3.22)$$

where r is in miles or kilometers and r_0 equals 1 mile or 1.6 km. γ can be expressed as a γth power in linear expression and a γ dB/dec in dB expression. The reason of taking a 1-mile point of interception is that within a 1-mile radius very few streets are available. Therefore we should avoid data with limited runs that do not provide the statistical mean. The adjustment factor is α_0. Equation (2.3.21) is a general formula that can be used for different frequency ranges as long as the frequency is above 30 MHz. The wavelength is usually smaller than the size of the built structure (in the frequencies above 30 MHz) so that the multipath reflection mechanism prevails.

For path-loss prediction we prefer that the received signal be predicted in an absolute power level (dBm) and be compared easily and directly with the experimental data. For this reason the following set of conditions is assumed:

Frequency f_0 = 900 MHz

Base-station antenna height = 30.48 m (100 ft)

Base-station power at the antenna = 10 watts

Base-station antenna gain = 6 dB above dipole gain

Mobile-unit antenna height = 3 m (10 ft)

Mobile-unit antenna gain = 0 dB above dipole gain

Then the adjustment factor is used for a different set of conditions

$$\alpha_1 = \left(\frac{\text{new base-station antenna height (m)}}{30.48 \text{ (m)}}\right)^2$$

$$= \left(\frac{\text{new base-station antenna height (ft)}}{100 \text{ (ft)}}\right)^2$$

$$\alpha_2 = \left(\frac{\text{new mobile-unit antenna height (m)}}{3}\right)^v$$

$$= \left(\frac{\text{new mobile-unit antenna height (ft)}}{10}\right)^v$$

$$\alpha_3 = \frac{\text{new transmitter power}}{10 \text{ watts}}$$

$$\alpha_4 = \frac{\text{new base-station antenna gain with respect to } \lambda/2 \text{ dipole}}{4}$$

α_5 = different antenna-gain correction factor at the mobile unit

There is a 2-dB signal gain received from an actual 4-dB gain antenna at the mobile unit in a suburban area, and less than 1-dB gain received from the same antenna in an urban area (see Chapter 3) for adjusting α_5. α_0 is expressed as

$$\alpha_0 = \alpha_1 \alpha_2 \alpha_3 \alpha_4 \alpha_5$$

where α_0 is a linear expression. Or α_0 can be expressed in dB

$$\alpha_0 = \sum_{i=1}^{5} \alpha_i \quad \text{dB} \tag{2.3.23}$$

where each α_i is converted into dB scale before summing. The parameters of γ and P_{ro} are found from the empirical data:

In free space

$$P_{ro} = 10^{-4.5} \text{ m watts}$$

$$\gamma = 2$$

$$P_{ro} = -45 \text{ dBm}$$

$$\gamma = 20 \text{ dB/dec}$$

In an open area[6]

$$P_{ro} = 10^{-4.9} \text{ m watts}$$

$$\gamma = 4.35$$

$$P_{ro} = -49 \text{ dBm}$$

$$\gamma = 43.5 \text{ dB/dec}$$

In suburban areas

Almost all suburban areas in the United States are alike. They consist of houses of ranch style, colonial style, etc., spread throughout a small town with two- to three-story buildings. Therefore the path-loss curve for suburban areas can be used in any suburb in the United States.[7]

$$P_{ro} = 10^{-6.17} \text{ m watts}$$

$$\gamma = 3.84$$

$$P_{ro} = -61.7 \text{ dBm}$$

$$\gamma = 38.4 \text{ dB/dec}$$

In Philadelphia (urban area)[8]

$$P_{ro} = 10^{-7} \text{ m watts}$$

$$\gamma = 3.68$$

$$P_{ro} = -70 \text{ dBm}$$

$$\gamma = 36.8 \text{ dB/dec}$$

(3/4 λ mobile antenna was used and has 1-dB gain in the urban area.)

In Newark (urban area)[9]

$$P_{ro} = 10^{-6.4} \text{ m watts}$$

$$\gamma = 4.31$$

$$P_{ro} = -64 \text{ dBm}$$

$$\gamma = 43.1 \text{ dB/dec}$$

In Tokyo, Japan (urban area)[10]

$$P_{ro} = 10^{-8.4}$$

$$\gamma = 3.05$$

$$P_{ro} = -84 \text{ dBm}$$

$$\gamma = 30.5 \text{ dB/dec}$$

Determine the Value of n in Eq. (2.3.22)

The value of n in Eq. (2.3.22) is found from empirical data. Okumura[10] indicates $n = 30$ dB/dec, and Young[11] indicates $n = 20$ dB/dec. Therefore

$$20 \text{ dB/dec} < n < 30 \text{ dB/dec} \tag{2.3.24}$$

when n is valid for the frequency range from 30 to 2000 MHz and the distance range from 2 to 30 km, or approximately 1.5 to 20 miles. The value n seems dependent on the geographical locations and the operating frequency ranges. In a suburban or open area with the operating frequency below 450 MHz, $n = 20$ dB/dec is recommended. In an urban area with the operating frequency above 450 MHz, $n = 30$ dB/dec is recommended.

Determine the Exponential Value of v in the Adjustment Factor α_2[10]

The value of v also can be found from the empirical data:

$$v \begin{cases} 2 & \text{for new mobile-unit antenna height} > 10 \text{ m (30 ft)} \\ 1 & \text{for new mobile-unit antenna height} < 3 \text{ m (10 ft)} \end{cases} \qquad (2.3.25)$$

A refined correction on the mobile unit antenna height appears in Eq. (2.3.29).

General Formula of the Model

$$P_r = -61.7 - 38.4 \log r - n \log\left(\frac{f}{900}\right) + \alpha_0 \qquad \text{dBm (suburban)}$$

$$= -70 - 36.8 \log r - n \log\left(\frac{f}{900}\right) + \alpha_0 \quad \text{dBm (Philadelphia)}$$

$$= -64 - 43.1 \log r - n \log\left(\frac{f}{900}\right) + \alpha_0 \quad \text{dBm (Newark)} \qquad (2.3.26)$$

$$\alpha_0 = 20 \log\left(\frac{h_1}{100'}\right) + 10 \log\left(\frac{P_t}{10\,w}\right) + (g_1 - 6) + g_2 + 10 \log\frac{h_2}{10'}$$

$$= 20 \log h_1 + 10 \log P_t + g_1 + g_2 + 10 \log h_2 - 64 \qquad (2.3.27)$$

where new values are P_t in watts, base antenna height h_1 and mobile antenna height h_2 in feet, base antenna gain g_1 and mobile antenna gain g_2 in dBd, r in miles and f in MHz. The three equations are shown in Fig. 2.8 with $\alpha_0 = 0$ dB, α_0 is a correction factor that consists of five subcorrection factors mentioned previously (α_0 expressed in Eq. (2.3.27) can be simply obtained by inserting new values of P_t, h_1, h_2, g_1 and g_2.). Equation (2.3.26) provides a median line. The standard deviation of predicted data would be 8 dB, either above or below the median line.

The 1-mile interception point P_{r_0} and the slope γ can be easily found by taking few field points at a 1-mile radius and at a 10-mile radius. Since we start our measurement at a distance of 1 mile from the base-station transmitter in an area with similar built structures, the different locations of the base-station antennas in that area will not affect the empirical path-loss curve. Therefore, if two respective median points are chosen from several measured data at two different distances (1 mile and 10 miles), and connected by a line between them, the 1-mile interception P_{r_0} and slope γ are obtained. If the terrain is different in the different directions from the base station, we may need a different 1-mile interception and different γ in each direction.

Drop Rate of Reception Level at the Boundary

We can try to figure out the reception level in dB drops as the normal-speed mobile unit moves out from its own coverage boundary, called a cell. Assume

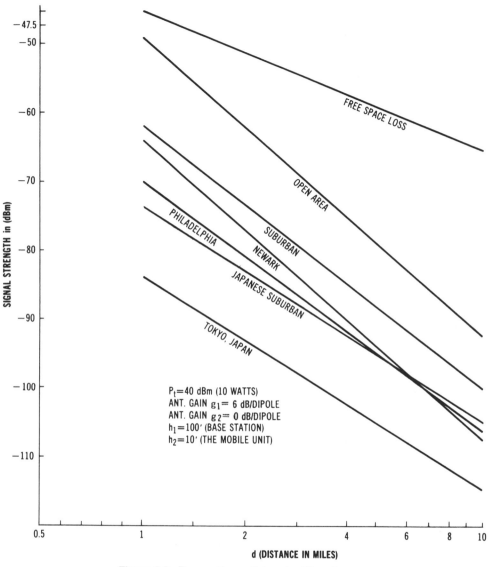

Figure 2.8. Propagation path loss in different areas.

that the unit moves at a speed of 15 mph in a 2-mile cell and a speed of 60 mph in an 8-mile cell. This is a reasonable assumption because a smaller cell indicates heavier traffic and slower speeds.

In this situation a mobile unit moves from a 2-mile cell boundary to 2.25 miles from the base station in a minute. The difference in reception levels from 2 miles to 2.25 miles (after a minute) can be found in Eq. (2.3.26).

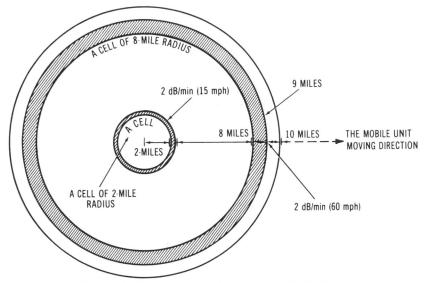

Figure 2.9. A rate of signal strength drop—2 dB/min rule.

Using the suburban prediction rate, the signal drop per minute is expressed as follows:

$$\Delta P = 38.4(\log 2 - \log 2.25) = 1.9 \sim 2 \text{ dB/min} \qquad (2.3.28a)$$

If in another situation a mobile unit moves from an 8-mile cell boundary to 9 miles from the base station in a minute (corresponding to a speed of 60 mph), the signal drop per minute is

$$\Delta P = -38.4(\log 8 - \log 9) = 1.9 \sim 2 \text{ dB/min} \qquad (2.3.28b)$$

Comparing Eqs. (2.3.28a) and (2.3.28b), we can quantitatively say that the field-strength level drops at a rate of 2 dB/min, as illustrated in Fig. 2.9. This can be a guideline for the action taken in the handoff process, which will be described in Chapter 6.

Comment on Selecting a Mobile Radio Path-Loss Model

A good model for predicting mobile radio propagation loss should be able to distinguish among open areas, suburban areas, and urban areas. The model described in this section has this feature. All urban areas, such as the hilly San Francisco area and the flat Chicago area, are unique in terrain, buildings, and street configurations. The model described in this section differentiate between urban areas. A good prediction model should follow the same guidelines, with no ambiguity, so that every user gets the same answer for given conditions.

2.3.7 The Model of Okumura et al.

Many models have been published in the past. We choose only one model in addition to the model described in the previous section. The model of Okumura and his colleagues can distinguish human-made structures and generate a complete set of empirical data.

The data for the Okumura model was collected in the Tokyo area. The Tokyo urban data were then used as a basic predictor for urban areas. The family of curves for a base-station antenna height of 200 m (656 ft) and a mobile-unit antenna height of 3 m (10 ft) are shown in Fig. 2.10. The correction factors for different antenna heights are as follows.

$$\text{Base-station antenna height correction} = 20 \log\left(\frac{h_1'}{200 \ m}\right)$$

$$h_1' > 10 \text{ m}$$

$$\text{Mobile-unit antenna height correction} = 10 \log\left(\frac{h_2'}{3 \ m}\right)$$

$$h_2' < 5 \text{ m} \tag{2.3.29}$$

$$\text{Mobile-unit antenna height correction} = 2h_2' \log\left(\frac{h_2'}{3 \ m}\right)$$

$$5 \text{ m} < h_2' < 10 \text{ m}$$

The correction factors for Japanese suburban areas do not match the U.S. suburban areas very well. The latter correspond better to the correction curve of the quasi-open area in Japan. This may be because houses in Japanese suburban areas are move densely packed than those in U.S. suburban areas. The correction factors for open areas, U.S. suburban areas, and Japanese suburban areas are shown in Fig. 2.11. The results from U.S. suburban areas on the whole agree with predictions from the new model in Section 2.3.6.

The data obtained from the Okumura model can be converted to the propagation–path-loss slopes; this is shown in Fig. 2.8. Each city has its own slope. Tokyo's slope is fairly flat, only 30 dB/dec, but the 1-mile intercept is 21.5 dB lower than that of U.S. suburban areas. This finding ensures that we need two parameters—a 1-mile intercept and the path-loss slope—to describe a general area as expressed by the previous model.

As the theoretical path-loss model described in Section 2.3.5 indicates, the path loss in a mobile radio environment is 40 dB/dec. The path-loss slopes in an open suburban area and in an urban area (Newark) are around 40 dB/dec. This is because all the base-station antennas stand clear from their sur-

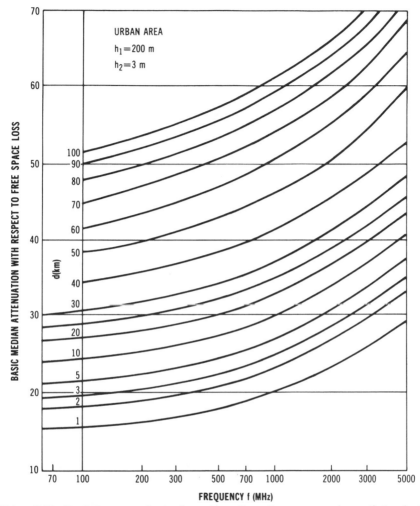

Figure 2.10. Prediction curve for basic median attenuation over quasi-smooth terrain in an urban area.

roundings so that the theoretical prediction of Eq. (2.3.20) is applied. In Philadelphia and Tokyo the base-station antennas almost certainly do not stand clear from their surroundings. The 1-mile intercepts are thus lower. Nevertheless, the path loss at 16 km (10 miles) would not be heavily affected by the surroundings at the base stations. As a result the path-loss slopes for Philadelphia and Tokyo are flatter.

2.3.8 A General Path-Loss Formula over Different Environments

More often the radio path between the base station and mobile unit may be propagated over more than one type of environment. The signal may be

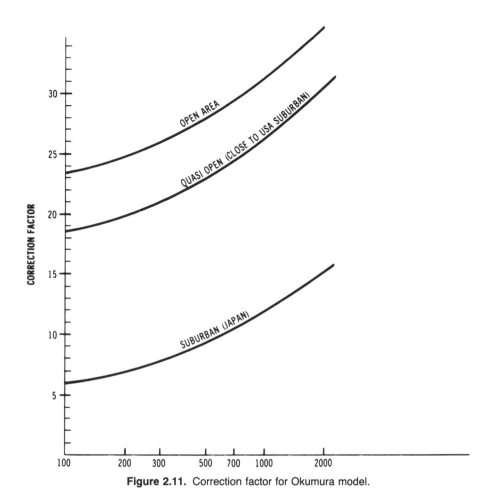

Figure 2.11. Correction factor for Okumura model.

transmitted over a suburban area where the base station sits, then be propagated into an urban area where the mobile unit travels. The slope γ in each
environment is different. The new model described in Section 2.3.6 can be
expanded to handle this situation.

The path-loss slope γ_1 is predicted in area A (suburban) and the path-loss
slope γ_2 is predicted in area B (urban). The predicted power level P_r received
in area B at distance r from the base station is

$$P_r = P_{r0}(r_1)^{-\gamma_1}\left(\frac{r}{r_1}\right)^{-\gamma_2}\alpha_0 \qquad r_1 \leq r \leq r_2 \qquad (2.3.30)$$

where r_1 is at the boundary of area A, r is within the boundary of area B,
and P_{r0} is the 1-mile interception point. Equation (2.3.30) is illustrated in Fig.
2.12. The reciprocity principle always holds. The same predicted reception

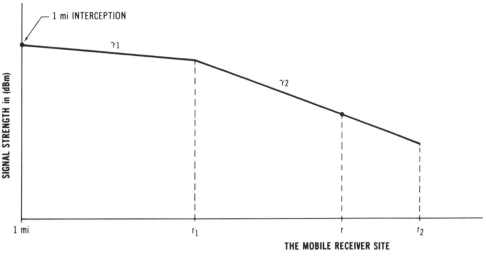

Figure 2.12. Propagation along two different environments.

level is found at the base station if the mobile unit is transmitting. However, if the base station and the mobile unit change locations between area A and area B, the result can be the same as long as both areas are flat. This statement is not true in the hilly area described in Section 2.5.

General Formula for Propagation through N Different Environments

The received signal level for a wave passing N different environments can be deduced from Eq. (2.3.30):

$$P_r = P_{r0}\alpha_0 \cdot (r_1)^{-\gamma_1}\left(\frac{r_2}{r_1}\right)^{-\gamma_2}\left(\frac{r_3}{r_2}\right)^{-\gamma_3}$$

$$\cdots \left(\frac{r}{r_{N-1}}\right)^{-\gamma_N} \quad r_{N-1} < r < r_N \quad (2.3.31)$$

The path loss $\mathscr{L}$ is defined as

$$\mathscr{L} = \frac{P_r}{P_t} = \frac{P_{r0}}{P_t}\,\alpha_0(r_1)^{-\gamma_1+\gamma_2}(r_2)^{-\gamma_2+\gamma_3}$$

$$\cdots (r_{N-1})^{-\gamma_{N-1}+\gamma_N} \cdot (r)^{-\gamma_N} \quad (2.3.32)$$

where P_t is the power delivered to the base station antenna and P_{r0} is the 1-mile intercept as shown in Eq. (2.3.31). The correction factor is α_0.

2.4   POINT-TO-POINT PREDICTION (PATH-LOSS PREDICTION OVER HILLY TERRAIN)[12,29]

In hilly terrain there are two situations: One is nonobstructed reception due to the flat terrain; the other is obstructed reception due to the hilly terrain.

2.4.1   Point-to-Point Prediction under Nonobstructive Conditions

Suppose that there are no obstacles between the base station and the mobile unit. The received signal consists of two types of waves, a direct wave and a reflected wave, when the actual path length is above the radio horizon. When the path length becomes longer in a mobile radio environment, the incidence angle becomes smaller; that is, the antenna heights at both ends (the base station and the mobile unit) are generally much shorter than the propagation path length. When the distance exceeds the radio horizon distance, there is an additional loss. For this reason the prediction is used here only for a propagation path length above the radio horizon. The maximum coverage of a base station is based on the radio horizon distance, which also can be called the *radius of coverage*

$$\text{Radius of coverage} \leq \sqrt{2h} \text{ miles}$$

$$\leq 2.9\sqrt{2h'} \text{ km} \qquad (2.4.1)$$

where h is the base station antenna height in feet, and h' is in meters. Beyond the radio horizon the signal attenuation will be greater than that of above the radio horizon. The signal attenuations are different at different frequencies as

$$22 \text{ dB/dec at } 43 \text{ MHz}$$

$$66 \text{ dB/dec at } 430 \text{ MHz}$$

$$79 \text{ dB/dec at } 850 \text{ MHz}$$

$$160 \text{ dB/dec at } 4.3 \text{ GHz}$$

Reflection Points on Hilly Terrain

Since the angle incidence on the ground for causing a reflected wave at the mobile unit is usually small, an approximate method of finding a reflection point on hilly terrain is shown in Fig. 2.13, which follows the same steps as shown in Fig. 2.4. Two types of transmission links are shown in Fig. 2.14 based on the different locations of reflection points. Type A has the base station situated on the flat ground and the mobile unit moving on a hilly slope. Type B has the base station situated on the top of the hill and the mobile unit traveling on flat ground.

For each of these two types, there are two kinds of reflected waves. One is called the *specular reflected wave;* the others are called *diffused reflected*

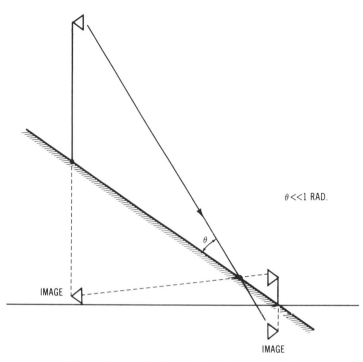

$\theta \ll 1$ RAD.

IMAGE

IMAGE

Figure 2.13. Reflection point on sloped ground.

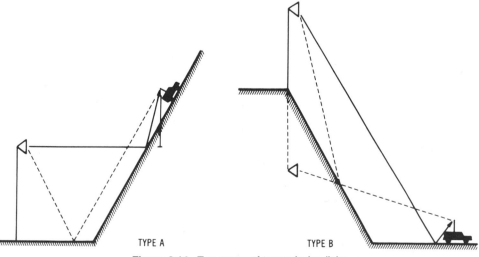

TYPE A

TYPE B

Figure 2.14. Two types of transmission links.

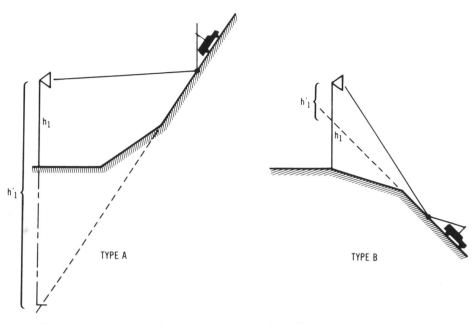

(A) Effective antenna height is greater   *(B) Effective antenna height is less*
than actual height.   *than actual height.*

Figure 2.15. Three-sloped ground planes.

waves. The specular reflected wave retains the major portion of reflected energy. It can be easily identified because its reflection point is always closer to the mobile unit than the other reflection points. Then the received signal becomes

$$P_r = P_t \left(\frac{h_1' h_2}{r^2} \right)^2 \tag{2.4.2}$$

where the antenna height of the base station is h_1', which is the effective antenna height. The measurement of an effective antenna height is as follows: Extend the sloped ground plane on which the specular reflection point lies to the base station site, and measure the antenna height from the antenna to the extension of the sloped ground plane, as shown in Fig. 2.15. The effective antenna height is greater than the actual antenna height, as case *A* in Fig. 2.15 demonstrates. The effective antenna height is less than the actual antenna height in case *B* of Fig. 2.15. The sloped ground plane actually always has a very small elevation angle. To show the reflection phenomenon, we enlarge the vertical (*y*-)axis so that the scales in the *y*-axis and in the *x*-axis are different. One is in feet (or meters); the other is in miles (or kilometers). The base-station antenna height (either actual height or effective height) is always measured upright (on the y-axis) and not perpendicular to the sloped ground plane (which in reality would have a very small tilt angle if the same scale applied to both x- and y-axes). The antenna height of the mobile unit h_2 is always its actual height. There is no effective height for the mobile antenna.

Two Special Considerations in Hilly Terrain

Scale Conversion. Plotting the propagation path on paper with the two different scales used in the x- and y-axes requires a certain amount of caution. Assume that a 100 ft : 1 mile scale is used for the two axes shown in Fig. 2.16. It means that the same length of 100 ft in the y-axis equals 1 mile in the x-axis. Then if the actual incidence angle is 0.5°, the angle plotted on the paper is 26.4° (= 0.5° × 5280/100), as is shown in the figure. This angle θ in line A is used for plotting the propagation path more easily.

Example 2.2. Assume that the base-station antenna height is 200 ft and the mobile-unit antenna height is 10 ft. The incidence angle is 0.5°. Find the location of the mobile unit and the reflection point on the graph (see Fig. 2.16). convert the angle of 0.5° to 26.4° in Fig. 2.16 and draw a line (line A) with the new angle. Use 200 ft as the base-station antenna height on the y-axis, and move horizontally to intercept a point on line A. The location of the reflection point is the same as the point Q on the x-axis. Also from point Q follow line A up 10 ft to point M. The location of the mobile unit is the same as M on the x-axis.

Use of Effective Base-Station Antenna Height. Find the effective base-station antenna height, and the distance r_1 between the specular reflection point and the mobile unit for two types.[12]

Since the scales for the x-axis and the y-axis are usually not the same when expressing hilly terrain, measuring a distance along the slope ground based on the x-axis scale is improper. If we use the same scales for both x-axis and y-axis, then the error between the actual distance measured along the sloped ground and the distance measured along the x-axis is negligible. Therefore, the distance between the base station and the mobile unit is measured on the x-axis (not along the slope ground)—a much simpler maneuver. Let R be the distance between the base-station site and the foot of the hill, r the distance between the base station and the mobile unit, and r_1 be the distance between the reflection point and the mobile unit. These parameters are shown in Fig. 2.17.

TYPE A (See Fig. 2.17)

$$h_{e_1} = h_1 + \frac{HR}{r - R} = h_1 + h_{11} \tag{2.4.3}$$

$$r_1 = \frac{h_2 r}{h_{e_1} + h_2} \tag{2.4.4}$$

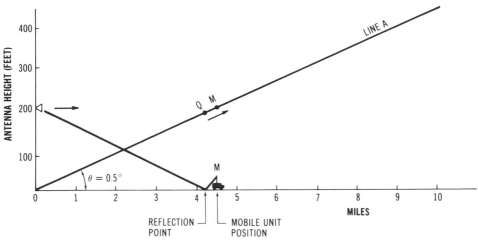

Figure 2.16. Conversion of scales, angles, and reflection points.

TYPE B (See Fig. 2.15(B))

$$h_{e_1} = h_1 + H \tag{2.4.5}$$

$$r_1 = \frac{h_2 r}{h_{e_1} + h_2} \tag{2.4.6}$$

From Eqs. (2.4.4) and (2.4.6), we can show that the location of the reflection point as well as the effective antenna height at the base station remain the same as shown in Fig. 2.17 provided that both the distance r and the mobile antenna height h_2 are the same.

TYPE A

$$h_{e_1} = h_1 + \frac{HR}{r - R} = A \quad \text{(a constant)} \tag{2.4.7}$$

TYPE B

$$h_{e_1} = h_1 + H = A \quad \text{(a constant)} \tag{2.4.8}$$

In the type A terrain situation, let R be fixed in Eq. (2.4.7); then let H and h_1 vary between cases 1 and 2 as shown in Fig. 2.17. Let h_1 be fixed in Eq. (2.4.8); then let H and R vary between case 3 and case 4. Let all R, H, and h_1 vary between cases 1 and 3. All four cases have the same values of r_1 and h_{e_1} mentioned previously.

In a type B terrain situation, let $h_1 = A - H$ in Eq. (2.4.8). As long as

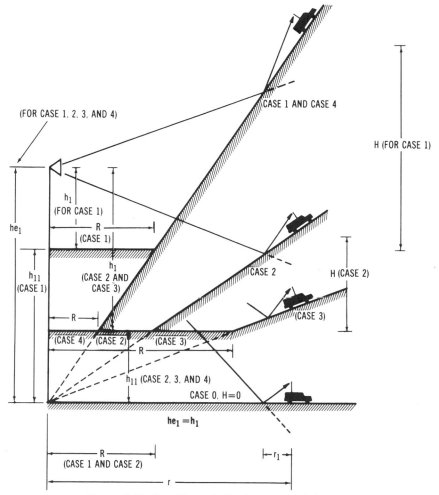

Figure 2.17. Conditions of effective antenna heights.

$H < A$, H can have any value, and h_1 is determined accordingly; the value of r_1 remains the same.

Reflection Points on Rolling Hills

In nature the ground does not form a straight line; a rolling hill is the more usual configuration. Determining the reflection points on a rolling hill can be done by applying the imaging method at the base station and at the mobile unit. Along rolling, hilly ground two reflection points obtained from two different image sites will not coincide. (See Fig. 2.18.) The reflection point that is closer to the mobile unit is the effective reflection point. The logic

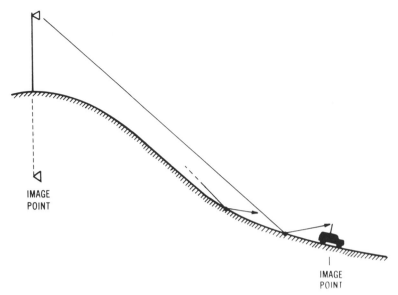

Figure 2.18. The reflection point on a rolling hill.

involved in this choice is the same as that used to choose the specular diffused reflection point as the effective point.

Application on Hilly Terrain

The prediction introduced here is a point-to-point prediction as described in Lee's paper.[12] It can predict a signal reception value much more accurately. Applying this prediction, we could reduce the range of 1-σ spread of 8 dB, derived from an area-to-area model (see Section 2.3.6), to 3 or 4 dB. In Lee's paper the 1-σ spread of his prediction in certain areas is within 1 dB. The reasoning behind the use of the point-to-point prediction is as follows:

1. The 8-dB spread obtained from an area-to-area model applied over flat terrain can be increased if the model is applied to rolling, hilly terrain. Signal reception varies in hilly terrain, resulting in a wide range of 1-σ spread obtained from the path-loss curve in that area. Then the value read off from the path-loss curve becomes meaningless because the predicted value and the measured value can be more than ± 8 dB, which is 16 dB apart.

2. Suppose that a drastic change in signal variation occurs in this area. Then before using many measured data points to generate a path-loss slope, we can use the effective antenna height-gain formula to correct the raw data as they were received from a smooth flat ground. The range of spread after correction therefore is drastically reduced. This is the way we use to find the mean path-loss slope and make an area-

to-area prediction first, and then use the antenna height-gain formula to adjust the path-loss value corresponding to that particular location. The point-to-point prediction is then obtained.

To illustrate the path-loss prediction in hilly terrain, the path-loss slope should first be obtained (as shown in Fig. 2.19B) from a particular terrain contour (as shown in Fig. 2-19A). (Assume it is in a suburban area.) In this particular run we plotted the mobile path in Fig. 2.19A. For simplicity assume that the radio path and the mobile path are the same in this illustration. Thus the elevation plot along the radio path would be as shown in Fig. 2.19A. The corrections—gains or losses due to the effective antenna height-gain formula—are listed in Table 2.2. The point-to-point predictions are plotted in Fig. 2.19B.

Effects of Terrain Roughness and Built Structures

We have to test the smoothness and roughness following the criterion shown in Section 2.3.2 (see Fig. 2.5): In a relatively flat or smooth ground surface area, the path loss can be obtained from the area-to-area prediction model presented in Section 2.3.6. In a rough surface the spacing between two humps S_R should be more elevated than the Rayleigh height H_R. In this area the path loss should be obtained from the point-to-point prediction.

In a human-made environment the predicted statistical values should be only applied in a randomly structured environment. If a high-rise building is located between a mobile unit and a base station, and no other scatterers are there, then no signal can be received by the mobile unit. This situation does not fit into our statistical model. For example, in New York City, depending on the location of the base-station antenna and the area in which the signal-strength data are collected by the mobile unit, the standard deviation or 1-σ spread could be 14 dB.

2.4.2   Point-to-Point Prediction under Obstructive Conditions—Shadow Loss

Shadow loss, predicted by the Fresnel-Kirchoff diffraction theory which was published in the midnineteenth century, is applied to optical- or ray-tracing techniques. In a mobile radio environment the hills, mountains, or any other obstructive objects are much larger than the wavelength that the knife-edge diffraction solution applies. Under these conditions no effective antenna height gain will be involved. Four parameters are required:

r_1 = distance from the knife edge to the base station
r_2 = distance from the knife edge to the mobile unit
h_p = height of the knife edge
λ = wavelength

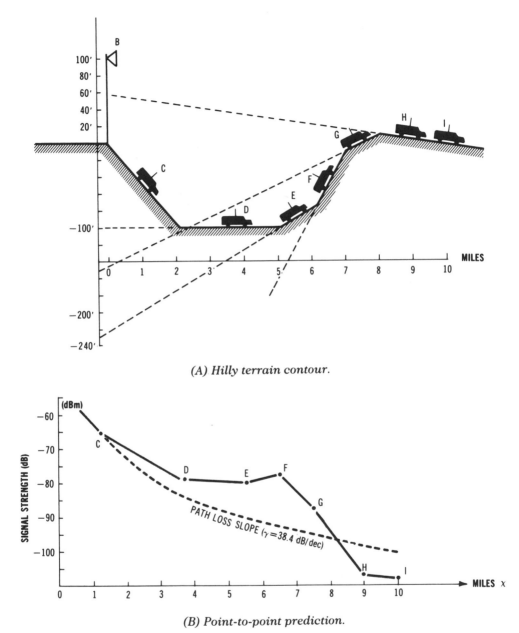

(A) Hilly terrain contour.

(B) Point-to-point prediction.

Figure 2.19. Illustration of the terrain effect on the effective antenna gain at each position.

TABLE 2.2 Gains or Losses Due to Effective Antenna Height Gain

Locations	B	C	D	E	F	G	H	I
Effective antenna h_e (ft)	100	100	200	330	620	250	40	40
Gain (loss) (20 log h_e/h_1) dB	0	0	6	10.37	15.85	8	(-7.6)	(-7.6)
Distance (miles)	0	1.15	3.75	5.5	6.5	7.5	9	10
Area-to-area path loss (dBm) (suburban area)	46	-63	-84	-90	-93	-95	-98	-100
Point-to-point path loss (dBm)	46	-63	-78	-79.63	-77.15	-87	-105.6	-107.6

A new parameter ν is used for plotting the shadow loss as[3]

$$\nu = -h_p\sqrt{\frac{2}{\lambda}\left(\frac{1}{r_1} + \frac{1}{r_2}\right)} \qquad (2.4.9)$$

which is shown in Fig. 2.20. When $h_p = 0$, $\nu = 0$, it indicates the condition of a 6-dB loss. The physical picture isshown in Fig. 2.21A with $h_p = 0$. Also two special cases are indicated:

1. If $r_1 + r_2 > r_1' + r_2'$, $h_p = h_p'$ and $\lambda = \lambda'$ (as shown in Fig. 2.21B), then the shadow loss of a long-distance propagation is less than that of a short-distance propagation. This is because, from Eq. (2.4.9),

$$\nu' > \nu, \qquad \mathcal{L}_L' > \mathcal{L}_L$$

   As can be observed from the equations above, the larger the diffraction angle, the greater the shadow loss.

2. If $r_1 \gg r_2$, then the shadow loss will be independent of r_1, as can be seen in Eq. (2.4.9)

$$\nu = -h_p\sqrt{\frac{2}{\lambda r_2}}, \qquad \mathcal{L}_L = f(h_p \cdot \lambda r_2)$$

The path-loss prediction from a double knife-edge diffraction case can be found in reference 13.

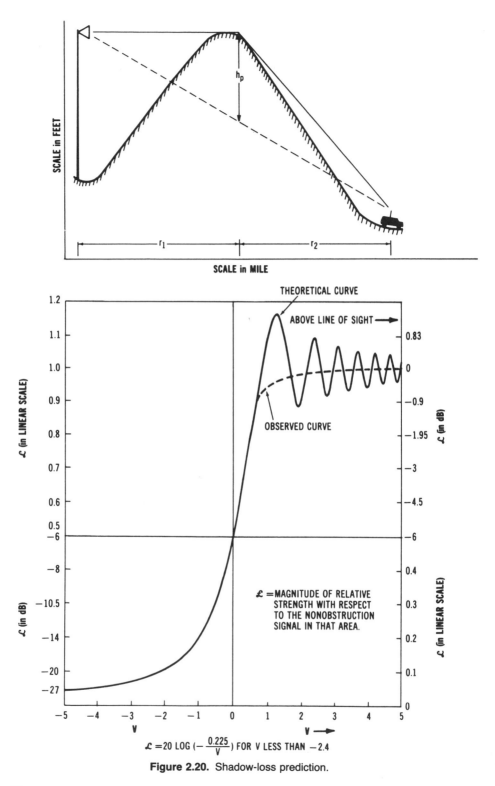

Figure 2.20. Shadow-loss prediction.

$$\mathcal{L} = 20 \text{ LOG} \left(-\frac{0.225}{V}\right) \text{ FOR V LESS THAN } -2.4$$

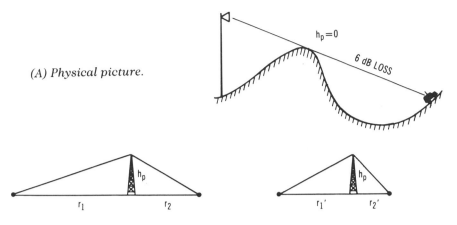

(A) Physical picture.

(B) Shadow loss comparison.
Figure 2.21. Shadow effect.

Example 2.3. Find the path loss at point A as the mobile unit travels on the island shown in Fig. E.2.3 with an operating frequency of 850 MHz.

The terrain contour is plotted in Fig. E.2.3. The three parameters r_1, r_2 and h_p are computed:

$r_1 = 5.28$ km (3.3 miles)
$r_2 = 2.88$ km (1.8 miles)
$h_p = 19.5$ m (65 ft)

The operating frequency f_c is given as $f_c = 850$ MHz (i.e., $\lambda = 1.17$ ft). Then the new parameter ν,

$$\nu = -h_p\sqrt{\frac{2}{\lambda}\left(\frac{1}{r_1} + \frac{1}{r_2}\right)} = -1.083.$$

The diffraction loss can be read from Fig. 2.20:

$$\mathscr{L}_L = 14 \text{ dB}$$

2.5 OTHER FACTORS

There are many other factors that can change the signal reception level. The foliage effect, the street channel effect, and the tunnel effect are the most important among these.

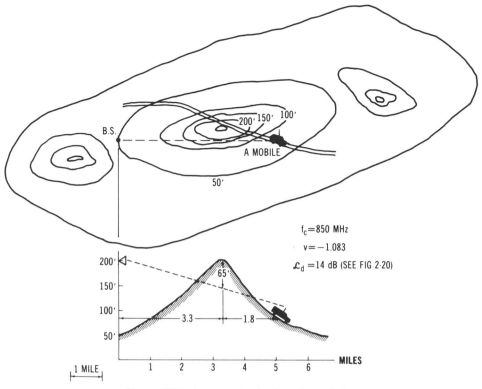

Figure E2.3. An example of calculating path loss.

2.5.1   Foliage Effects

There are two considerations when dealing with the foliage effect. In the Northern Hemisphere the leaves fall during the winter and grow back during the summer. The leaves of the oak, maple hickory, or similar trees often cause an additional signal loss in the operating frequencies above 400 MHz. In the summer this additional loss depends on the type of leaves and the operating frequency. A study could be made that considers such factors as the diversity of leaves, trunks, and branches; the height of the trees; and the thickness of the foliage. Foliage loss itself could be a major research task. Here we only use the information for design purposes. In the winter the leaves fall; the signal received is stronger than in the summer time.

Near the equator the leaves never fall, and the shapes of the leaves, such as palm tree leaves, have a different leaf structure compared to those of the northern states. The signal attenuation due to palm tree leaves is different from that due to northern leaves.

In a tropical rain forest area[14,15,16,17,18] experiments were carried out at frequencies of 50 to 800 MHz, and at distances of 40 m to 4 km (131 ft to

2.5 miles). All the receiving and transmitting terminals were surrounded by jungle. The results are as follows:

1. The loss increases rather linearly in log scale, while increasing in distance. Loss rules of 40 dB/dec at 800 MHz and 35 dB/dec at 50 MHz can be found. Therefore, if there is foliage loss and the mobile radio loss occurs at a frequency of 800 MHz, then a path-loss rule of more than 4 dB/dec should be expected. The actual path-loss slope, however, varies in a real situation.

2. The loss increases exponentially with frequencies in a log scale. At a distance of 4 km (2.5 miles), the difference in loss between 80 and 800 MHz is 20 dB for vertical polarization and 35 dB for horizontal polarization. The foliage loss with respect to the frequency of fourth power ($\sim f^4$) found by Tamir's theoretical prediction agrees with the loss of horizontal polarization.

3. The difference in loss due to two polarizations is 8 to 25 dB at 50 MHz and 1 to 2 dB at 800 MHz. Horizontal polarization has less loss than the vertical polarization in general.

4. The rate of foliage attenuation over the frequency range of 50 to 800 MHz is:

$$0.005 \text{ dB/m} - 0.3 \text{ dB/m} \qquad \text{(for horizontal polarization)}$$

$$0.005 \text{ dB/m} - 0.51 \text{ dB/m} \qquad \text{(for vertical polarization)}$$

5. A delay spread of 0.2 μs is observed[19,20] from the following setup: The receiving antenna is 8 m (26 ft) above the tree tops, the transmitting antenna is located in the trees, and the propagation distance is 160 m (525 ft).

From a design point of view, at least in the Northern Hemisphere, if there is a heavy forest in the area, then a 10-dB allowance[21,22,23] in the signal reception level should be added to the value obtained from the propagation–path-loss model to compensate for the summer foliage loss. However, dense foliage areas are rarely found in suburban and urban areas, so no foliage loss need be considered in these areas. In a jungle area, though, a special study should be made of the foliage situation following the path-loss rules stated previously. In the southern states the foliage attenuation on the signal reception should be estimated by the rules stated previously.

2.5.2 Street Orientation Channel Effect

When the mobile unit is closer to the base station, within 1.6 or 3.2 km (1 or 2 miles), for instance, the receiving signal is greatly affected by the structures and buildings around the base station and the base-station antenna

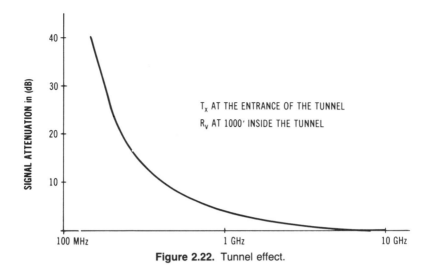

Figure 2.22. Tunnel effect.

height at the site. In general, the difference between two signal strengths, one signal received from a street in line with the base station and the other from a street perpendicular to the base station, is about 10 dB.[24] This phenomenon diminishes at a distance of 8 km (5 miles) or more. When a test is run, therefore, it is best to either avoid in-line and perpendicular streets or sample equal numbers of these two types of streets within a 3.2-km (2-mile) radius. This is to create an unbiased average path-loss slope for use in design.

2.5.3 The Tunnel and Underpass Effects

Tunnel Effect
The signal is attenuated by tunnel size,[25,26] in one experiment approximately 4.8 m (16 ft) high and 6 m (20 ft) wide. A transmitter is situated at the entrance of the tunnel; the receiver is 305 m (1000 ft) inside the tunnel. As shown in the graph in Fig. 2.22, at 1 GHz, a 4-dB loss will be observed 305 m (1000 ft) inside the tunnel.

Underpass Effect
When the mobile unit drives through an underpass 6 to 15 m (20 to 50 ft) long, the signal drops 5 to 10 dB. The period of signal attenuation depends on the speed of the vehicle. At a speed of 24 km/h (15 mph) (22 ft/s), it takes two seconds to pass the underpass. Normally the voice channel will not be affected. When the traffic is heavy and many vehicles are stopped under the underpass, if those vehicles are making mobile telephone calls, the signals can be lost.

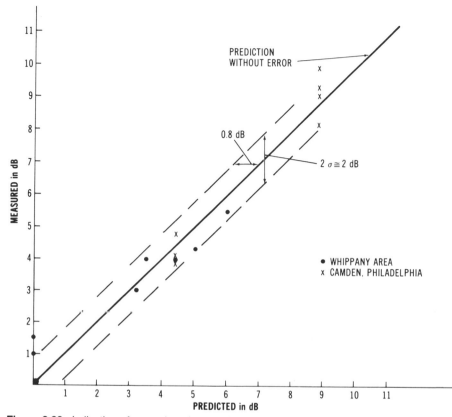

Figure 2.23. Indication of errors in point-to-point predictions under nonobstruction conditions.

2.6 THE MERIT OF POINT-TO-POINT PREDICTION

The area-to-area model usually only provides an accuracy of prediction within a standard deviation of 8 dB. This means that 68% of the actual path-loss data is within ±8 dB of the predicted value. This uncertainty range is too large. Point-to-point prediction reduces the uncertainty range by applying the detailed terrain contour information to the path-loss predictions.

Comparisons between the predicted values and the measured ones, using point-to-point prediction, were made in many areas.[27] Differences were compared in Whippany, New Jersey, and the Camden-Philadelphia area. First, points were plotted with predicted values at the x-axis and measured values at the y-axis, as shown in Fig. 2.23. The 45° line is the line of prediction without error. The dotted points are from the Whippany area and the cross points are from the Camden-Philadelphia area. Most of them are close to the line of prediction without error and the mean value of all the data is directly on the line of prediction without error. The predicted value deviated only 0.8 dB from the measured one.

   In other areas slightly larger differences were found. However, the largest difference between the predicted value and the measured one was roughly 3 dB. This range of accuracy is much less when compared to 8 dB from the area-to-area model.

   Point-to-point prediction is very useful in mobile cellular system design (see Section 5.5), where the radius of each cell is 10 miles or less. It can provide information to insure uniform coverage and avoidance of co-channel interference. Moreover the occurrence of handoff in the cellular system can be predicted more accurately. More information on point-to-point prediction can be found in references 28 to 30 which describe Lee's Model.

2.7   MICROCELL PREDICTION MODEL[30]

When the size of the cells is small, less than 1 km, the street orientation and individual blocks of buildings make a difference in signal reception, as mentioned previously. Those street orientations and individual blocks of buildings do not make any noticeable difference in reception when this signal is well attenuated at a distance over 1 km. Over a large distance the relatively great mobile radio propagation loss of 40 dB/dec is due to the situation that two waves, direct and reflected, are more or less equal in strength. The local scatterers (buildings surrounding the mobile unit) reflect this signal causing only the multipath fading not the path-loss at the mobile unit. When the cells are small, the signal arriving at the mobile unit is blocked by the individual buildings; this weakens the signal strength and is considered as part of the path loss. Therefore we have to take another approach in our prediction, to be described in this section. In small cells we are calculating the loss based on the dimensions of the building blocks. Since the ground incident angles of the reflected waves are, in general, small due to the low antenna heights used in small cells, the exact height of buildings in the middle of the propagation paths is not important, as shown in Fig. 2.24. Therefore only a two-dimension photo map is used. Although the strong received signal at the mobile unit comes from the multipath reflected waves not from the waves penetrating through the buildings, there is a correlation between the attenuation of the signal and the total building blocks, along the radio path. The larger the building blocks, the higher the signal attenuation. We can use an aerial photograph (see Fig. 2.25) to calculate the proportional length of the direct wave path being attenuated by the building blocks. When the wave is not being blocked by the building it is a line-of-sight condition. From the measurement data along the streets in an open line-of-sight condition, we formulate the line-of-sight signal reception curve P_{los}. Also, from the measured signal P_{os} along the streets in out-of-sight conditions within the cells, we formulate the additional signal attenuation α_B curve due to the portion of building blocks over the direct path by subtracting the received signal from P_{los}. The steps for forming an additional signal attenuation formula α_B are as follows:

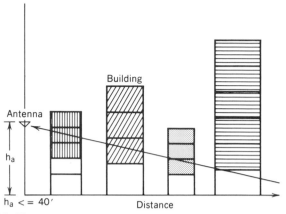

Figure 2.24. The propagation mechanics of low-antenna height at the cell site.

Figure 2.25. A sample aerial photograph.

1. Calculate the total blockage length B by adding the individual building blocks. For example, $B = a + b + c$ at point A shown in Fig. 2.26.
2. Measure the signal strength P_{los} for line-of-sight conditions.
3. Measure the signal strength P_{os} for out-of-sight conditions.
4. The local mean at point A is P_{os} (at A). The distance from the base to the mobile unit is d_A. The blockage length B at point A is $B = a + b + c$. Then the value of α_B for a blockage of B can be expressed as

$$\alpha_B(B = a + b + c) = P_{los}(d = d_A) - P_{os} \quad \text{(at } d_A)$$

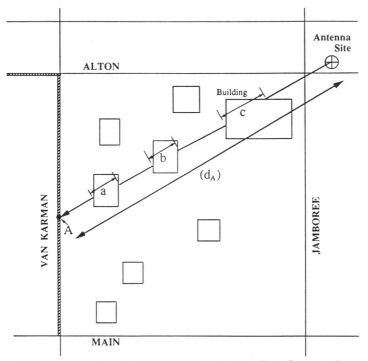

Figure 2.26. Building block occupancy at location A. Then $B = a + b + c$.

Then the additional signal attenuation curve based on the building blockage is found experimentally as shown in Fig. 2.27. The α_B curve was obtained at Irvine, California. The curve shows the rapid attenuation occurred while B is less than 500 ft. When B is greater than 1000 ft, a nearly constant value of 20 dB attenuation is observed. It can be explained by the street corner phenomenon as shown in Fig. 2.28. The rapid attenuation was seen on the mobile signal during the turning from one street to another as B starting from 0 ft and increasing. After B reaches 500 ft the received signal strength P_{os} will remain 18 dB below the P_{los} as the distance d increases. The path losses due to a line-of-sight condition for a series of antenna heights have been measured along many streets. The 9 dB/oct (or 30 dB/dec) antenna height gain over an antenna height change is usually observed in a small cell, as shown in Fig. 2.29. It is due to the fact that the incident angle in the small cell is usually larger than 10°. In the small-cell prediction model we use the two curves, P_{los} and α_B to predict the received signal strength. Therefore the microcell (small cell) model can be formed as

$$P_r = P_{los} - \alpha_B \qquad (2.7.1)$$

where P_{los} is the line-of-sight pathloss (measured) and α_B is the additional loss due to the length of the total building blocks B along the paths. In Fig. 2.27, the cell site ERP is 1 watt and the antenna height is 20 ft.

Microcell Prediction

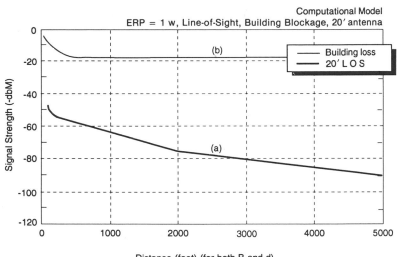

Figure 2.27. Microcell prediction parameters: (a) Line-of-sight P_{los}; (b) the α_B due to building blockage.

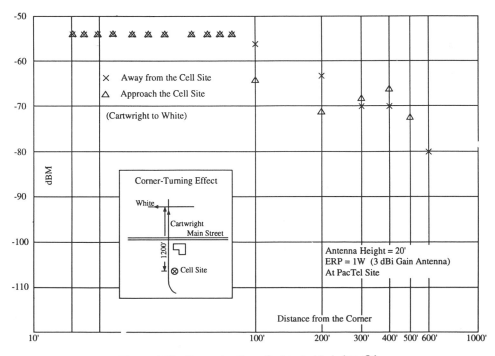

Figure 2.28. Corner turning effect tested in Irvine, CA.

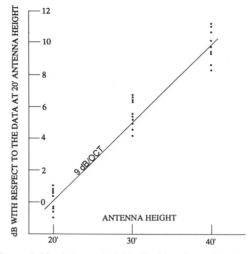

Figure 2.29. Antenna height effect in microcell systems.

The expressions to be evaluated are (from Fig. 2.27)

$$P_{los} = P_t - 77\,\text{dBm} - 21.5\log\frac{d}{100'} + 30\log\frac{h_1}{20} \quad 100' \le d < 200'$$

$$= P_t - 83.5\,\text{dBm} - 14\log\frac{d}{200'} + 30\log\frac{h_1}{20} \quad 200' \le d < 1000' \quad (2.7.2)$$

$$= P_t - 93.3\,\text{dBm} - 36.5\log\frac{d}{1000'} + 30\log\frac{h_1}{20} \quad 1000' \le d < 5000'$$

(use the Macrocell model prediction for $d > 5000'$)

and

$$\alpha_B = 0 \qquad\qquad\qquad 1' \le B$$

$$= 1 + 0.5\log(B/10) \qquad 1' \le B < 25' \qquad (2.7.3)$$

$$= 1.2 + 12.5\log(B/25) \qquad 25 \le B < 600'$$

$$= 17.95 + 3\log(B/600') \qquad 600' \le B < 3000'$$

$$= 20\,\text{dB} \qquad\qquad\qquad 3000' \le B$$

where P_t is the ERP in dBm, d is the total distance in feet, h, is the antenna height in feet. B is the length of blocking. Substitute Eq. (2.7.2) and Eq. (2.7.3) into Eq. (2.7.1), the predicted received signal P_r is obtained.

$$P_r = P_o + \gamma_o \log\frac{d_1}{d_o} - \gamma_1 \log B \qquad (2.7.4)$$

This microcell model has been verified in the areas of Irvine and San Diego, California, with good results, as shown in Figs. 2.30 and 2.31.

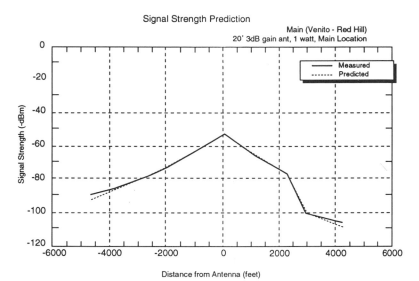

Figure 2.30. Comparison of the measured data and the predicted curve at Main Street, Irvine, CA.

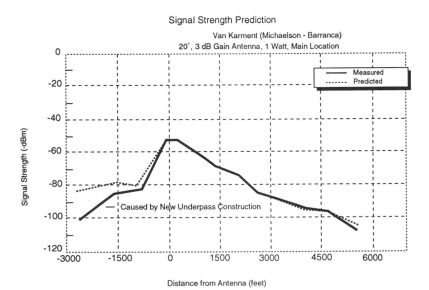

Figure 2.31. Comparison of the measured data and the predicted curve at Von Karmen, Irvine, CA.

In a hilly area Eq. (2.7.1) can be modified by adding the term antenna height gain obtained from Fig. 2.29 as

$$P_r = P_{los} - \alpha_B + 30 \log \frac{h_e}{h_a} \qquad (2.7.5)$$

The prediction from the microcell model is not as accurate as that from the macro-cell model. This is due to the fact that we are using a statistical prediction tool to predict the received signal indirectly by correlating the building blockage with the multipath-wave reception in more or less deterministic conditions where the propagation distance is short.

REFERENCES

1. Lee, W. C. Y., and Y. S. Yeh, "On the Estimation of the Second-Order Statistics of Log-Normal Fading in Mobile Radio Environment," *IEEE Trans. Commun.* Com-22: 6 (June 1974): 869–873.

2. Lee, W. C. Y., "Estimate of Local Average Power of a Mobile Radio Signal," *IEEE Trans. Veh. Tech.* VT-34: 1 (Feb. 1985): 22–27.

3. Attwood, S. S., ed., "The Propagation of Radio Waves through the Standard Atmosphere," *Summary Technical Report of the Committee on Propagation* 3 (Washington, DC: Reports and Documents, 1946): 250.

4. Beckmann, P., and A. Spizzichino, *The Scattering of Electromagnetic Waves from Rough Surfaces* (Macmillan, 1963): 20.

5. Lee, W. C. Y., *Mobile Communications Engineering* (McGraw-Hill, 1982): 107.

6. Bell System Practices *Public Land Mobile and UHF Maritime Systems Estimates of Expected Coverage* (Radio Systems General, July 1963).

7. Kelley, K. K., II, "Flat Suburban Area Propagation of 821 MHz," *IEEE Trans. Veh. Tech.* 27 (Nov. 1978): 198–204.

8. Ott, G. D., and A. Plitkins, "Urban Path-Loss Characteristics at 820 MHz," *IEEE Trans. Veh. Tech.* 27 (Nov. 1978): 189–197.

9. AT&T to FCC, "Advanced Mobile Phone Service—Development System Report," no. 5 (June 5, 1978).

10. Okumura, Y., E. Ohmori, T. Kawano, and K. Fukuda, "Field Strength and Its Variability in VHF and UHF Land Mobile Service," *Rev. Elec. Comm. Lab* 16 (Sept.–Oct. 1968): 825–873; also reprinted in *IEEE*.

11. Young, W. R., "Mobile Radio Transmission Compared at 150 to 3700 MC," *Bell Sys. Tech. J.* 31 (Nov. 1952): 1068–1085.

12. Lee, W. C. Y., "Studies of Base-Station Antenna Height Effects on Mobile Radio," *IEEE Trans. Veh. Tech.* VT-29: 2 (May 1980): 252–260.

13. Lee, W. C. Y., *Mobile Communications Engineering* (McGraw-Hill, 1982): 126.

14. Swarup, S., and R. K. Tewari, "Propagation Characteristics of VHF/UHF Signals in Tropical Moist Deciduous Forest," *J. Instn. Electronics Telecom. Engr.* 21: 3 (1975): 123–125.

15. Swarup, S., and R. K. Tewari, "Depolarization of Radio Waves in a Jungle

Environment," *IEEE Trans. Antenna Propagation* AP-27: 1 (Jan. 1979): 113–116.

16. Vincent, W. R., and G. H. Hagn, "Comments on the Performance of VHF Vehicular Radio Sets in Tropical Forests," *IEEE Trans. Veh. Tech.* VT-18: 2 (Aug. 1969): 61–65.

17. Tamir, T., "On Radio-Wave Propagation in Forest Environments," *IEEE Trans. Antenna Propagation* 15 (Nov. 1967): 806–817.

18. Tamir, T., "On Radio-Wave Propagation along Mixed Paths in Forest Environments," *IEEE Trans. Antenna Propagation* AP-25 (July 1971): 471–477.

19. Sass, P. F., "Propagation Measurements for UHF Spread Spectrum Mobile Communications," *IEEE Trans. Veh. Tech.* 32 (May 1983): 168–176.

20. Hufford, G. A., R. W. Hubbard, L. E. Patt, J. E. Adams, S. J. Paulson, and P. F. Sass, *Wideband Propagation Measurements in the Presence of Forests* (Fort Monmouth, NJ: U.S. Army Communications Electronics Command Jan., 1982): ADA113698.

21. Reudink, D. O., and M. F. Wazzowicz, "Some Propagation Experiments Relating to Foliage and Diffraction Loss at X-band and UHF Frequencies," *IEEE Trans. Commun.* 21 (Nov. 1973): 1198–1206.

22. Barsis, A. P., M. E. Johnson, and M. J. Miles, "Analysis of Propagation Measurements over Irregular Terrain in the 96- to 9200-MHz Range," *ESSA Tech. Rep.* (Boulder, CO: U.S. Dept. of Commerce March, 1969): ERL 114-ITS 82.

23. Lee, W. C. Y., *Mobile Communications Engineering*, 134.

24. Ibid., 133.

25. Reudink, D. O., "Mobile Radio Propagation in Tunnels," *IEEE Trans. Veh. Tech.* (Group Conference, San Francisco, CA, Dec. 2–4, 1968).

26. Emslie, A. G., R. L. Lagace, and P. F. Strong, "Theory of the Propagation of UHF Radio Waves in Coal Mine Tunnels," *IEEE Trans. Antenna Propagation* 23 (March 1975): 192–205.

27. Lee, W. C. Y., "Base-Station Antenna Height," 252–260.

28. Lee, W. C. Y., "A New Propagation Path-Loss Prediction Model for Military Mobile Access," *IEEE Milcom.* 85: 2, Boston, MA (Oct. 1985): 19.2.1–19.2.10.

29. IEEE VTS Committee on Radio Propagation "Lee's Model," *IEEE Trans. on Veh. Tech.*, Feb. 1988, pp. 68–70.

30. Lee, W. C. Y., "Lee's Model," *IEEE VTS 42nd Conference Proceedings*, Denver, CO, May 10–13, 1992, pp. 343–348.

ADDITIONAL REFERENCES

1. Bullington, K., "Radio Propagation for Vehicular Communications," *IEEE Trans. Veh. Tech.* VT-26: 4 (Nov. 1977): 295–308.

2. Dadson, Clifford E., "Radio Propagation Terrain Factors; Mobile Radio Field Strength Prediction and Frequency Assignment; Computer Methods," *IEEE Trans. Veh. Tech.* 24 (Feb. 1975): 1–8.

3. Forrest, Robert T., "Land Mobile Radio, Propagation Measurements for System Design," *IEEE Trans. Veh. Tech.* VT-24 (Nov. 1975): 46–53.

4. Hagn, G., "Radio System Performance Model for Predicting Communications Operational Ranges in Irregular Terrain," *Proc. 29th IEEE Vehicular Technology Conference Record* (1979): 322–330.

5. Jensen, Robert, "900 MHz Mobile Radio Propagation in the Copenhagen Area," *IEEE Trans. Veh. Tech.* VT-26 (Nov. 1977).

6. Turin, G. L., "Simulation of Urban Location Systems," *Proc. 21st IEEE Vehicular Technology Conference Record* (1970).

7. Wait, James R., "Radiowave Propagation; Hills and Knife-Edge Obstacles; Diffraction Losses," *IEEE Trans. Antenna Propagation* 15 (Nov. 1968): 700.

8. Nielson, D. L., "Microwave Propagation Measurements for Mobile Digital Radio Applications," *IEEE Trans. Veh. Tech.* VT-27 (Aug. 1978): 117–132.

9. French, R. C., "Radio Propagation in London," *Radio Electronic Engr.* 46 (July 1976): 333–336.

10. Young, W. R., "Comparison of Mobile Radio Transmissions at 150, 450, 900 and 3700 MHz," *Bell Sys. Tech. J.* 31 (Nov. 1952): 1068–1085.

11. Graziano, V., "Propagation Correlations at 900 MHz," *IEEE Trans. Veh. Tech.* VT-27 (Nov. 1978): 182–188.

12. Reudink, D. O., "Properties of Mobile Radio Propagation above 400 MHz," *IEEE Trans. Veh. Tech.* VT-23 (Nov. 1974): 143–160.

13. Hata, M., "Empirical Formula for Propagation Loss in Land Mobile Radio Services," *IEEE Trans. Veh. Tech.* VT-29 (1980): 317–325.

14. Akeyama, A., Nagatsu, T., and Ebine, Y., "Mobile Radio Propagation Characteristics and Radio Zone Design Method in Local Cities," *Rev. Elec. Comm. Lab.* 30: 2 (1982): 308–317.

15. Longley, A. G., and Rice, P. L. "Prediction of Tropospheric Radio Transmission Loss over Irregular Terrain, a Computer Method-1968," *ESSA Tech. Report ERL 79-ITS 67* NTIS 676874 (1968).

16. Barsis, A. P., "Radio Wave Propagation over Irregular Terrain in the 76- to 9200-MHz Frequency Range," *IEEE Trans. Veh. Tech.* VT-20: 2 (1971): 41–62.

17. Durkin, J., "Computer Prediction of Service Areas for VHF and UHF Land Mobile Radio Services," *IEEE Trans. Veh. Tech.* VT-26: 4 (1977): 323–327.

18. Palmer, F. H., "The CRC VHF/UHF Propagation Prediction Program: Description and Comparison with Field Measurements," *AGARD Conference Proc. 238* (Canada, Nov. 1978): 49-1–49-15.

19. Egli, J. J., "Radio Propagation above 40 MHz over Irregular Terrain," *Proc. IRE* 45 (Oct. 1975): 1382–1391.

20. Murphy, J. P., "Statistical Propagation Model for Irregular Terrain Paths between Transportable and Mobile Antennas," *AGARD Conf. Proc.* 70 (1970): 49-1–49-20.

21. Allsebrook, K., and Parsons, J. D., "Mobile Radio Propagation in British Cities at Frequencies in the VHF and UHF Bands," *IEEE Trans. Veh. Tech.* VT-26: 4 (1977): 313–323.

22. Ibrahim, M. F., and Parsons, J. D., "Urban Mobile Radio Propagation at 900 MHz," *IEEE Elec. Letters* 18: 3 (1982): 113–115.

PROBLEMS

2.1 How can the length $2L$ which is equivalent to equal 20–40 wavelengths be reached? Explain why the local mean has to obtain signal strength data with a window of $2L$?

2.2 How many number of samples are needed for a 90% C/I = 2 dB?

2.3 A local terrain variation height of 50 ft. is observed. Will rough terrain be considered if the operation frequency is 850 MHz and the incident angle is 0.5° (or 0.0087 rad.)? Keeping the same incident angle and changing the operation frequency to 85 MHz, will rough terrain be considered?

2.4 Verify from the reflection coefficients, a_n and a_v in Eq. (2.3.7) and Eq. (2.3-8) respectively, that the wave reflection coefficient always approaches minus one regardless of the type of terrain.

2.5 If the antenna height is changed from 135 ft to 532 ft, what will be the antenna-height gain? (Note: the antenna-height gain should be obtained using the effective antenna height.)

2.6 By comparing the two curves shown in Fig. 2.19, explain why the area-to-area prediction would have a standard deviation of 8 dB whereas the point-to-point prediction would have a standard deviation of 2–3 dB.

2.7 Calculate the diffraction loss in Fig. P2.1. Assume a suburban area condition.

2.8 Two transmitting antennas are shown in Fig. P2.2 with power P_1 and P_2 respectively. At a distance of one mile, two signals $S_1 = S_2$, calculate the two signals S_1 and S_2 received at a distance of 1.5 miles. (Apply 40 dB/dec rule for path loss)

2.9 Determine the effective antenna height in Figure P2.3. If the system was designed based on the actual antenna height, what will the antenna-height gain (loss) be at the mobile location due to the effective antenna height? If the mobile is to gain 6 dB over the actual antenna height, how high should the new antenna height be?

2.10 The base-station antenna height and location, and the mobile location shown in Figs. P2.3 and P2.4 are the same. The terrain between the base station and the mobile is different. Find the effective antenna heights in Figs. P2.3 and P2.4. What is the antenna-height gain (loss) at the mobile location due to the difference between the effective antenna height and the actual antenna height? If the mobile is to gain 6 dB over the actual antenna height, how high should the new antenna height be?

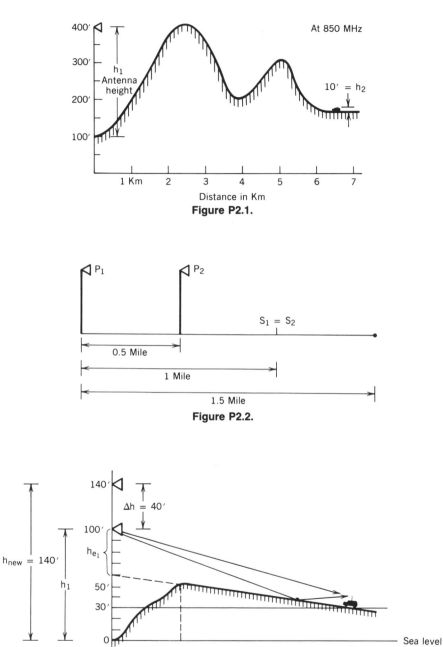

Figure P2.1.

Figure P2.2.

Figure P2.3.

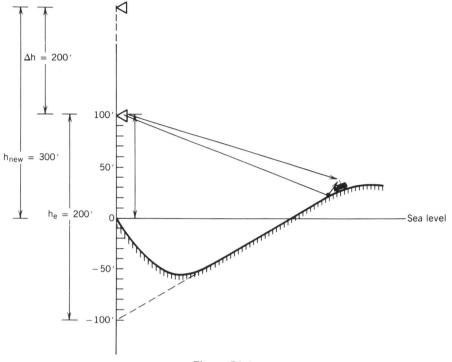

$\Delta h = 200'$

$h_{new} = 300'$

$h_e = 200'$

100'

50'

0

−50'

−100'

Sea level

Figure P2.4.

CALCULATION OF FADES AND METHODS OF REDUCING FADES

3.1 AMPLITUDE FADES

The probability density function (pdf) and the cumulative probability distribution (CPD) are both first-order statistics. By definition, they are not functions of time. The pdf and CPD of Rayleigh (related to nondirect wave components) and Rician distributions (related to nondirect waves plus a direct wave component) were shown in Sections 1.5.2 and 1.5.3, respectively. This chapter concentrates on second-order statistics on fades; second-order statistics are functions of time, such as level-crossing rates, average duration of fades, and the distribution of the duration of fades. A general formula for these second-order statistics appeared in Section 1.5.4. Having nondirect wave components at reception, a case of Rayleigh fading, is the worst type of fade in the mobile radio environment. This case will be thoroughly considered. In mobile radio communications, the radios are in motion, thus the received signal strength varies; this is called a short-term fading signal. The short-term fading is a function of time or vehicle speed, so the properties of short-term fading are second-order statistics. Level-crossing rates, average duration of fades, and the distribution of duration of fades of Rayleigh fading are as follows.

3.1.1 Level-Crossing Rates

The level-crossing rate (lcr) is

$$n(R) = n_0 \cdot n_R \qquad (3.1.1)$$

where n_R is independent of frequency and velocity and only a function of signal amplitude expressed as

$$n_{R_1} = R_1 e^{-R_1^2} \qquad \text{(for } E\text{-field)} \tag{3.1.2}$$

$$n_{R_2} = \sqrt{1 - \frac{1}{2}\cos 2\alpha} \cdot R_2 \cdot \exp(-R_2^2) \qquad \text{(for } H_x\text{-field)} \tag{3.1.3}$$

$$n_{R_3} = \sqrt{1 + \frac{1}{2}\cos 2\alpha} \cdot R_3 \cdot \exp(-R_3^2) \qquad \text{(for } H_y\text{-field)} \tag{3.1.4}$$

where R_1 is the envelope of E with respect to its rms value ($R_1 = r_1/\sqrt{\overline{r_1^2}}$), R_2 is the envelope of H_x with respect to its rms value ($R_2 = r_2/\sqrt{\overline{r_2^2}}$), and R_3 is the envelope of H_y with respect to its rms value ($R_3 = r_3/\sqrt{\overline{r_3^2}}$). The relationship among three rms values is $\overline{r_1^2} = 2\overline{r_2^2} = 2\overline{r_3^2}$. The direction of the vehicle's motion is α. The coordinates of these equations are shown in Fig. 3.1. n_R is closely related to the pdf of Rayleigh fading as shown in Eq. (1.5.10).

$$n_{R_1} = \frac{1}{2}p(R) \tag{3.1.5}$$

and the normalization factor n_0 of Eq. (3.1.1) is

$$n_0 = \frac{\beta V}{\sqrt{2\pi}} = \sqrt{2\pi}\frac{V}{\lambda} = 2.5 \cdot \frac{V}{\lambda} \tag{3.1.6}$$

which is a function of frequency and vehicle speed. The previous equations are plotted in Fig. 3.2. The theoretical level-crossing rate can be obtained easily. It agrees closely with the experimental values.[1]

Given frequency and vehicle speed, the theoretical level-crossing rates at different levels R can be obtained from Fig. 3.2. The experimental level-crossing rate can be obtained by counting the crossings as described in Section 1.5.4. The maximum level-crossing rate always occurs at 3 dB below the average power level. It can be proved by differentiating Eq. (3.1.2) or Eq. (3.1.5) with respect to R_1:

$$\frac{d}{dR_1}(n_{R_1}) = 0$$

This equation implies that the maximum of R_1 is at $R = 1/\sqrt{2}$, which is -3 dB with respect to its rms value.

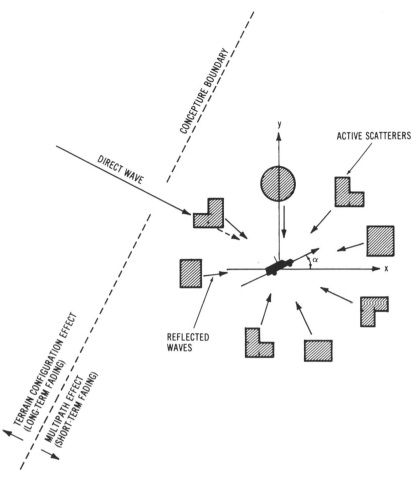

Figure 3.1. The coordinates of the multipath model.

Example 3.1. A signal at 850 MHz is received by a dipole at a mobile unit traveling at 24 km/h (15 mph). What would be the expected level-crossing rate at 10 dB below its average power level? From Fig. 3.2

$$R_1 = -10 \text{ dB}$$

$$n_0 = 47$$

$$n_{R_1} = 0.284$$

Then $n(R_1) = 47 \times 0.284 = 13.35$ crossings per second.

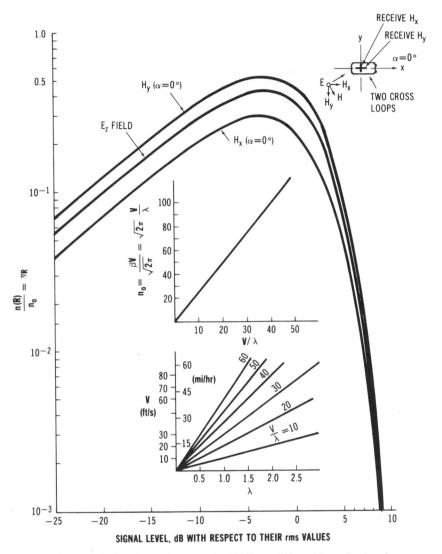

Figure 3.2. Level-crossing rates for E_z, H_x, and H_y mobile radio signals.

To obtain the experimental level-crossing rate, first calculate the average power level:

$$\text{Average power} = (\text{rms})^2 = \frac{\sum_1^N r_i^2}{N} \qquad (r_i \text{ in volts in digital data})$$

$$= \frac{1}{\tau} \int_0^\tau r^2\, dr \qquad (r \text{ in volts in analog data}) \qquad (3.1.7)$$

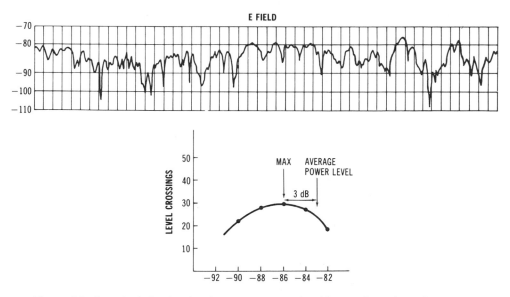

Figure 3.3. A method of estimating the average power level from a piece of raw data.

Then calculate the level-crossing rate at any power level with respect to this level.

Example 3.2. Sometimes it is easy to effectively find the power level from a piece of measured data, especially if a computer facility is not available. Assuming that the measured data comes from the conditions described in Example 3.1, the steps to obtain the average power level of this piece of data (shown in Fig. 3.3) are described as follows:

1. First, draw five lines at 2-dB incremental levels (in dBm or dB referred to any power level) across the received data in the vicinity of the average power level; use eyeball accuracy, as shown in Fig. 3.3.
2. Count the positive slope of crossings in each level over a length of data. Divide the level-crossing count by the length of time. The result is the level-crossing rate at each amplitude level.
3. Plot the level-crossing rates against the amplitude levels.
4. Draw a curve fitting all five level-crossing-rate points, and pick the maximum position.
5. The average power level will be 3 db above the level of maximum level crossing.

Once the average power level is found for that particular piece of data, any other level can be identified based on the average power level (or so-called rms value level).

Example 3.3. Compare the experimental lcr with the theoretical prediction. From the raw data shown in Fig. 3.3, find the 10-dB level below the average power level first. Then count the number of level crossings in a period of 5 sec as 73. Convert the level crossings to a level-crossing rate 14.6, which is very close to the expected lcr of 13.35 shown in Example 3.1.

3.1.2 Average Duration of Fades[2,3]

The average duration of fade (adf), $\bar{t}(R)$ is

$$\bar{t}(R) = \frac{\text{CPD}}{\text{lcr}} = \frac{P(r \le R)}{n(R)} \tag{3.1.8}$$

$$\bar{t}(R) = \left(\frac{1}{n_0}\right)\frac{P(r \le R)}{n_R} = \bar{t}_0 \cdot \bar{t}_R \tag{3.1.9}$$

See Eq. (1.5.27). Equation (3.1.9) is plotted in Fig. 3.4.

Example 3.4. Under the same conditions as Example 3.1, what are the theoretical predictions of $\bar{t}_0$ and $\bar{t}_R$ for an E-field signal?
Solution:

$$\bar{t}_0 = \frac{1}{47} = 0.0213$$

$$\bar{t}_R = 0.352$$

$$\bar{t}(R) = \bar{t}_0 \cdot \bar{t}_R = 0.00749 \text{ sec}$$

Example 3.5. From the measured data (and without computer assistance) calculate an adf with the same conditions described in Example 3.1. It is hard to obtain the adf directly. Therefore obtain the CPD function and lcr first, gaining the adf of Eq. (3.1.8) in the process.

In a Rayleigh-fading environment the CPD always follows the Rayleigh curve shown in Fig. 1.12. At the -10-dB level, $P\ (x \le -10 \text{ dB}) = 0.09$. The lcr is obtained by experiment from Example 3.3, as $n(R) = 14.6$. Then an adf is obtained from Eq. (3.1.8)

$$t(R) = \frac{0.09}{14.7} = 0.00612 \text{ sec}$$

The results obtained from Example 3.4 and from this example closely agree.

3.1.3 Distribution of Duration of Fades[4]

Deriving the distribution of duration of fades is very complicated. However, it is quite useful in designing the signaling format in a fading environment.

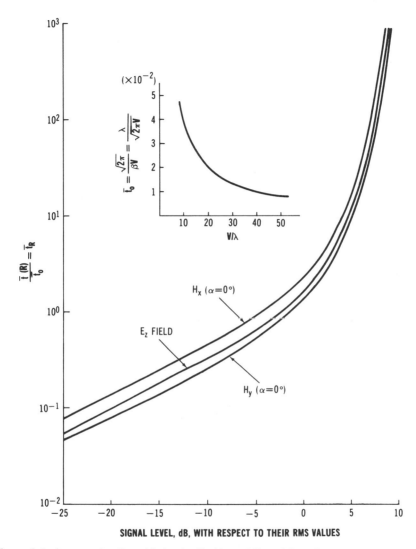

Figure 3.4. Average duration of fades for E_z, H_x, and H_y mobile radio signals.

S. O. Rice calculated the distribution of duration of fades $F_\tau (u, R)$ shown in Fig. 3.5, where $u = \tau/\bar{t}$. The probability that $R(t) < R$ for an interval lasting longer than τ is indicated. The average duration of fades is $\bar{t}$, and R is the envelope with respect to its rms values.

Example 3.6. What is the probability of a fade occurring, whose duration is greater than twice that of adf at a level of 3 dB below its rms value?

From Fig. 3.5, we find that the probability of occurrence is 10%.

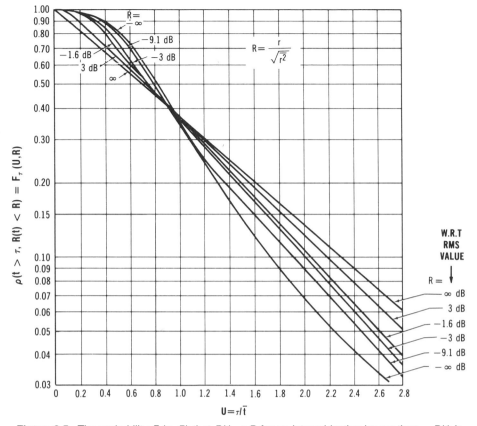

Figure 3.5. The probability $F_r(u, R)$ that $R(t) < R$ for an interval lasting longer than τ. $R(t)$ is the envelope of narrowband Gaussian noise.

3.1.4 Envelope Correlation between Two Closely Spaced Antennas at the Mobile Unit[5]

The correlation coefficient between two received signals in frequency and time separation at a mobile unit while it is traveling with a velocity V can be expressed in two cases:

1. Between two E-fields (or two H-fields)

$$\rho_r(\Delta\omega, \tau) = \frac{J_0^2(\beta V\tau)}{1 + (\Delta\omega)^2\Delta^2} \qquad \text{(frequency and time separation)}$$

$$(3.1.10)$$

or

$$\rho_r(\Delta\omega, \Delta d) = \frac{J_0^2(\beta \cdot \Delta d)}{1 + (\Delta\omega)^2 \Delta^2} \qquad \text{(frequency and space separation)}$$

$$(3.1.11)$$

where $J_0(\cdot)$ is the Bessel function of first kind and zero order. The speed of the vehicle is V, and τ is the time separation, which corresponds to space separation

$$\Delta d = V\tau$$

the frequency separation is $(\Delta\omega)$, and Δ is the delay spread described in Section 1.5.6.

2. Between E- and H-fields, since the correlation coefficient between E and H is[6]

$$\rho_{eh}(\Delta d) = J_1^2(\beta \cdot \Delta d)$$

Then following the same format as Eq. (3.1.1), we obtain

$$\rho_{eh}(\Delta\omega, \Delta d) = \frac{J_1^2(\beta \cdot \Delta d)}{1 + (\Delta\omega)^2 \cdot \Delta^2} \qquad (3.1.12)$$

and

$$\rho_{eh}(\Delta\omega, 0) = 0$$

where $J_1(\cdot)$ is the Bessel function of first kind and first order.

Equations (3.1.10) and (3.1.12) are plotted in Fig. 3.6. In the case of $\Delta d = 0$ (or $\tau = 0$), it may be interpreted that only one antenna is in use at the mobile unit. In the case of $\Delta\omega = 0$, it means that only one frequency is used. In the case of both $\Delta\omega$ and $\Delta d \neq 0$, it means that two kinds of diversity— frequency and space—are implemented at the same time. It can be called a *hybrid diversity system*. In practice we use either space diversity ($\Delta\omega = 0$) or frequency diversity ($\Delta d = 0$). The former saves the frequency spectrum, the latter requires only one antenna. Sometimes the performance of a hybrid diversity system is better than a four-branch diversity of either one.

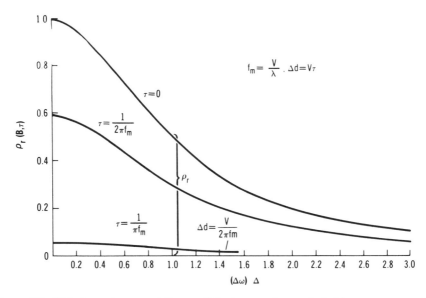

Figure 3.6. Plot of envelope correlation coefficient versus the product of frequency separation $\Delta\omega = \omega_2 - \omega_1$ and time delay spread Δ.

3.1.5 Power Spectrum

At IF

The power spectrum $S_{ez}(f)$, $S_{hy}(f)$, and $S_{hx}(f)$ of three field components at IF (intermediate frequency) are[7]

$$S_{ez}(f) = \frac{3}{2\pi \sqrt{f_m^2 - f^2}} \tag{3.1.13}$$

$$S_{hx}(f) = \frac{3}{2\pi f_m^2} \sqrt{f_m^2 - f^2} \tag{3.1.14}$$

$$S_{hy}(f) = \frac{3f^2}{2\pi f_m^2 \sqrt{f_m^2 - f^2}} \tag{3.1.15}$$

The three preceding equations are plotted in Fig. 3.7A. Equation (3.1.13) represents a typical FM power spectrum. Equation (3.1.14) indicates that the arrival of waves from the car's sides contributes the most to the power spectrum. However, Eq. (3.1.15) indicates that the front and rear wave arrival contribute the most.

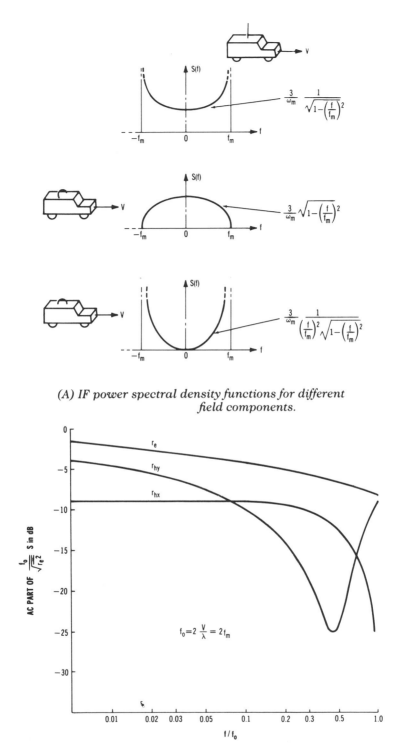

(A) IF power spectral density functions for different field components.

(B) Baseband spectra density functions of the field envelopes.

Figure 3.7. Power spectra of the envelopes of three field components.

At Baseband Frequency

The power spectrum $S_{re}(f)$, $S_{hy}(f)$, and $S_{hx}(f)$ of three field components at the baseband frequency are[8,9]

$$\frac{1}{\sqrt{r_e^2}} \cdot S_{re}(f) = \frac{\pi}{4} \delta(f) + 2\left(1 - \frac{\pi}{4}\right)\frac{K(\sqrt{1 - (f/f_0)^2})}{\pi^2 f_0} \qquad (3.1.16)$$

$$\frac{1}{\sqrt{r_e^2}} \cdot S_{rhx}(f) = \frac{\pi}{8} \delta(f)$$

$$+ \frac{1}{3\pi f_0}\left\{\left[1 + \left(\frac{f}{f_0}\right)^2\right] \cdot E\left[\sqrt{1 - \left(\frac{f}{f_0}\right)^2}\right]\right.$$

$$\left. - 2\left(\frac{f}{f_0}\right)^2 \cdot K\left[\sqrt{1 - \left(\frac{f}{f_0}\right)^2}\right]\right\} \qquad (3.1.17)$$

$$\frac{1}{\sqrt{r_e^2}} \cdot S_{rhy}(f) = \frac{\pi}{8} \delta(f)$$

$$+ \frac{1}{4\pi f_0}\left\{\left[1 + \frac{4}{3}\left(\frac{f}{f_0}\right)^2\right]K\left[\sqrt{1 + \left(\frac{f}{f_0}\right)^2}\right]\right.$$

$$\left. - \frac{8}{3}\left[1 - \frac{1}{2}\left(\frac{f}{f_0}\right)^2\right] \cdot E\left[\sqrt{1 - \left(\frac{f}{f_0}\right)^2}\right]\right\} \qquad (3.1.18)$$

where the delta function $\delta(f)$ represents a dc power, and $E(\cdot)$ and $K(\cdot)$ are the elliptic integral and complete elliptic integrals of the first kind of functions, respectively. Equations (3.1.16), (3.1.17), and (3.1.18) are plotted in Fig. 3.7B. The cutoff frequency f_0 at baseband is twice that of the fading frequency ($f_0 = 2f_m$).

3.2 RANDOM PM AND RANDOM FM

A signal $s(t)$ whose amplitude term is constant A, and whose phase term contains a message $\psi_s(t)$, can be expressed as

$$s(t) = Ae^{j(\omega t + \psi_s(t))} \qquad (3.2.1)$$

where ω is the angular frequency, while the signal $s_0(t)$ received at the mobile unit can be represented by

$$s_0(t) = A \cdot r(t) \cdot e^{j(\psi_s(t) + \psi_r(t))} e^{j\omega t}$$

$$= A \cdot m(t) \cdot r_0(t) e^{j(\omega t + \psi_s(t) + \psi_r(t))} \qquad (3.2.2)$$

where $r(t)$ is the envelope of received signal. Thus $r(t)$ can be artificially separated into two parts, $m(t)$ and $r_0(t)$, due to an understanding of the mobile radio environment, as described in Section 1.3. We introduce an additional phase term $\psi_r(t)$ to include the multipath effect; $\psi_r(t)$ is also a random variable.

3.2.1 Random Phase $\psi_r(t)$

Random phase $\psi_r(t)$ is known as uniformly distributed. It can be any phase angle between $0°$ and 2π with equal probability.

$$\rho(\psi_r) = \begin{cases} \dfrac{1}{2\pi} & 0 < \psi_r < 2\pi \\ 0 \end{cases} \qquad (3.2.3)$$

Otherwise, the probability (CPD) that ψ_r is less than a given phase Ψ is

$$\rho(\psi_r \le \Psi) = \int_0^{\Psi} \rho(\psi_r) \, d\psi_r = \frac{\Psi}{2\pi} \qquad (3.2.4)$$

Equations (3.2.3) and (3.2.4) are plotted in Fig. 3.8A. The CPD of ψ_r is independent of the vehicle speed.

3.2.2 Random RM $\dot{\psi}_r(t)$

When we deal with an FM modulation, the random phase $\psi_r(t)$ becomes its derivative $\dot{\psi}_r(t)$, a random variable in FM called *random FM*. The characteristics[10] of the pdf and CPD of $\dot{\psi}_r(t)$ are shown in Fig. 3.8B. They are functions of vehicle speed V or the fading frequency, V/λ, as shown in Fig. 3.8B. The random FM variable $\dot{\psi}_r(t)$ is randomly but symmetrically spread in frequencies around the carrier frequency. The CPD of $\dot{\psi}_r(t)$ is a nonlinear function and is shown in Fig. 3.8B. The power spectrum of $\dot{\psi}_r(t)$ has a sharp cutoff when the frequency exceeds the value of $2V/\lambda$ shown in Fig. 3.8C. In the mobile radio environment the vehicle could reach 105 km/h (70 mph). Also if we assume that the operating frequency is 1 GHz, then approximately $1\lambda \doteq 0.3$ m (1 ft). Under these conditions the effective random FM will have only a slight effect when its frequency is above $2V/\lambda$ or 176 Hz because of the steep dropping slope shown in Fig. 3.8C. It means

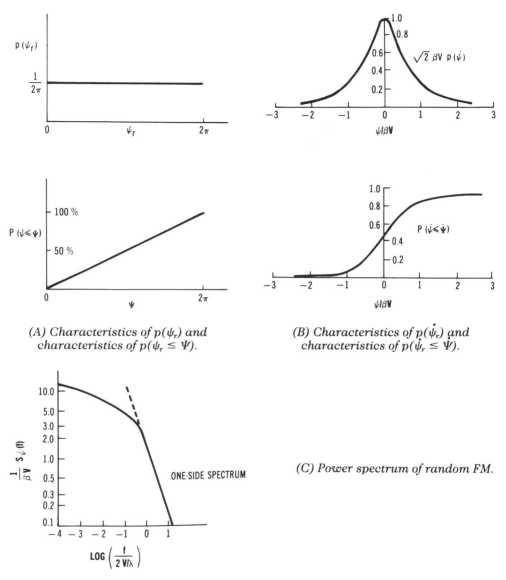

(A) *Characteristics of $p(\psi_r)$ and characteristics of $p(\psi_r \leq \Psi)$.*

(B) *Characteristics of $p(\dot{\psi}_r)$ and characteristics of $p(\dot{\psi}_r \leq \dot{\Psi})$.*

(C) *Power spectrum of random FM.*

Figure 3.8. Characteristics of random phase and random FM.

that if the lower band of a desired signal is designed to be higher than 176 Hz, the random FM will not affect the desired signal. Since the frequency of the human voice is higher than 300 Hz, a bandpass filter could filter out the random FM. In other words, the random FM does not affect mobile radio communication of voice.

In data communications it is critical to use the proper signaling waveform to avoid the energy concentrated in the random FM range. The Manchester code is a good code for this application because its power is not concentrated at the zero frequency.[11]

3.3 SELECTIVE FADING AND SELECTIVE RANDOM FM

3.3.1 Selective Fading

Selective fading usually means frequency-selective fading. That is, if two different frequencies separated by a finite frequency range propagating in a medium do not observe the same fading. Selective fading is closely related to the time-delay spread Δ. If the time-delay spread equals zero, no selective fading exists. However, in a mobile radio environment, the multipath fading results delay spread, as mentioned in Section 1.5.6. Fades are selective and dependent on the coherence bandwidth B_c. Equation (1.5.58) can be derived from Eq. (3.1.11). For a single antenna case, let $\Delta d = 0$ in Eq. (3.1.11), and let the correlation coefficient $\rho = 0.5$ as a criterion for determining the coherence bandwidth. If the separation in frequency $B = \Delta\omega/2\pi$ is so large that $\rho < 0.5$, the two frequencies are not in the coherence bandwidth $B > B_c$. If $\rho > 0.5$, the two frequencies separated by B are within their coherent bandwidth. Using $\rho_c = 0.5$ for this criterion, Eq. (3.1.11) becomes

$$\rho_c = 0.5 = \frac{1}{1 + (\Delta\omega_c)^2 \cdot \Delta^2}$$

or

$$B_c = \frac{\Delta\omega_c}{2\pi} = \frac{1}{2\pi\Delta} \tag{3.3.1}$$

Equation (3.3.1) is defined as a coherence bandwidth between two amplitude fading signals. If we substitute into the equation the data of delay spread listed in Section 1.5.6, then the coherence bandwidths in different kinds of manmade environments are as follows:

$$B_c = 300 \text{ kHz} \quad \text{for } \Delta = 0.5 \text{ } \mu\text{s in suburban areas}$$

$$B_c = 50 \text{ kHz} \quad \Delta = 3 \text{ } \mu\text{s in urban areas}$$

$$B_c = 0.8 \text{ MHz} \quad \Delta = 0.2 \text{ } \mu\text{s in open areas}$$

The coherence bandwidth in a suburban area is greater than in an urban area. In an urban area uncorrelated signal fadings will occur if two frequencies are separated by more than 50 kHz. In a suburban area a separation of 300 kHz will cause uncorrelated fades; if in an open area, 0.8 MHz. If a frequency diversity is implemented in the mobile radio environment, the required fre-

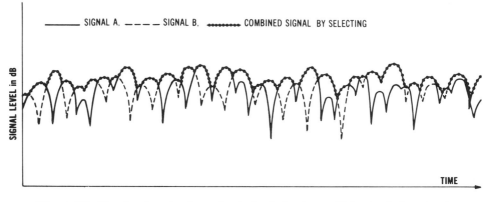

Figure 3.9. Showing the advantage of reducing fading by combining two fading signals.

quency separation should be 300 kHz. The reason is stated in the frequency diversity scheme of Section 3.4.2.

3.3.2 Selective Random FM

A coherence bandwidth for two different random FM also can be obtained, but its derivation is tedious and lengthy. We only express its simple formula as follows:[12]

$$B'_c = \frac{1}{4\pi\Delta} \tag{3.3.2}$$

In this case the coherent bandwidth of selective random FM is half as wide as the coherence bandwidth of selective fading. In section 3.5 we will apply the coherence bandwidth B'_c to a system design.

3.4 DIVERSITY SCHEMES

"Diversity schemes" provide two or more inputs at the mobile reception unit so that the fading phenomena among those inputs are uncorrelated. We have to be cautioned that the "correlation" we mean only deals with two signal-fading channels. The message carried by two signal channels should always be the same. This section will discuss the methods that can create the least correlation between two fading signals. After reception they can be combined and the fades smoothed out before the message is detected. An illustration is provided in Fig. 3.9. There are two kinds of fading: long-term fading and short-term fading. To reduce long-term fading, we need to use macroscopic diversity, and to reduce short-term fading, we need to use microscopic diversity.

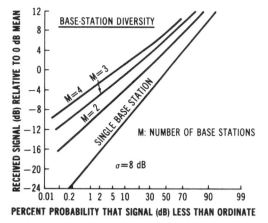

Figure 3.10. Performance of selective combining in macroscopic diversity.

3.4.1 Macroscopic Diversity (Apply on Separated Antenna Sites)

The variation of local means is due to the mutability of terrain contours. If only one antenna site is used, the traveling mobile unit may not be able to receive the signal at certain geographical locations due to terrain variations such as hills or mountains. Therefore two separated antenna sites must be used to transmit or receive two signals and to combine them to reduce long-term fading. As described in Section 2.3, long-term fading follows a lognormal distribution with a standard deviation, whose value depends on terrain variations. Figure 3.10 shows the combining of two to four long-term fading signals with the same standard deviation of 8 dB. The selective combining technique is a recommended technique in the macroscopic diversity scheme. Selective combining means always selecting the strongest from the two fading signals in real time.

3.4.2 Microscopic Diversity (Apply on Co-located Antenna Site)

In a fading environment the local average signal received at the mobile unit suffers with an increase in distance; it also varies due to the terrain contour along the radio path. This local average signal is called the *local mean* (or *long-term fading signal* as described in Section 2.2). The multipath phenomenon results in a Rayleigh fading whose amplitude dynamic variation is about 40 dB along the local mean.

A reduction in long-term and Rayleigh fading is desirable. In Section 3.4.1 macroscopic diversity is used to reduce the long-term fading. In Section 3.4.2 six microscopic diversity schemes are introduced. They will all reduce Rayleigh fading, and they all require two or more antennas or two frequencies at the same antenna site (co-located site). After the creation of diversity branches, there are ways to combine them all. The combining techniques appear in Section 3.5. The six diversity schemes are space, frequency, polarization, field component, angle, and time.

Space Diversity

Two antennas separated physically by a distance d can provide two signals with low correlation among their fadings. The separation d in general varies with the antenna height h. The separation will be specified in Chapter 6 for the base-station antenna height[13] and in Chapter 7 for the mobile-unit antenna height.

Frequency Diversity

When two frequencies separated by a bandwidth B_c are such that two received fading signals on two different frequencies are uncorrelated. The value of B_c can be determined from Section 3.1. Equation (3.3.1) shows that a value of B_c larger than 50 kHz should be used for urban areas, and larger than 300 kHz for suburban areas. In open areas the value of B_c should be greater than 0.8 MHz. However, in open areas no severe fading is observed, so no diversity is needed. To use frequency diversity in both urban and suburban areas, B_c has to be 300 kHz or greater. The frequency separation required to reduce fading in a suburban area will also reduce fading in an urban area. The coherence bandwidth remains the same at any carrier frequency as long as the carrier frequency ranges from 30 MHz and up; therefore this frequency separation remains constant. The frequency separation also remains the same when diversity is provided either at the mobile unit or at the base station.

Polarization Diversity

Two polarization components E_x and E_y transmitted by two polarized antennas at the base station and received by two polarized antennas at the mobile unit, and vice versa, can provide two uncorrelated signal fadings.[14] This has been shown theoretically and proved by experiments in the mobile radio environment. The drawback of using polarization diversity is the 3-dB power reduction at the transmitting site due to the splitting of power into two different polarized antennas.

Field Component Diversity

The idea of using field components is based on the electromagnetic theory that when an E-field is propagating, the H-field is always associated with it. Both E and H carry the same information message. If there are no scatterers, two components cannot be distinguished from each other. Suppose that these components are bounced back and forth in a multipath environment. Then the reflection mechanisms for E and H would be different. A simple example is to show that the patterns of the standing waves reflecting E and H waves from a scatterer are 90° phase apart. When E is a maximum, H is a minimum. In a mobile radio environment, we can sum up the many pairs of both E-field and H-field standing waves. The results are predictable: all the components E_z, H_x, and H_y are uncorrelated in a mobile radio environment.[15,16] It is also proved in Section 3.1. This scheme does not require physical separation between antennas. The advantage of using this diversity scheme is for systems operating at lower operating frequencies, such as below 100 MHz. When an operating frequency is higher—above 1 GHz—space diversity can be easily implemented physically; there is no need for field component diversity. Both this scheme and space diversity are always preferable to polarization diversity because they do not have the 3-dB transmitted power reduction like polarization. Field component diversity antennas will be described in Section 7.8.

Angle Diversity
When an operating frequency is 10 GHz or higher, two or more directional antennas can be pointed at different directions at the receiving site. This scheme is more effective at the mobile unit than at the base station.

Time Diversity
Time diversity means transmitting identical messages in different time slots. This yields two uncorrelated fading signals at the receiving end. Time diversity is a good scheme for reducing intermodulation at a multichannel site. But in a mobile radio environment a mobile unit may be at a standstill at any location that has a weak local mean or is caught in a deep fade. In either situation time diversity would not reduce the fades.

3.5   COMBINING TECHNIQUES

3.5.1   Combining Techniques on Diversity Schemes

There are four major combining techniques: selective, switched, maximal-ratio, and equal-gain. Every diversity scheme can be applied to one of these combining techniques. They are shown in Fig. 3.11 with a two-branch diversity receiver.

Selective Combining
By selecting the strongest signal among M diversity branches, we can reduce log-normal fading by macroscopic diversity with M separated antenna sites, as shown in Fig. 3.10. Reducing Rayleigh fading by microscopic diversity with M co-sited antennas is shown in Fig. 3.12. The figures show reduced fading as M increases. A two-branch selective combining receiver always needs two receiving front ends—one for receiving the maximal signal and the other for monitoring purposes.

Switched Combining
Switched combining is different from selective combining. In switched combining, two diversity signals are selected, based on a given threshold level in

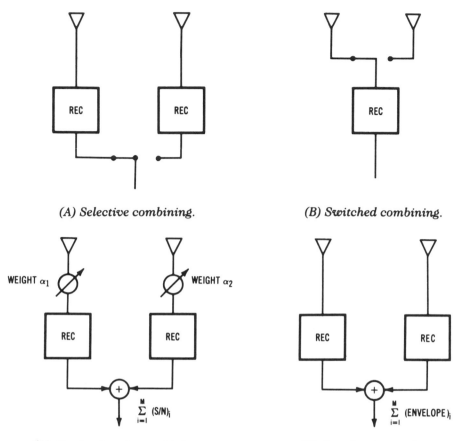

(A) Selective combining.   *(B) Switched combining.*

(C) Maximal-ratio combining.   *(D) Equal-gain combining.*

Figure 3.11. Four different diversity combiners.

one receiver. If signal A above a threshold L is selected to receive, it receives until it falls below the level L. Then the receiver switches to signal B regardless of whether signal B is above L or below. Signal B should be above L, but if it is below, then depending on the switching algorithm of the receiver, it can switch back to signal A or stay in signal B until signal B rises above level L. Switched combining performance (shown in Fig. 3.13) is not always as good as that of selective combining. Since switched combining needs only one receiver, it is less costly and may be used on mobile units. However, the performance is greatly affected by threshold level and switching noise. To improve this scheme, the threshold level L needs to be changed dynamically in real time based on the present signal level received in the environment. Switching noise due to switching impulses can be eliminated by a blank-and-hold device. Research still needs to be done on this combining technique to justify the cost of improving it over other combining techniques.

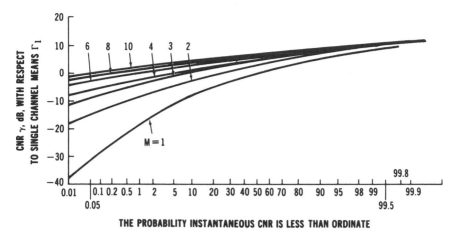

Figure 3.12. Performance curves for selectively combined microscopic-diversity Rayleigh-fading signals.

Maximal-Ratio Combining

The term "maximal ratio" refers to maximal signal-to-noise ratio. This is the best combining technique, as has been proved by mathematics. Each branch needs proper weights, as shown in Fig. 3.11C. At a baseband the combined signal is the sum of the instantaneous SNRs (γ_i) on the individual branches.

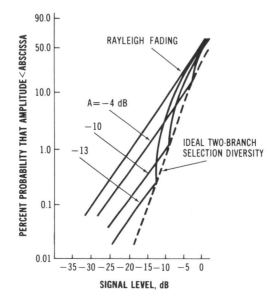

Figure 3.13. Performance of a two-branch switched-combined signal with various threshold levels.

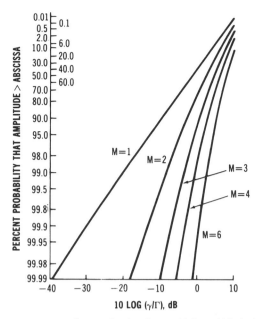

Figure 3.14. Performance curves for maximal-ratio combining within independent channels.

$$\gamma = \sum_{i=1}^{M} \gamma_i \tag{3.5.1}$$

The performance of maximal-ratio combining is shown in Fig. 3.14 where Γ is the average SNR (sometimes called carrier-to-noise ratio, CNR) of a single channel, $\Gamma = \langle \gamma_1 \rangle$. This combining technique for a two-branch diversity needs two receivers; the circuit is very complicated.

Equal-Gain Combining

Equal-gain combining is a co-phase combining that brings all phases to a common point and combines them. The combined signal is the sum of the instantaneous fading envelopes of the individual branches:

$$r = \sum_{i=1}^{M} r_i \tag{3.5.2}$$

The performance of equal-gain combining is shown in Fig. 3.15 where $\langle r^2 \rangle / \langle r_1^2 \rangle = \gamma / \Gamma$. $\langle r_1^2 \rangle$ is the average power of a single branch. Γ is the average SNR of a single branch. Equal-gain combining has only one dB degradation as compared to maximal-ratio combining. Because of this, and because the circuit is relatively simple, equal-gain combining is usually used at base stations.

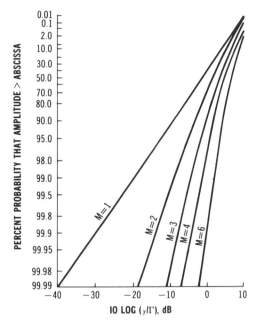

Figure 3.15. Cumulative probability distribution for equal-gain combined branches.

3.5.2 Combining Techniques for Reducing Random Phase

The following combining techniques are used for each branch. Their purpose is to reduce the random phase (described in Section 3.3) in each branch while the signal is being received.

Feedforward Combining

A technique to reduce the random phase by using two mixers, M_1 and M_2, is shown in Fig. 3.16A. The output at S_8 is trying to reduce the random phase component. But due to this feedforward combining the front-end noise of the receiver is still retained.

Feedback Combining (Granlund)

Feedback combining is similar to feedforward combining, but it has a feedback loop, as shown in Fig. 3.16B. Reducing random phase by using feedback combining is always more effective than using feedforward combining. The price for this type of combining is the necessity to design proper filters. Both feedforward combining and feedback combining can be used as co-phase combining of M branches. Eliminating the random phase in each branch, and combining them, is equivalent to an equal-gain combiner. Therefore a combination of feedback combining to reduce random phase and equal-gain combining to reduce multipath fading has been suggested for base-station receivers.

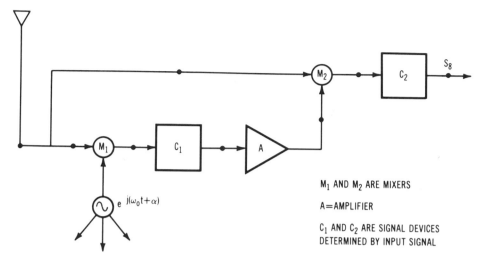

M$_1$ AND M$_2$ ARE MIXERS

A=AMPLIFIER

C$_1$ AND C$_2$ ARE SIGNAL DEVICES
DETERMINED BY INPUT SIGNAL

(A) Typical feedforward-combining circuit.

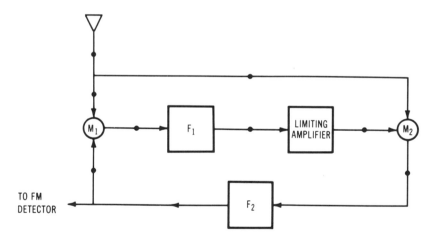

(B) Granlund combiner.

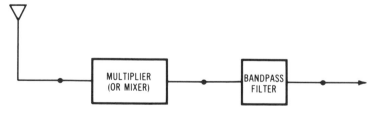

(C) Model circuit for the alternate-pilot-signal method.
Figure 3.16. Reducing random FM techniques.

Combining with a Pilot Tone

By transmitting a pilot tone very close to the desired signal carrier, we are able to filter it after reception. The pilot tone usually only transmits a little energy. At the receiving end both the desired signal and the pilot tone carry the random phase information. They will carry the same random phase if the frequency separation between them is not far. After going through a mixer (see Fig. 3.16C), the random phase is canceled out. The mixer helps to reduce front-end noise, but it does have one drawback in that extra use is made of the frequency spectrum. In order to keep the same random phase, we use a frequency separation B' which should be within the coherence bandwidth B'_c shown in Eq. (3.3.2):

$$B' < B'_c = \frac{1}{4\pi\Delta} \tag{3.5.3}$$

Suppose that in a suburban area $\Delta = 0.5 \ \mu s$; then $B'_c = 150$ MHz. Now suppose that in an urban area $\Delta = 3 \ \mu s$; then $B'_c = 25$ MHz. Since B' should always be less than B'_c, then B' will be less than 25 kHz. This takes care of the random phase reduction in the urban area of our example.

3.6 BIT-ERROR RATE AND WORD-ERROR RATE IN FADING ENVIRONMENT

3.6.1 In the Gaussian Noise Environment

In the Gaussian noise environment the bit-error rate (BER) is a function of signal level. In digital modulation the binary waveform is superimposed on a carrier, and then the phase modulation and frequency modulation are commonly employed. Because of the two-level nature of the carrier modulating signal, phase modulation is referred to as *phase shift keying* (PSK), and frequency modulation is referred to as frequency shift keying (FSK). Synchronous demodulation of an FSK signal is called *coherent FSK*. Nonsynchronous demodulation of an FSK signal is called *noncoherent FSK*. Differential phase-shift keying (DPSK) is a modification of PSK that avoids the necessity of providing the synchronous carrier required at the receiver for demodulating a PSK signal. For example, the BER of four kinds of binary (0 and 1) digital modulation schemes are listed as follows:

$$Pe = \frac{1}{2} e^{-\gamma} \quad \text{(DPSK)} \tag{3.6.1}$$

$$Pe = \frac{1}{2} e^{-\gamma/2} \quad \text{(noncoherent FSK)} \tag{3.6.2}$$

$$Pe = \frac{1}{2} \text{erfc}(\sqrt{\gamma}) \quad \text{(PSK)} \tag{3.6.3}$$

$$Pe = \frac{1}{2} \text{erfc}\left(\sqrt{\frac{\gamma}{2}}\right) \quad \text{(coherent FSK)} \tag{3.6.4}$$

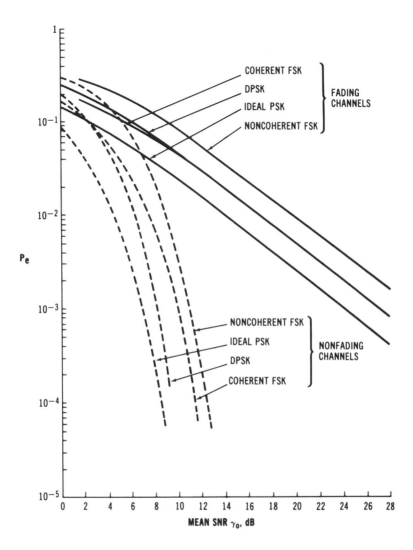

Figure 3.17. Probability of error for several systems in Rayleigh fading.

where γ is the carrier-to-noise ratio C/N, and $\mathrm{erfc}(\cdot)$* is the complementary error function. Equations (3.6.1) through (3.6.4) are plotted in Fig. 3.17. In a Gaussian noise environment the bit-error rate depends only on C/N. The word-error rate P_{ew} can be obtained by assuming that the error of considering each bit independently is the same as the error of considering each bit with its adjacent bits. It is an independent bit condition. The word-error rate of

$$*\mathrm{erfc}(x) = 1 - (2/\sqrt{\pi}) \int_0^x e^{-t^2}\, dt = (2/\sqrt{\pi}) \int_x^\infty e^{-t^2}\, dt$$

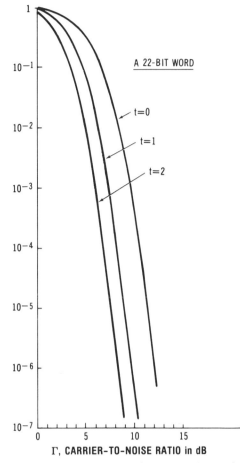

Figure 3.18. Word-error rate in a Gaussian noise environment.

an N-bit word after correcting t bits is

$$p_{ew} = 1 - p(N, 0) - \sum_{m=1}^{t} p(N, m) \qquad (3.6.5)$$

where $p(N, m)$ is the probability that, given a word length of N bits, an m bit in error can be expressed as

$$p(N, m) = \frac{N!}{m!(N - m)!} (1 - P_e)^{N-m} P_e^m$$

where P_e is one of the bit-error rates shown in Eqs. (3.6.1) through (3.6.4). The word-error rate for DPSK is plotted in Fig. 3.18 with t-bit correction. A detailed description of the word-error rate in different word lengths, environments, and vehicle speeds will appear in Section 8.1.

3.6.2   In a Rayleigh Fading Environment

In a Rayleigh fading environment, however, the received C/N varies due to multipath fading. Hence the bit-error rate cannot be based on one constant C/N level γ. An average BER is used in the Rayleigh fading case.

$$\langle Pe \rangle = \int_0^\infty Pe \cdot p(\gamma) \, d\gamma \tag{3.6.6}$$

where $p(\gamma)$ is the pdf of Rayleigh fading and is derived from Eq. (1.5.10) as

$$p(\gamma) = \frac{1}{\Gamma} \exp\left(-\frac{\gamma}{\Gamma}\right) \tag{3.6.7}$$

and Γ is the average C/N ratio of a single branch (channel) receiver. Substituting Eqs. (3.6.1) through (3.6.4) into Eq. (3.6.6) yields

$$\langle Pe \rangle = \frac{1}{2 + 2\Gamma} \quad \text{(DPSK)} \tag{3.6.8}$$

$$\langle Pe \rangle = \frac{1}{2 + \Gamma} \quad \text{(noncoherent FSK)} \tag{3.6.9}$$

$$\langle Pe \rangle = \frac{1}{2}\left[1 - \sqrt{\frac{\Gamma}{\Gamma + 1}}\right] \quad \text{(PSK)} \tag{3.6.10}$$

$$\langle Pe \rangle = \frac{1}{2}\left[1 - \sqrt{\frac{\Gamma}{\Gamma + 2}}\right] \quad \text{(coherent FSK)} \tag{3.6.11}$$

The preceding four equations are plotted in Fig. 3.17. Also for comparison with $\Gamma \gg 1$, the $\langle Pe \rangle$ of both DPSK and coherent FSK coincides and follows the slope of $1/(2\Gamma)$. In general, in a Rayleigh fading environment the BER of digital systems is as follows:

$$\langle Pe \rangle \propto \frac{1}{\Gamma} \quad \text{when } \Gamma \text{ is larger}$$

The BER also increases greatly when there is Rayleigh fading, in contrast to its behavior in a Gaussian environment. For this reason multipath fading in the received mobile radio signal, which results in a higher BER, must be reduced. The following section discusses schemes to reduce fading.

3.6.3 Diversity Transmission for Error Reduction

Diversity schemes described in Section 3.5 are always used to reduce signal fading. Since reducing fading, they reduce the BER. The average bit-error rates for different systems with a maximal-ratio combiner operating in a mobile fading environment are

$$\langle Pe \rangle = \frac{1}{2}\left(\frac{1}{\Gamma + 1}\right)^M \quad \text{(DPSK)} \tag{3.6.12}$$

$$\langle Pe \rangle = \frac{1}{2}\left(\frac{1}{\frac{1}{2}\Gamma + 1}\right)^M \quad \text{(noncoherent FSK)} \tag{3.6.13}$$

The following average bit-error rates are only applied to cases of expecting low bit-error rates:

$$\langle Pe \rangle = \frac{1}{2\sqrt{\pi}}\frac{1}{\Gamma^M}\frac{\left(M - \frac{1}{2}\right)!}{M!} \quad \langle Pe \rangle \ll 1 \text{ (PSK)} \tag{3.6.14}$$

$$\langle Pe \rangle = \frac{1}{2\sqrt{\pi}\left(\frac{1}{2}\Gamma\right)^M} \cdot \frac{\left(M - \frac{1}{2}\right)!}{M!}$$

$$\langle Pe \rangle \ll 1 \text{ (coherent FSK)} \tag{3.6.15}$$

Equations (3.6.12) and (3.6.13) are plotted in Fig. 3.19, and Eqs. (3.6.14) and (3.6.15) in Fig. 3.20. The average bit-error rates decrease as M increases, as shown in all curves. To keep the same average bit-error rates, a noncoherent FSK system needs 3 dB more signal strength than the average bit-error rate received from a coherent FSK system.

3.6.4 Irreducible Bit-Error Rate

In a mobile radio environment the bit-error rate is a function of average signal level and coherence bandwidth based on delay spread. The bit-error rate can be reduced as the average signal level increases. When the average signal level reaches a certain point, the BER will remain constant as the average

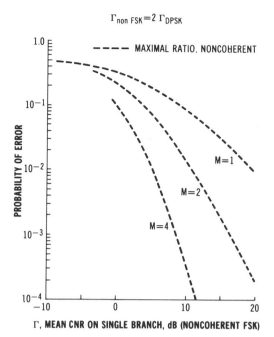

Figure 3.19. Error rates for noncoherent FSK with maximal-ratio combining.

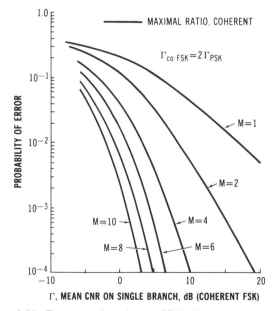

Figure 3.20. Error rates for coherent FSK with maximal-ratio combining.

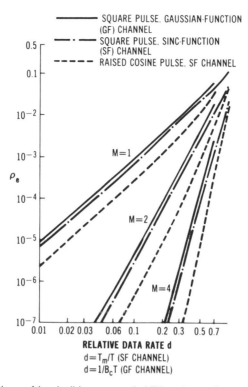

Figure 3.21. Comparison of irreducible-error probabilities due to frequency-selective fadings for the three channel-signal combinations investigated.

signal level continues to increase. This bit-error rate is called an *irreducible bit-error rate*. The irreducible bit-error rate of a DPSK system is shown in Fig. 3.21. The rate of an FSK system is shown in Fig. 3.22. The irreducible bit-error rate can be decreased by increasing the coherence bandwidth, or by using diversity schemes. As Figs. 3.21 and 3.22 indicate, diversity schemes can further lower the irreducible BER.

3.6.5 Overall Bit-Error Rate

The total BER can be summed by three individual BERs based on two different transmission rates, R_{Rfm} and R_Δ. Transmission rate R_{Rfm} is the rate below which random FM errors will occur, and R_Δ is the rate above which intersymbol interference will occur:

$$Pe = Pe_1 + Pe_2 + Pe_3$$

where Pe_1 is the BER due to Rayleigh fading, Pe_2 is the BER due to random FM, and Pe_3 is the irreducible BER due to frequency-selective fading. All three cases of Pe_1, Pe_2, and Pe_3 are shown in Fig. 3.23A where R_t is the transmission rate, f_d the fading frequency, and Δ the time delay spread.

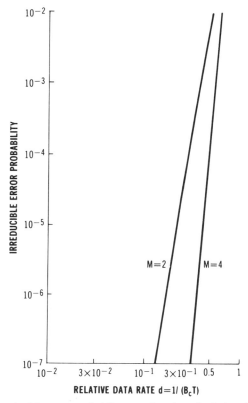

Figure 3.22. Irreducible error probability as a function of relative data rate for FSK.

Usually

$$Pe = Pe_1 + Pe_2 \qquad \text{(when transmission rate is low, } R_t < R_{Rfm})$$

$$Pe = Pe_1 + Pe_3 \qquad \text{(when transmission rate is high, } R_t > R_\Delta)$$

$$Pe = Pe_1 \qquad\qquad \text{(when transmission rate is } R_{Rfm} < R_t < R_\Delta)$$

The total BER is shown in Fig. 3.23B for a DPSK two-branch diversity combining system.

3.7   CALCULATION OF SIGNAL STRENGTH ABOVE A LEVEL IN A CELL (FOR A STATIONARY MOBILE UNIT)

If a mobile unit is stationary within a cell, then what is the chance that it can receive a signal arbitrarily located in that cell? Suppose that the cell radius is 16 km (10 miles), the average power received at 10 miles is P_0, and the CNR at 10 miles is 18 dB. Let us set up a threshold level 6 dB below the average power—in other words, an CNR above 12 dB will be acceptable.

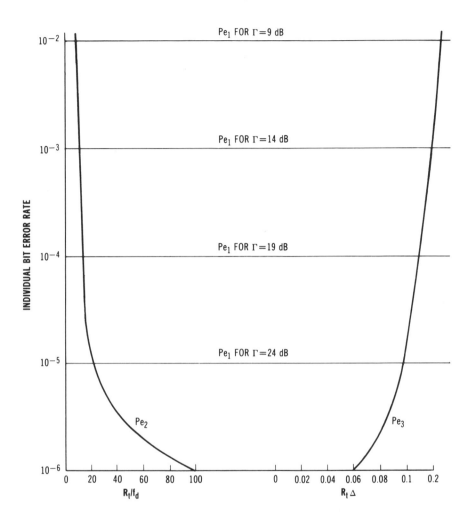

(A) Bit-error rate in a two-branch diversity combining DPSK system subject to envelope fading, random FM, and frequency selective fading. (From Ref. 17)

Figure 3.23. The total bit-error rate in a two-branch diversity combining DPSK system subject to envelope fading, random FM, and frequency-selective fading.

At the cell boundary the probability that a signal exceeds a level of aP_0, where $0 < a < 1$ and P_0 is the average power, is

$$\text{Prob}_{r = 10 \text{ miles}}(P > aP_0) = \int_{aP_0}^{\infty} \frac{1}{P_0} e^{-P/P_0} \, dP$$

$$= \int_{a}^{\infty} e^{-x} \, dx$$

$$= e^{-a} \qquad (3.7.1)$$

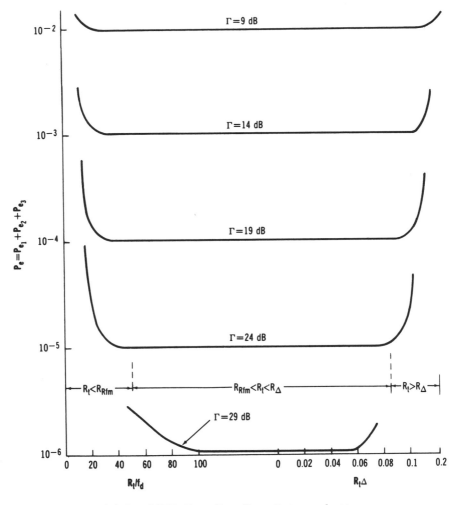

(B) Total BER, $P_e = P_1 + P_2 + P_3$ in a selective Rayleigh fading environment.

Figure 3.23. *(Continued)*

Referring to Fig. 3.24, if the mobile unit is at a distance r_1 from the central transmitting site with $r_1 = 16$ km (10 miles), then the chance of successful signal reception is still based on a level of aP_0, but the average power P_1 is now higher than P_0. Equation (3.7.1) is then modified as follows:

$$\text{Prob}_{r=r_1}(P > aP_0) = \int_{aP_0}^{\infty} \frac{1}{P_1} e^{-P/P_1} \, dP$$

$$= e^{-a(P_0/P_1)} \tag{3.7.2}$$

Suppose that the propagation loss at a distance r_1 is

$$P_1 = kr_1^{-4} \tag{3.7.3}$$

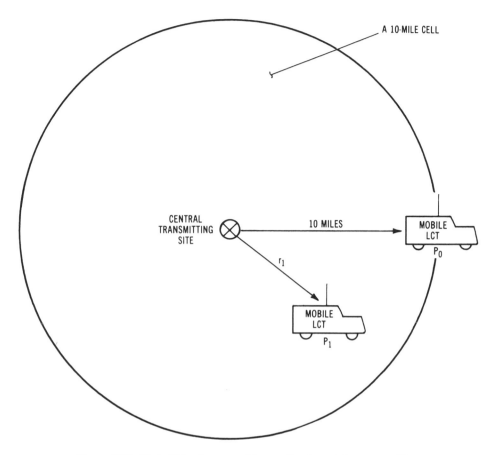

Figure 3.24. Probability of successful operation in a stationary condition.

where k is a constant (see Section 2.3.5). When $r_1 = 10$ miles, $P_1 = P_0$, and Eq. (3.7.3) becomes

$$P_0 = k10^{-4} \qquad (3.7.4)$$

Combining Eqs. (3.7.3) and (3.7.4) and eliminating the factor k, we obtain

$$P_1 = \left(\frac{10}{r_1}\right)^4 P_0 \qquad (3.7.5)$$

Substituting Eq. (3.7.5) for Eq. (3.7.2) yields

$$\text{Prob}_{r=r_1}(P > aP_0) = e^{-a(r_1/10)^4} \qquad (3.7.6)$$

If we let $a = 0.25$ (the threshold level is 6 dB below the average power),

Eq. (3.7.6) can be interpreted as the fraction of the rim of a radius r_1 with received CNR $\geq$ 12 dB.

The area of the portion of the cell that has CNR $\geq$ 12 dB is

$$A_{\text{CNR} \geq 12\text{dB}} = \int_0^{10} e^{-0.25(r_1/10)^4} 2\pi r_1 \, dr_1 \qquad (3.7.7)$$

To find the percentage of the cell that has CNR $\geq$ 12 dB the total cell area is divided by $\pi(10)^2 = 100\pi$:

$$\frac{A_{\text{CNR} \geq 12\text{dB}}}{A_{\text{Total}}} = \frac{2\pi \int_0^{10} e^{-0.25(r_1/10)^4} r_1 \, dr_1}{100\pi} \qquad (3.7.8)$$

Equation (3.7.8) was evaluated by numerical integration, and the fractional area of the cell having CNR $\geq$ 12 dB was found to be 92.25%. This means that a stationary mobile unit can communicate successfully in 92.25% of the area of the cell. In the remaining 7.75% of the cell area, slight vehicle movement would probably move the cellular radio antenna out of the deep null for successful communication.

If a mobile unit does not have a proper signaling format to combat, or reduce, multipath fading, then the severe fading will cause the drop calls while the mobile is moving during the calling. For example, a mobile unit that has almost no protection against fading on its signaling performance must remain stationary at a strong signal spot while calling.

3.8 SINGLE-SIDEBAND (SSB) MODULATION

Single-sideband modulation can be operated with diversity scheme that can help reduce signal fading. Single-sideband modulation is always an attractive modulation in view of spectrum efficiency. A 5-kHz channel bandwidth is used to carry a voice channel in a nonfading environment. Nevertheless, SSB alone cannot be used in a fading environment because the signal (voice or data) is modulated on the amplitude of the carrier. Signal fading is also shown on the amplitude of the carrier due to the transmission medium. Therefore the signal is impaired by the signal fading, especially when severe fading occurs. A two-branch diversity scheme may not be effectively used in this modulation.

In 1978 Lusigman[18] proposed a scheme of using both amplitude and frequency compandings in an SSB modulation for mobile uses—frequency companding to reduce the bandwidth, and amplitude companding to enhance the signal-to-voice performance. This scheme, however, cannot combat severe fading. An additional noise—signal suppression noise—is induced into the signal. A natural approach to reduce the signal suppression noise in SSB modulation in a fading environment is to use a pilot signal as an AGC (automatic gain control). Still we cannot send a pilot signal with a modulation signal such that the frequency separation between them is within the coherent bandwidth criterion for reducing amplitude fading effectively, [shown in Eq. (3.3.1)].

Since the coherent bandwidth criterion [see Eq. (3.3.1)] is derived for angle modulation (FM or PM), not amplitude modulation, the coherent bandwidth does not need to be narrow. The two envelopes of signal fadings do not carry the signal information (voice or data). On the contrary, in SSB modulation, the required frequency separation between the pilot and the modulated carrier needs to be very narrow to achieve the objective of reducing fading effectively. The required envelope correlation coefficient ρ_r to obtain an output signal to signal-suppression-noise ratio (S/N$_s$) of 20 dB is 0.9998.[19] In this condition the pilot AGC limits at -30 dB with respect to the mean power. When the correlation coefficient approaches one, the frequency of the pilot signal should be very close to that of the modulated signal. We can find the separated pilot frequency from Eq. (3.1.11) if we assume that $\Delta d = 0$ and that only one antenna is in use.

$$\rho_r = \frac{1}{1 + (\Delta\omega)^2\Delta^2} \tag{3.8.1}$$

Then

$$\Delta\omega = \frac{1}{\Delta}\sqrt{\frac{1 - \rho_r}{\rho_r}} \tag{3.8.2}$$

or

$$\Delta f = \frac{1}{2\pi\Delta}\sqrt{\frac{1 - \rho_r}{\rho_r}} \tag{3.8.3}$$

where Δf is the frequency separation, Δ is the delay spread, and ρ_r is the envelope correlation coefficient.

Example 3.5. Calculate the two required frequency separations for a pilot signal in a suburban area and an urban area.

Let $\Delta = 0.5$ μs as obtained in a suburban area, and let $\Delta = 3$ μs as in an urban area. Also let $\rho_r = 0.9998$ as mentioned previously. Then the required separation Δf for a pilot can be obtained from Eq. (3.8.3) as

$$f = 4.5 \text{ kHz} \qquad \text{(suburban)}$$
$$f = 0.75 \text{ kHz} \qquad \text{(urban)}$$

This means that Δf has to be equal to or smaller than 4.5 kHz in suburban and 0.75 kHz in urban areas.

McGeehan[20] realized the fact shown in Example 3.5 and came up with an in-band pilot scheme. He separated the voice band and created a gap in the middle of the voice band where the pilot is put before the signal is sent out. At the receiving end, the voice band closes up after the pilot is filtered out. The pilot signal then is effectively used as AGC to combat fading. The practical performance of this as a future system remains to be seen.

The in-band pilot scheme could be less effective in combating rapid fading in urban areas because of the circuit delay in canceling fading. Since the pilot AGC can be effectively changed over a 30-dB dynamic range, a high received carrier-to-noise ratio, say, 30 dB, could be designed for this scheme. However, if the carrier-to-noise ratio at the receiver is less than 30 dB, some voice portions cannot be recovered because of human-made noise level introduced at the receiver.

In general, SSB can be used for fixed point-to-point communications and can take advantage of spectrum efficiency. In a cellular mobile system (see Section 5.5), however, the frequency reuse scheme is used. Since some frequencies will be transmitted simultaneously in more than two different areas, co-channel interference poses a major problem. An analysis of this problem was made by comparing spectrum efficiency between today's existing FM and ideal SSB in cellular mobile systems[20] under the assumption that SSB can completely remove the Rayleigh fading. The conclusion was that FM permitted larger cells while SSB required smaller cells to provide the same voice quality for the same area. Therefore, in a cellular mobile radio environment, FM requires fewer larger cells with less separation between co-channel cells, and SSB requires more smaller cells with greater separation between co-channel cells.

The carrier-to-noise ratio level, 18 dB for FM, used in the analysis is just for establishing a reference level. The conclusion will remain the same if the realistic carrier-to-noise ratio is other than 18 dB. The detail of the analysis will be found in reference 21.

REFERENCES

1. Lee, W. C. Y., "Statistical Analysis of the Level Crossings and Duration of Fades of the Signal from an Energy Density Mobile Radio Antenna," *Bell Sys. Tech. J.* 46 (Feb. 1967): 417–448.
   This was the first in the literature to detail derivations of the level-crossing rates and average deviation of fades of a mobile radio signal received by a whip antenna, a loop antenna, and an energy-density antenna.

2. Ibid.

3. Lee, W. C. Y., *Mobile Communication Engineering* (McGraw Hill, 1982): 189.

4. Rice, S. O., "Distribution of the Duration of Fades in Radio Transmission," *Bell Sys. Tech. J.* 37 (May 1958): 581–635.

5. Lee, W. C. Y., *Mobile Communication Engineering*, 198.

6. Gilbert, E. N., "Energy Reception for Mobile Radio," *Bell Sys. Tech. J.* 44 (Oct. 1965): 1779–1803.

7. Gans, M. J., "A Power-Spectral Theory of Propagation in the Mobile Radio Environment," *IEEE Trans. Veh. Tech.* 21: 1 (Feb. 1972): 27–38.

8. Jakes, W. C., *Microwave Mobile Communications* (Wiley): 29.

9. Lee, W. C. Y., "Comparison of an Energy Density Antenna System with Pre-detection Combining Systems for Mobile Radio," *IEEE Trans. Commun.* 17 (April 1969): 277–284.

10. Gans, M. J., "Propagation," 27–38.

11. Lee, W. C. Y., *Mobile Communication Engineering*, 343.

12. Lee, W. C. Y., Ibid., 219.

13. Lee, W. C. Y., "Mobile Radio Signal Correlation versus Antenna Height and Spacing," *IEEE Trans. Veh. Tech.* 25: 4 (Aug. 1977): 290–292.

14. Lee, W. C. Y., and Y. S. Yeh, "Polarization Diversity System for Mobile Radio," *IEEE Trans. Commun.* Com-20 (Oct. 1972): 912–923.

15. Lee, W. C. Y., "Level Crossings," 417–448.

16. Gilbert, E. N., "Energy Reception," 1779–1803.

17. Jakes, W. C., *Microwave Mobile Communications*, 530.

18. Wilmott, R. M., and B. B. Lusigman, "Spectrum Efficiency Technology for Voice Communications," *UHF Task Force Report FCC/OPP UTF 78-01 (PB 278340) FCC* (Feb. 1978).

19. Gams, M. J., and Y. S. Yeh, "Modulation, Noise and Interference," *Microwave Mobile Communications*, ed. W. C. Jakes (Wiley, 1974): 206, ch. 4.

20. McGeehan, J. P., and A. J. Bateman. "Theoretical and Experimental Investigation of Feedforward Signal Regeneration as a Means of Combating Multipath Propagation Effects in Pilot-Based SSB Mobile Radio Systems," *IEEE Trans. Veh. Tech.* 32 (Feb. 1983): 106–120.

21. Lee, W. C. Y., "Spectrum Efficiency: A Comparison between FM and SSB in Cellular Mobile Systems" (Presented at Office of Science and Technology, FCC, Aug. 2, 1985. Also in *Telephony*, Nov. 1985).

PROBLEMS

3.1   When a vehicle travels at a speed of 35 mph and the signal is received at a frequency of 850 MHz, prove that the level-crossing rate (LCR) at a level of 10 dB below the average power level is approximately 35. What is the LCR when the vehicle speed is at 50 mph? What is the LCR when the vehicle speed is at 80 Kmph?

3.3   When the threshold level is 3 dB below the average power, what is the probability that the fading interval is longer than 1.5 times the average duration of fades $\bar{t}$?

3.4   Let $P_r = 0.5$ and $\tau = 0$. Using Eq. (3.1.11), find the requirement of frequency separation for frequency diversity in suburban and urban areas.

3.5   Simply adding two antenna inputs at the mobile unit cannot reduce the signal fading. Why?

3.6   Explain why in FM the required correlation coefficient between the desired signal and the pilot signal can be 0.5, whereas in SSB the correlation coefficient between pilot signal and the desired signal has to be 0.9998 in order to remove the necessary fading situation by using the pilot.

3.7   The random FM energy will reside at a low frequency, namely, below $f_{rfm} = 2V/\lambda$. If the vehicle is traveling 65 mph, what would be the random FM frequency above which the energy can be neglected? Assume that 850 MHz is the carrier frequency.

3.8   When the mobile unit is moving, both Rayleigh fading and random FM will affect the received signal at the mobile unit. When the mobile unit is standing still, will the Rayleigh fading and random FM still affect the received signal?

3.9   If the data transmission is designed for the mobile unit, what would be the minimum data transmission rate in order to avoid the random FM? Assume that the vehicle is traveling at 65 mph.

3.10   How large must a human-made structure be to be considered a radio scatterer in terms of wavelength?

4

MOBILE RADIO INTERFERENCE

4.1 NOISE-LIMITED AND INTERFERENCE-LIMITED ENVIRONMENT

4.1.1 Noise-Limited Environment

If there are only two transceiver (R/T) communication units in the field in a point-to-point communication, then human-made noise will dominate the performance. Assume that one or both of two units are in motion, then multipath fading also affects the performance. The BER of the modulation schemes; the noncoherent FSK; and the coherent FSK, PSK, and DPSK are described in Section 3.6. The bandwidth requirement for these modulation schemes can be reduced to a relatively low value, depending on the transmitted power or the range-of-communication link since propagation path loss is the only concern in this noise-limited environment.

4.1.2 Interference-Limited Environment

The interference-limited environment is formed only when there are many R/T communication units in the field, when some users are using the same channels, and while others are using adjacent (nearby) channels. Assume that all the situations described in the noise-limited environment are applicable in this environment with additional interference due to co-channel and adjacent-channel communications. This chapter only deals with the interference-limited environment that has always been a factor in designing a system. Traffic conditions are the major concern in this environment, and they will be described in Chapter 8.

4.2 CO-CHANNEL AND ADJACENT-CHANNEL INTERFERENCE

4.2.1 Co-channel Interference

A co-channel arrangement is when two or more communication channels are assigned to the same frequency. The purpose of doing this is to increase the spectrum utilization. In a co-channel environment, two or more co-channel communications are on the air; even the presence of a large deviation in PM or FM (meaning a wideband modulation) does not help reduce the interference in a mobile radio environment.[1,2] Suppose that each antenna have signal coverage in its own cell of radius R and that the distance between two co-channel cells is D. Then the ratio of D/R is used as a key parameter in dealing with co-channel interference:

$$a = \frac{D}{R} \tag{4.2.1}$$

The value a (called the *co-channel reduction factor*) can be determined for any level of signal-to-interference ratio that is required. Therefore a good multiantenna site configuration for a large area should be based on the co-channel reduction factor.

Now we determine the value a from a co-channel interference environment, in which the carrier-to-co-channel interference ratio is greater than 18 dB, or equivalent to 63.1 as

$$\frac{C}{N_0 + I} = \frac{C}{N_0 + \sum\limits_{i=1}^{M} I_i} = 63.1 \tag{4.2.2}$$

where M is the number of interferers. We assume that $M = 6$ and a path loss of 40 dB/dec (i.e., loss is proportional to R^{-4}) are used in a mobile radio environment, as shown in Fig. 4.1. The carrier-to-interference ratio received at a desired cell site (base station) is noted in Fig. 4.1A. The ratio at a desired mobile-unit site is noted in Fig. 4.1B. C_b and C_m are carrier levels, N_b and N_m are noise levels, and I_i and I'_i are interference levels. The letter b stands for the base unit and the letter m for the mobile unit. Carrier levels C_b and C_m should be equal because the reciprocity holds. Noise levels N_b and N_m may be different by 1 to 2 dB (see Section 6.4.1). Since there is only a slight difference between case 1 and case 2, there is no need to distinguish these two cases in terms of their carrier-to-interference ratios. For simplifying the calculation, we assume that the interferers are equidistant from the desired cell. Usually the interference level is much stronger than the local noise level, so local noise is negligible. Then Eq. (4.2.2) becomes

$$\frac{C}{I} = \frac{C}{\sum\limits_{i=1}^{6} I_i} = \frac{R^{-4}}{6D^{-4}} = \frac{a^4}{6} \geq 63.1 \tag{4.2.3}$$

or

$$a = 4.4$$

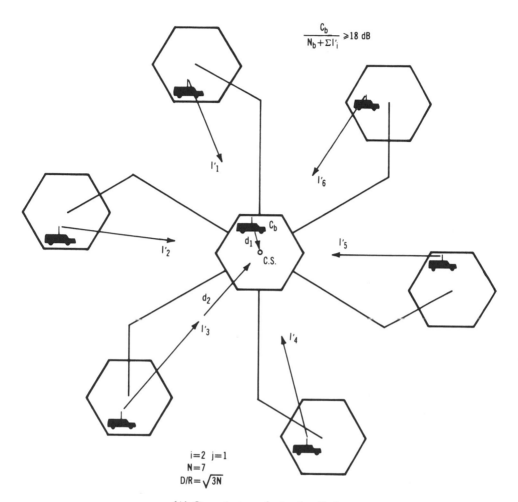

(A) Case 1: At a desired cell site.

Figure 4.1. Co-channel interference in a mobile radio environment with six interferers.

In the preceding equation the factor a is independent of transmitted power as long as the transmitted power is above a minimum value. The determination of minimal transmitted power is based on the carrier-to-noise ratio in a noise-limited environment. Factor a is dependent on the number of interferers. Space separation based on $a = 4.4$ means that when $R = 13$ km (8 miles), $D = 56$ km (35 miles); and when $R = 6$ km (4 miles), $D = 28$ km (17.6 miles). As long as the transmitted power at every co-channel cell site is the same, any transmitted power value can be used without changing the carrier-to-co-channel interference ratio. This is a very important concept.

Co-channel interference can be reduced by other means such as directional antennas, tilted antenna beams, lowered antenna height, and appropriately

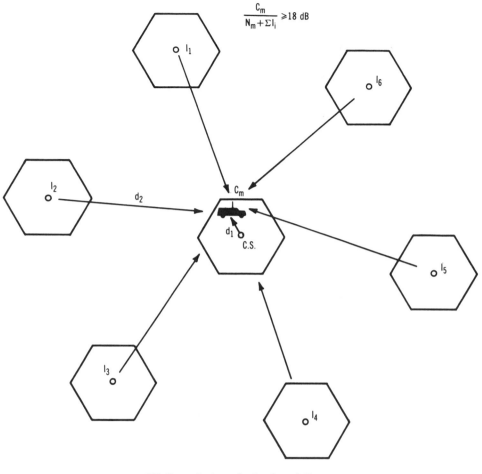

$$\frac{C_m}{N_m + \Sigma I_i} \geqslant 18 \text{ dB}$$

(B) Case 2: At a desired mobile unit.

Figure 4.1. *(Continued)*

chosen site. Directional antennas and the tilting of antenna beams are described in Section 6.3.

4.2.2 Adjacent-Channel Interference

Adjacent-channel interference has slightly better control than co-channel interference. In addition to the other means described in Section 4.2.1, the filter characteristics can help reduce the interference. There is in-band and out-of-band adjacent-channel interference. The former is similar to co-channel interference, and it cannot be filtered.

Out-of-band interference is the adjacent-channel interference that is being considered here. Assume that the filter has a slope of 6 dB/oct, and that the bandwidth of each channel is 30 kHz. The frequency at the edge of the channel is 15 kHz from the center frequency (carrier frequency). Starting from the edge of the channel, we may follow the 6-dB/oct loss (or isolation) and carry it over the frequency range. If a frequency is 240 kHz away from the center of the desired channel, then the loss (or isolation) can be found by letting $f_1 = 15$ kHz, $f_2 = 240$ kHz, and $K = 6$ in Eq. (4.2.4)

$$\text{Loss} = K \log_2\left(\frac{f_2}{f_1}\right) = \frac{K}{0.3} \log_{10}\left(\frac{f_2}{f_1}\right) \quad \text{(dB)} \quad (4.2.4)$$

The result is

$$\text{Loss} = \frac{6}{0.3} \times \log_{10}\frac{240}{15} = 24 \text{ dB}$$

This means that there will be a loss of 24 dB if the power is received at a frequency separated by 240 kHz. Of course the losses due to geographical separation and antenna direction will be added up, along with the loss due to frequency separation. The additional path loss is shown in Eq. (2.3.20) of Section 2.3.5. If the desired source is further away from the receiver than the interference source, then

$$\text{Additional path loss} = 40 \log_{10}\left(\frac{d_1}{d_2}\right) \quad (4.2.5)$$

where d_1 and d_2 are the distances from two co-channel sources to the base-station receiver in a cell, and $d_1 > d_2$, as shown in Fig. 4.2. By setting Eq. (4.2.4) equal to Eq. (4.2.5), we get

$$\left(\frac{d_1}{d_2}\right)^4 = \left(\frac{f_2}{f_1}\right)^{K/3}$$

or

$$f_2 = f_1\left(\frac{d_1}{d_2}\right)^{12/K} \quad (4.2.6)$$

Equation (4.2.6) is illustrated in Fig. 4.2 for different values of K. Once the frequency f_2 is obtained, the number of channel separations is found from

$$\text{Channel separation} = \frac{|f_2 - f_1|}{B} \quad (4.2.7)$$

where B is the channel bandwidth. Equation (4.2.6) indicates that interference

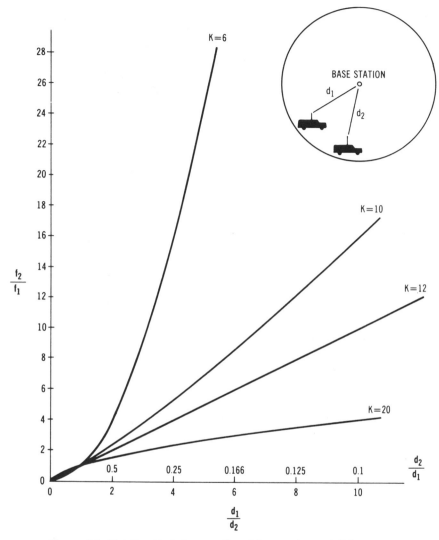

Figure 4.2. Relationship between ratios of frequencies and distances.

due to geographical position can be reduced by using frequency channel separation. Here the definition of "adjacent channel" needs clarification. There are actually two definitions, one for naturally adjacent channels and the other for system-adjacent channels. The naturally adjacent channels are the channels next to each other in the frequency spectrum. These natural adjacent channels affect the system's design so we try to separate them from system adjacent channels. The system-adjacent channels are those channel frequencies closest to each other among a set of discreet channels. For example, if there are 10 channels, each channel has 30 kHz bandwidth, and the separation between the two closed channels (adjacent channels) is 500 kHz, then the total channel bandwidth is 5 MHz. The system-adjacent channels are the channels that need enough frequency separation to avoid adjacent-channel interference. In other words, adjacent-channel interference can be reduced by considering the losses from the following conditions:

$$D = \text{geographical separation, } D/R$$

$$|f_1 - f_2| = \text{frequency separation, } \Delta f = |f_1 - f_2|$$

as shown in Eq. (4.2.7). Other ways to reduce interference are by using directional antenna beams pointing in different horizontal directions (Section 6.3) and at different angles of elevation (Section 6.3). Reducing adjacent-channel interference by selective modulation schemes is not effective unless a spread spectrum system is used. The pros and cons of using a spread spectrum system will appear in chapter 9.

4.3   INTERMODULATION (IM)

Intermodulation (IM) occurs because of a nonlinear process, as when an input signal is the sum of N frequencies. Passing a signal through a power amplifier or a hard limiter will produce IM terms.

4.3.1   Through a Power Amplifier

There are two cases to be considered:

CASE 1

Effect on an Angle-Modulated Signal. Assume that an angle-modulated signal is

$$e_i = A_c \cos[2\pi f_c t + \phi_s(t)] \tag{4.3.1}$$

where A_c and f_c are the amplitude and frequency, respectively, and $\phi_s(t)$ is the transmitted signal. Also assume that the power amplifier has the following nonlinear process

$$e_0 = a_1 e_i + a_2 e_i^2 + a_3 e_i^3 \tag{4.3.2}$$

where a_1, a_2, and a_3 are constants. Substituting Eq. (4.3.1) into Eq. (4.3.2) yields

$$e_0 = \frac{1}{2} a_2 A_c^2 + \left(a_1 A_c + \frac{3}{4} a_3 A_c^3 \right) \cos(2\pi f_c t + \phi_s(t))$$

$$+ \text{ higher-order terms} \tag{4.3.3}$$

A filter can be used to extract the angle-modulated signal centered at f_c. The amplitude of the output e_0' of the filter is

$$|e_0'| = a_1 A_c + \frac{3}{4} a_3 A_c^3 \tag{4.3.4}$$

The nonlinear characteristic has done nothing more than modify the gain $a_1 A_c$ with an additional term in Eq. (4.3.4). The signal information retained in the phase of the angle modulation remains unchanged. This is an important difference between using AM and angle modulation, and it is a primary reason why angle modulation is used in mobile radio systems where the use of nonlinear power amplifiers cannot be avoided.

CASE 2

Effect on N Signal Inputs. Assume that an input signal is the sum of three individual sinusoids ($N = 3$) as follows:

$$e_i = A \cos \alpha t + B \cos \beta t + C \cos \gamma t \tag{4.3.5}$$

where α, β, and γ are the angular frequencies. Substituting Eq. (4.3.5) into Eq. (4.3.2) yields

e_0 = dc (3 terms) + first order (3 desired terms plus 9 undesired terms)

+ second order (9 IM terms) + third order (19 IM terms)   (4.3.6)

Each order is listed in Table 4.1.

If all amplitudes A, B, and C are equal, then the power contained in the $\alpha \pm \beta$ products is 6 dB higher than the products 2α, 2β, or 2γ, as can be seen from Table 4.1:

$$P_{\alpha \pm \beta} = P_{2\alpha} + 6 \text{ (dBm)} \tag{4.3.7}$$

Similarly

$$P_{2\alpha \pm \beta} = P_{3\alpha} + 9.6 \quad \text{(dBm)} \tag{4.3.8}$$

$$P_{\alpha \pm \beta \pm \gamma} = P_{3\alpha} + 15.6 \quad \text{(dBm)} \tag{4.3.9}$$

All 28 IM terms will contaminate the medium while being transmitted from the power amplifier. If this contamination cannot be properly suppressed, the spurious noise level in the medium will rise.

Besides the consideration of IM reduction, the three frequencies (α, β, γ)

TABLE 4.1 Frequencies and Relative Magnitudes To Be Found in Output, $e_0 = a_1 e_i^1 + a_2 e_i^2 + a_3 e_i^3$, from Applied Signal, $e_i = A \cos \alpha t + B \cos \beta t + C \cos \gamma t$

	Term 1	Term 2	Term 3
d–c		$\tfrac{1}{2} a_2(A^2 + B^2 + C^2)$	
First order	$a_1 A \cos \alpha t + a_1 B \cos \beta t$ $+ a_1 C \cos \gamma t$		$\tfrac{3}{4} a_3 A(A^2 + 2B^2 + 2C^2)\cos \alpha t$ $+ \tfrac{3}{4} a_3 B(B^2 + 2C^2 + 2A^2)\cos \beta t$ $+ \tfrac{3}{4} a_3 C(C^2 + 2A^2 + 2B^2)\cos \gamma t$
Second order		$\tfrac{1}{2} a_2(A^2 \cos 2\alpha t + B^2 \cos 2\beta t$ $+ C^2 \cos 2\gamma t)$ $+ a_2 AB[\cos(\alpha + \beta)t + \cos(\alpha - \beta)t]$ $+ a_2 BC[\cos(\beta + \gamma)t + \cos(\beta - \gamma)t]$ $+ a_2 AC[\cos(\alpha + \gamma)t + \cos(\alpha - \gamma)t]$	
Third order			$\tfrac{1}{4} a_3(A^3 \cos 3\alpha t + B^3 \cos 3\beta t + C^3 \cos 3\gamma t)$ $+ \tfrac{3}{4} a_3 \left\{ \begin{array}{l} A^2 B[\cos(2\alpha + \beta)t + \cos(2\alpha - \beta)t] \\ A^2 C[\cos(2\alpha + \gamma)t + \cos(2\alpha - \gamma)t] \\ B^2 A[\cos(2\beta + \alpha)t + \cos(2\beta - \alpha)t] \\ B^2 C[\cos(2\beta + \gamma)t + \cos(2\beta - \gamma)t] \\ C^2 A[\cos(2\gamma + \alpha)t + \cos(2\gamma - \alpha)t] \\ C^2 B[\cos(2\gamma + \beta)t + \cos(2\gamma - \beta)t] \end{array} \right.$ $+ \tfrac{3}{2} a_3 ABC[\cos(\alpha + \beta + \gamma)t + \cos(\alpha + \beta - \gamma)t$ $+ \cos(\alpha - \beta + \gamma)t + \cos(\alpha - \beta - \gamma)t]$

TABLE 4.2   Number of IM Products of Various Orders N Input Sinusoids

Im Form	Order m	Number of Frequencies L in IM	Total Number of IM Products
$2\alpha - \beta$	3	2	$N(N - 1)$
$^a\alpha + \beta - \gamma$	3	3	$T_3 \triangleq N(N - 1)(N - 2)/2$
$3\alpha - \beta - \gamma$	5	3	$N(N - 1)(N - 2)/2$
$\alpha + 2\beta - 2\gamma$	5	3	$N(N - 1)(N - 2)$
$^a\alpha + \beta + \gamma - \sigma - \eta$	5	5	$T_5 \triangleq N(N - 1(N - 2)$ $(N - 3)(N - 4)/12$

Source: From reference 3.
Note: The notation α, β, γ, σ, or η represents any of the input frequencies but $\alpha \neq \beta \neq \gamma \neq \sigma \neq \eta$.
[a]Indicates the dominant form of the third- and fifth-order cross products.

should be carefully assigned so that the following three relationships hold and spurious noise is avoided:

$$\gamma \neq \alpha \pm \beta$$

$$\gamma \neq 2\alpha \pm \beta \qquad (4.3.10)$$

$$2\gamma \neq \alpha \pm \beta$$

The spurious noise is caused by the IM product falling into the desired frequencies.

4.3.2   Through a Hard Limiter

Assume that an input signal e_i is a set of five angle-modulated signals:

$$e_i = A \cos(\alpha t) + B \cos(\beta t) + C \cos(\sigma t)$$

$$+ D \cos(\sigma t) + E \cos(\eta t) \qquad (4.3.11)$$

Let e_i pass through a hard limiter

$$e_0 = sgn \ e_i = \begin{cases} +1 & e_i > 0 \\ -1 & e_i < 0 \end{cases} \qquad (4.3.12)$$

The symbol "*sgn*" is used to represent Eq. (4.3.12) for the limiting condition. The number of IM products of various orders are listed in Table 4.2.[3] For

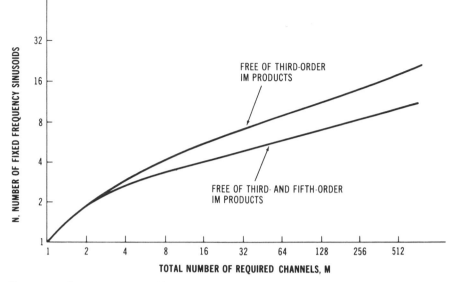

Figure 4.3. Required number of frequency channels versus number of frequency for a nonlinear transponder.

the order $m \geq 3$ listed in Table 4.2, the IM power P_m can be obtained approximately,

$$P_m \approx \frac{P}{m^2} \qquad (4.3.13)$$

where P is the desired signal power.

Free of IM Products

Suppose that N frequency sinusoids subchannels are to be packed in a combined channel of bandwidth MB, where B is the bandwidth of subchannels and M is the total number of channels required to avoid IM among N subchannels. We can compute the minimum bandwidth MB required to avoid third-order, or third- and fifth-order, IM products for a given N, as shown in Fig. 4.3. If $N = 8$, then $M = 48$—or a total channel of 48 B is needed to be free of third-order IM only among 8 subchannels. If the same $N = 8$, then $M = 150$—or a channel of 150 B is needed to be free of both third- and fifth-order IM products among those 8 subchannels. To be free of IM products, a wide RF bandwidth must be used for just a few channels.

Frequency Assignments Specifics

In Table 4.3 several specific channel frequency assignments are shown[4] for two sets of circumstances: no IM product spreading, or IM product spreading

TABLE 4.3 Frequency Plans to Avoid Third-Order IM Products with and without Spreading

Signal Channels	N	Total Channels M	Frequencies F_i
IM product	3	4	1, 2, 4
spreading	4	7	1, 2, 5, 7
	5	12	1, 2, 5, 10, 12
	6	18	1, 2, 5, 11, 13, 18
	7	26	1, 2, 5, 11, 19, 24, 26
	8	35	1, 2, 5, 10, 16, 23, 33, 35
	9	46	1, 2, 5, 14, 25, 31, 39, 41, 46
	10	ว2	1, 2, 8, 12, 27, 46, 48, 57, 60, 62
No IM	3	7	1, 3, 7
product spreading	4	15	1, 3, 7, 15

Source: From reference 4.

limited to 3 B. For example, in an IM product-spreading case four different carrier frequencies would take a total bandwidth of 7 B. The same four carrier frequencies used in the no-IM product-spreading case would take a total bandwidth of 15 B. Therefore a trade-off is necessary: Either IM products must be spread or a large frequency spectrum must be sacrificed.

4.4 NEAR-END-TO-FAR-END RATIO

Geographical distance separation always helps reduce signal interference. However, there is one case where mobile radio communication may be hurt by geographical distance separation. Consider that every mobile unit moves within the coverage of a base station. Some mobile units are always closer to the base station than others. Imagine that two mobile units simultaneously transmit signals to the base station. The signal received from the mobile unit closer to the base station is stronger than the signal received from the mobile unit farther away. The stronger received signal will mask the weaker one. The degree of masking depends on the difference in distance to the base station. The power difference due to the path loss between the receiving location and the two transmitters is called *near-end-to-far-end ratio* interference.[5]

$$\text{Near-end-to-far-end ratio} = \frac{\text{path loss due to } d_2 \text{ (near end)}}{\text{path loss due to } d_1 \text{ (far end)}} \quad (4.4.1)$$

An illustration of near-end-to-far-end ratio appears in Fig. 4.4. Let the transmitting power of two mobile units be the same, let Tx_1 be the desired signal at a distance $d_1 = 16$ km (10 miles) from the base station, and let Tx_2

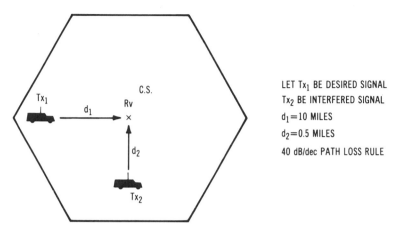

Figure 4.4. Illustration of the near-end-to-far-end ratio situation.

be the interferring signal at a distance d_2 = 0.8 km (0.5 mile). Use a 40 dB/dec path-loss rule. Then the near-end-to-far-end power ratio is

$$\text{Near-end-to-far-end ratio} = 40 \log_{10} \frac{d_2}{d_1}$$

$$= 40 \log_{20} = 52 \text{ dB} \qquad (4.4.2)$$

Equation (4.4.2) shows that the signal received at the base station from the far-end mobile unit is 52 dB weaker than the signal from the near-end mobile unit.

To reduce this far-end-to-near-end ratio interference for mobile unit Tx_1, two signals must be transmitted in two different frequencies that have enough separation dependent on the filter characteristics of the response shape. If a filter characteristic of 12 dB/oct is used, the frequency separation between two signals that will achieve isolation can be found by converting Eq. (4.2.4) as follows:

$$\frac{f_2}{f_1} = 10^{\eta} \qquad (4.4.3)$$

where

$$\eta = \frac{\text{Loss} \times 0.3}{K} \qquad (4.4.4)$$

All the parameters are described in Eq. (4.2.4). A frequency separation based on an isolation of 52 dB can be found by applying Eq. (4.4.3). Let loss be 52 dB and $K = 12$, then

$$\eta = \frac{52 \times 0.3}{12} = 1.3$$

and

$$\frac{f_2}{f_1} = 10^{1.3} = 20 \qquad (4.4.5)$$

If the bandwidth of a desired signal is 30 kHz, then $f_1 = 15$ kHz and $f_2 = 300$ kHz. The ratio $f_2/f_1 = 20$ also indicates that a separation of 20 channels is needed, regardless of the bandwidth of each channel.

The near-end-to-far-end ratio can exist in two systems: mobile to base and mobile to mobile. The former always uses the base station as a relay to connect to another mobile unit or to wireline telephones. (This was described in the early part of this section.) The latter does not involve a base station, and it is the system usually used by the military. The frequency separation requirement for reducing the near-end-to-far-end ratio is the same.

The near-end-to-far-end ratio interference is also commonly called the near-far interference. The other method of reducing this interference is to use power control to reduce the transmit power of the near-end mobile unit.

4.5 INTERSYMBOL INTERFERENCE

Intersymbol interference is due to a relatively large delay spread in a multipath medium or a relatively high transmission bit rate. Assume that 1 bps requires 1 Hz for a binary PSK system; then the transmission bit rate R_t can be determined by

$$R_t < \frac{1}{\Delta} \qquad (4.5.1)$$

In Section 1.5.6 we saw that if the delay spread Δ in an urban area is 3 μs, then R_t in the urban area cannot exceed

$$R_t < \frac{1}{3 \times 10^{-6}} = 3.33 \times 10^5 \text{ bps}$$

Tighter parameters may be needed by the coherence bandwidth criterion shown in Eq. (3.3.1). If two adjacent bits read in two time slots within a time separation of $2\pi\Delta$ tentatively interfere with each other and if two bits read in two time slots separated by more than $2\pi\Delta$ are not interfered with by adjacent bits, then the transmission bit rate in a Rayleigh fading environment would be

$$R_t < \frac{1}{2\pi\Delta} \qquad (4.5.2)$$

Equation (3.3.1) is based on a correlation coefficient between two consecutive bits within 0.5 as a criterion for the coherence bandwidth. The transmission rate R_t in Eq. (4.5.2) is also based on the criterion. Transmission rate R_t in an urban area is

$$R_t < \frac{1}{2\pi \times 3 \times 10^{-6}} = 5.3 \times 10^4 \text{ bps} \qquad (4.5.3)$$

An upper-level transmission rate in an urban area can also be determined more accurately by setting a specified bit-error rate that is affected by the delay spread, as described in Section 3.6 and shown in Fig. 3.23. When an irreducible bit-error rate is set at 10^{-4}, the value of d read from the curve is

$$d = \frac{R_t}{B_c} = 0.06$$

Assume that an urban area is being considered, then

$$R_t = 0.06B_c = \frac{0.06}{2\pi\Delta} = 3183 \text{ bps}$$

The value R_t can be increased by adding *diversity,* as shown in Figs. 3.21 and 3.22. R_t is limited by the delay spread of the medium at the high end and the random FM at the low end. When R_t exceeds the high end, intersymbol interference occurs.

4.6 SIMULCAST INTERFERENCE

Audio information transmitted over two or more transmitters operating on the same *rf* carrier frequency is called simulcast. Two cases are shown in Fig. 4.5. Two relays carry the same signal to the mobile unit in case 1 and three transmitters carry the same signal in case 2. The advantage of using simulcast in areawide coverage is to simplify dispatching, or to simplify area-wide, mobile-to-mobile communications.[6] Simulcast can sometimes improve the coverage in a mobile radio system. A simulcast is considered to create intentional multipaths in the environment. The receiver has to detect the sum of the two transmitted signals from two transmitters as

$$s_r = A \sin(\omega_a t + \phi_a) + B \sin(\omega_b t + \phi_b) \qquad (4.6.1)$$

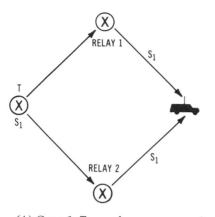 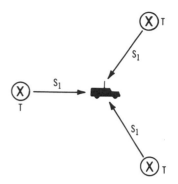

(A) Case 1: Two relays carry on the same signal.   *(B) Case 2: Three transmitters carry on the same signal.*

Figure 4.5. A simulcast environment.

Thus A and B are *rf* amplitudes, ω_a and ω_b are carrier frequencies, ϕ_a and ϕ_b are angle modulation (FM or PM), and ϕ_a and ϕ_b can be expanded as

$$\phi_a = \phi_{ma} \sin(\omega_{ma}t + \theta_a) + \phi_{ca} \tag{4.6.2}$$

$$\phi_b = \phi_{mb} \sin(\omega_{mb} + \theta_b) + \phi_{cb} \tag{4.6.3}$$

where ϕ_{ma} and ϕ_{mb} are peak deviations, ω_{ma} and ω_{mb} are the modulating audio frequencies, θ_a and θ_b are the audio phase delay, and ϕ_{ca} and ϕ_{cb} are the *rf* carrier phase delay from transmitter A and transmitter B, respectively. Ideally Eqs. (4.6.2) and (4.6.3) should be identical in order to eliminate simulcast interference provided $\omega_a = \omega_b$.

In a real mobile radio environment, the following situations can occur:

1. $\omega_a \neq \omega_b$: This situation causes a beat frequency with its harmonics due to the FM detection in the receiver.
2. $\omega_{ma} \neq \omega_{mb}$: Audio wires leading to each transmitter can cause a frequency shift. Audio distortion results.
3. $\phi_{ca} \neq \phi_{cb}$: This phase difference between carrier frequencies causes *rf* standing waves to occur. The *rf* amplitude level will vary. This situation always occurs in the mobile radio environment.
4. $\theta_a \neq \theta_b$: A phase difference between two audio-modulating signals results in harmonics and loss of detected signal.
5. $\phi_{ma} \neq \phi_{mb}$: An amplitude difference between two audio-modulating signals results in the same kind of disturbance as caused by the audio phase difference.

Simulcast is technically challenging. Good synchronization and good re-

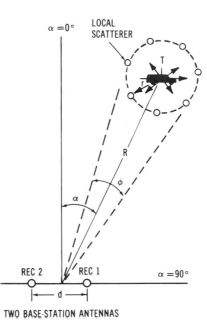

Figure 4.6. A model for determining the radius of local scatterers.

duction of audio phase and level differences between transmitters are the primary requirements. In a mobile radio telephone system, audio and data messages are not sent on the same frequency but assigned to different transmitters. Therefore no simulcast interference exists. However, the simulcast interference can be occurred in a paging system when one or more repeaters are used in the paging system.

4.7 RADIUS OF LOCAL SCATTERERS[7]

The terrain configuration usually dominates the propagation path loss in that area, and the local scatterers surrounding the mobile unit cause short-term fading. Short-term fading is one kind of interference. The local scatterers are so named if two requirements are met: (1) the sizes of the scatterers are greater than the operating wavelength, and (2) the heights of the scatterers are higher than the mobile antenna height. Naturally the surrounding houses and buildings around the mobile unit meet the two requirements and are local scatterers. Now the question is raised, How large is the area in which the effective scatterers are surrounding the mobile unit? The radius of a group of active local scatterers cannot be actually measured. However, we can obtain it indirectly by comparing the measured data to a theoretical model described in reference 8.

First a theoretical derivation has been based on a model shown in Fig. 4.6. Assume that there is no direct path from the mobile transmitter to the base-

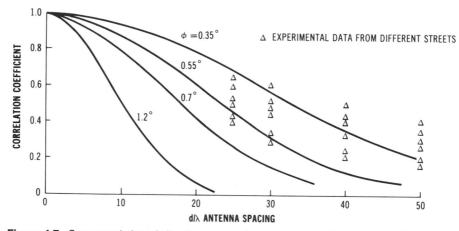

Figure 4.7. Crosscorrelation of signal envelopes from two base-station antennas: Theoretical and experimental correlations versus antenna spacing for the broadside propagation case.

station receiver. The distance between the base station and the mobile unit is R, and the radius of scatterers surrounding the mobile transmitter is r. The radius of scatterers r is defined as all active scatterers within the radius r. The objects outside the radius r will not be considered as scatterers because their scattering energy has only a negligible effect on the received signal at the base station. The angular sector ϕ of the signal arrival can be expressed as

$$\phi = \frac{R}{2r}$$

The theoretical correlation coefficients between two signals received by two base-station antennas with different values of angular sector ϕ are plotted against the base-station antenna spacing d, as shown in Fig. 4.7 for the $\alpha = 0°$ case. A collection of experimental correlation data taken 5 km (3 miles) away from the base station in a suburban area is also plotted in Fig. 4.7. The experimental data match the theoretical curve closely at an angular sector of 0.5°. Then the radius of local scatterers can be roughly estimated by

$$r = \frac{R\phi}{2} = \frac{(3 \text{ miles})(0.5°/57.3^{\text{red}})}{2} = 69 \text{ ft} \qquad \text{(suburban area)}$$

This result can be converted to wavelengths as 69λ at 850 MHz. Since the experimental data were taken in a suburban area where the houses, buildings, and other built structures were within 21 m (69 ft) of the mobile transmitter, the radius of scatterers of 69 ft indicates that the physical objects closer to the mobile transmitter are the active ones. It also indicates that secondary reflections due to the houses or buildings further away than 69 ft do not interfere with the received signal at the base station. Although the radius of scatterers is mainly dependent on the built environment, it is also affected by the wavelength. When the operating frequency is lower, the propagation loss is smaller. The radius of scatterers should actually be slightly larger. Therefore the radius of scatterers in the mobile radio environment is around 15 to 30 m (50 to 100 ft) for the frequencies around 850 MHz. Since the antenna spacing d at the base station is measured in wavelengths, we convert r from its physical length into the wavelengths

$$r = 50\lambda \text{ to } 100\lambda \qquad \text{(suburban area)}$$

for all the frequencies from 30 MHz to 10 GHz.

The radius of scatterers r is different in physical length with different wavelengths. This is why the required antenna spacing for a given correlation coefficient between two base-station signals is dependent on the wavelength and not the physical length.

REFERENCES

1. Lee, W. C. Y., *Mobile Communication Engineering* (McGraw-Hill, 1982): 369.
2. Prabhu, V. K., and H. E. Rowe, "Spectral Density Bounds of a PM Wave," *Bell Sys. Tech. J.* 48 (March 1969): 789–811.
3. Spilker, J. J., Jr., *Digital Communicating by Satellite* (Prentice Hall, 1977): 243.
4. Spilker, J. J., Jr., *Digital Communicating,* 252.
5. Lee, W. C. Y., "Elements of Mobile Cellular System," will be published by *IEEE Transactions on Vehicular Tech.,* May 1986.
6. Ade, John E., "Some Aspects of the Theory of Simulcast" (Paper presented at the IEEE 32nd Vehicular Technology Conference, San Diego, CA, May 1982): 133–139.
7. Lee, W. C. Y., "Effects on Correlation between Two Mobile Radio Base-Station Antennas," *IEEE Trans. Commun.* 21 (Nov. 1973): 1214–1224.
8. Lee, W. C. Y., "Antenna Spacing Requirement for a Mobile Radio Base-Station Diversity," *Bell Sys. Tech. J.* 50: 6 (July–Aug. 1971): 1859–1876.

PROBLEMS

4.1   In the equal strength contour of a cell shown in Fig. P4.1, find the virtual center and the average radius. How is the D/R ratio measured?

4.2   Co-channel interference suppression is based on C/I equal or greater than 18 dB. If the noise level is 20 dB below the signal (i.e., $C/N = 20$ dB), then what is $C/(N + I)$?

4.3 Filter A has a slope of 24 dB/oct, and filter B has a slope of 15 dB/oct. The f_2 is three times the channel distance away from f_1. What would be the distance separation between a mobile unit with f_2 and a mobile unit with f_1 within a cell if the mobile unit with f_1 is at the boundary of a cell whose radius is R? First assume filter A is used then filter B is used in both mobile units.

4.4 Find the IM power through a hard limiter for the third order and the fifth order. What is the signal-to-IM (third and fifth) noise?

4.5 If in Fig. 4.1, $C_b = -100$ dBm, $N_b = -119$ dBm, and each of six interferers $I_i' = -121$ dBm, find the value of

$$\frac{C_b}{\left(N_b + \sum_6^6 I_i\right)}$$

4.6 If $C_b = -100$ dBm, $I_i' = -121$ dBm, and $C_b/(N_b + \sum_1^6 I_i) = 18$ dB, what will be the level of N_b?

4.7 If without interference, $C/N = 18$ dB and with interference $C/(N + I) = 15$ dB, what is C/I?

4.8 If $C/N = 18$ dB and $C/I = 16$ dB, what is $C/(N + I)$?

4.9 Given $C/(N + I) = 20$ dB and $I/N = 4$ dB, find the level of C if N is -120 dBm.

4.10 In an interference environment $C/(N + I) = 20$ dB. Given $N = -115$ dBm and $I = -113$ dBm, what is the value of C?

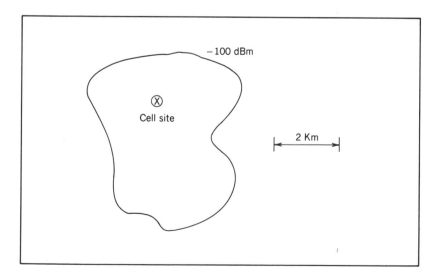

Figure P4.1

5

FREQUENCY PLANS AND THEIR
ASSOCIATED SCHEMES

As described in Chapter 4, a proper frequency plan is needed to reduce adjacent-channel interference and intermodulation. This chapter covers many frequency plans that maximize spectrum utilization.

5.1 CHANNELIZED SCHEMES AND FREQUENCY REUSE

5.1.1 Channelized Schemes

At the base station there are N channels associated with either N individual power amplifiers or one shared power amplifier, as shown in Fig. 5.1. The former is called a *channelized scheme;* the latter, a *multiplexed scheme*. This section covers the channelized scheme, and the multiplexed scheme is described in the following section.

In channelized schemes each channel has its own power amplifier, and the intermodulation (IM) products will not be produced by nonlinear processing since there is only one input. However, the power combiner combines all individual channels and sends them to the transmitting antenna. This power combiner cannot be the perfect impedance match device; the power at the combiner can leak from one port to another, feed backward to the power amplifier, and cause IM. A typical power combiner[1] at 1 GHz has 16 cavity resonators, each functioning as a narrowband filter feeding a common load— a transmitting antenna. This combiner can have a maximum of 3-dB loss per channel and a minimum channel-to-channel isolation of 18 dB. To meet the 3-dB loss per channel, the channels are spaced 630 kHz or 21 channels (21 × 30 kHz) apart; the filter cutoff edge can be broadened to reduce losses. Thus IM is controlled by ferrite isolators, providing 30-dB reverse losses. A diagram of this particular combiner is shown in Fig. 5.2. With three channels simultaneously power excited, the measured intermodulation products (with channel spacing of 21 B, where B is the frequency bandwidth of each channel)

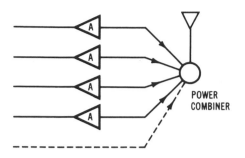
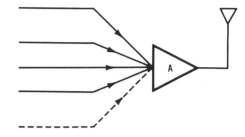

(A) Channelizing scheme.

(B) Frequency divisions multiplexing (FDM).

Figure 5.1. Comparison of channelizing scheme and FDM.

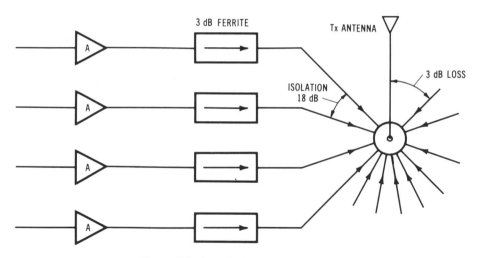

Figure 5.2. A particular power combiner.

are down at least 55 dB from the desired signals. For frequencies lower than 150 MHz, the physical size of the waveguide becomes impractical. Then the combiner may be designed using coaxial cables or LC type circuits. The principal considerations remain the same.

5.1.2 Frequency Reuse

Since the IM products level can be controlled by channelized schemes, the same channel can be reused, at a certain distance governed by the co-channel reduction factor as $(a = D/R)$ stated in Section 4.2.1. The value of a will be increased if the number of co-channel sites increases. For the same number of co-channel sites, there can be two cases: an average case and a worst cast.

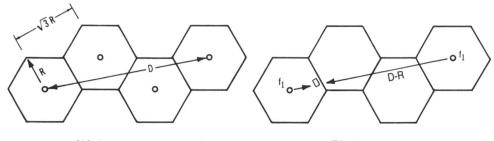

(A) An average case.   (B) A worst case.

Figure 5.3. Estimate of co-channel interference.

Consider an average case. Let a_1 be the co-channel interference reduction factor CIRF of an average case. (See Fig. 5.3A.)

$$a_1 = \frac{D}{R} \tag{5.1.1}$$

Then the new parameter a_1' is defined as follows:[2]

$$a_1' = \frac{D - R}{R}$$

$$= a_1 - 1 \tag{5.1.2}$$

For the average case (Fig. 5.3A), Section 4.2.1 shows that the value of a has to be equal or greater than 4.4 in order to have a C/I ratio of 18 dB based on six equidistant co-channel interferers.

If the worst case is considered, as shown in Fig. 5.3B for one co-channel cell, the distance between the co-channel cell site and the mobile unit is $D - R$. In reality, due to the imperfect site locations and the geographical shadowing, we may assume that the distance between the mobile unit and all co-channel interferers is $D - R$ for the worst situation. The carrier-to-interference ratio $(C/I)_w$, with six co-channel interferers, can be expressed as

$$\left(\frac{C}{I}\right)_w = \frac{C}{\sum_{i=1}^{6} I_i} = \frac{R^{-4}}{6(D - R)^{-4}} = \frac{a_1'^4}{6} \tag{5.1.3}$$

Let C/I be 18 dB or above (i.e., a value of 63). This means the carrier is 63 times stronger than the interferers. In Eq. (5.1.3),

$$\left(\frac{C}{I}\right)_w = \frac{a_1'^4}{6} \geq 63 \tag{5.1.4}$$

In Eq. (5.1.4), $a_1' = 4.4$. Then the co-channel interference reduction factor, a_1 from Eq. (5.1.2) becomes

$$a_1 = a_1' + 1 = 5.4$$

For example, when a worst case is considered a 13-km (8-mile) cell needs a reuse distance of 13 × 5.4, which is 70 km (43.2 miles).

An average case is generally used to estimate the values of all parameters. In designing a system, all parameter values should be determined for a worst case to ensure system performance. See reference 2 for how to compute accurately the worst case.

5.2 FREQUENCY-DIVISION MULTIPLEXING (FDM)

A given frequency band is divided into many frequency channels. Each signal is assigned to a separate non-overlapping frequency channel. All the signals are combined through a common power amplifier. Intermodulation products through a common power amplifier are either accepted or minimized by appropriate frequency selection and/or reduction of input power levels to permit quasi-linear operation.

The channel assignment format of FDM depends on signal distortion, adjacent-channel interference, and IM effects.[3] Guard bands are usually used for adjacent frequency channels.

5.2.1 FDM Signal Suppression

By Bandpass Nonlinearities (A General Consideration)

Consider a fixed envelope sinusoid with an amplitude B, received by the mobile unit with a large number of other sinusoidal signals that form a time-varying Gaussian interference.

The envelope A of the time-varying Gaussian interference is Rayleigh. Its probability density of A can be expressed

$$p(A) = \frac{A}{\sigma^2} e^{-A^2/2\sigma^2}$$

and its average power is

$$2\sigma^2 = \int_0^\infty A^2 p(A) \, dA$$

Now show the signal suppression of an input signal B when signal B and the rest of the interference pass through nonlinear devices. The sum of signal B and the combined interference A at the input of a nonlinear device is

$$s_i = Ae^{j\alpha} + Be^{j\beta} \qquad (5.2.1)$$

where α is a random phase resulting from combining the interfering signals, and β is the desired signal phase. Assume that interference A is much stronger than signal B; then Eq. (5.2.1) becomes

$$s_i = [A + B \cos(\beta - \alpha)]e^{j\alpha} + jB \sin(\beta - \alpha)e^{j\alpha}$$

Also assume that the nonlinear device is well behaved. Its envelope output $g(A)$ can be represented by a Taylor's series about A if $B \ll A$, as

$$g(A + B \cos(\beta - \alpha)) \simeq g(A) + Bg'(A)\cos(\beta - \alpha) \qquad (5.2.2)$$

The derivation of course is tedious. The output signal-to-interference ratio after the nonlinear device is[4]

$$\left(\frac{S}{I}\right)_o = \left(\frac{S}{I}\right)_{in} \times R \qquad (5.2.3)$$

where

$$\left(\frac{S}{I}\right)_{in} = \frac{B^2}{2\sigma^2}$$

and R is called the *signal-suppression ratio* expressed as

$$R = \frac{\left[\int_0^\infty Ag(A)p(A)\, dA\right]^2}{\left[\int_0^\infty g^2(A)p(A)\, dA\right]\left[\int_0^\infty A^2 p(A)\, dA\right]}$$

$$= \frac{\langle Ag \rangle^2}{\langle g^2 \rangle \langle A^2 \rangle} \qquad (5.2.4)$$

Applying the Schwartz inequality $\langle A \cdot g \rangle^2 \le \langle A \cdot A \rangle \langle g \cdot g \rangle$, we find that the value R is always less than unity no matter what $g(A)$ is used. Equation (5.2.3) becomes

$$\left(\frac{S}{I}\right)_o \le \left(\frac{S}{I}\right)_{in} \qquad (5.2.5)$$

for a strong Gaussian interference. Therefore no signal enhancement of signal B can be made through a nonlinear device when the Gaussian interference is strong.

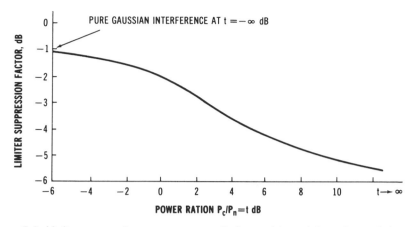

Figure 5.4. Limiter suppression versus power ratio for a mixture of Gaussian and sinusoidal interference. (From reference 5.)

By Bandpass Hard Limits

Consider an input environment consisting of a sinusoidal input signal plus a Rician interference. The Rician distribution is shown in Eq. (1.5.13) with a ratio $t = i/n$, a fixed envelope interference sinusoid to the Gaussian noise. The effective signal suppression factor for a weak sinusoid in the presence of a strong Rician interference with a hard-limiting band pass nolinearity[5] is shown in Fig. 5.4.

In Fig. 5.4, when $t = 0$, a pure Gaussian interference results; the suppression factor is 1 dB. When $t = \infty$, a pure sinusoidal interference results; the suppression factor is 6 dB.

5.2.2   FDM Signal Distortion

By Amplitude Nonlinearity

Amplifier nonlinearities not only cause signal suppression but also signal distortion. The ratio of signal-spectral density to intermodulation density represents the true signal-to-distortion power ratio. Assume that an input signal to the nonlinear device has a Gaussian power-spectral density with a rms value of σ. The output signal-to-intermodulation-distortion ratio, S/IM, at the center frequency f_0 versus the ratio c/σ, is plotted in Fig. 5.5, where c is the clipping level. When the hard-limiting level c decreases as $c/\sigma \to 0$, the output S/IM decreases to a minimum of approximately 9 dB. The predominant effect is caused by third- and fifth-order IM distortion. However, for a 3-dB output power backoff (which reduces the output power drive by 3 dB) the S/IM increases to 17 dB.

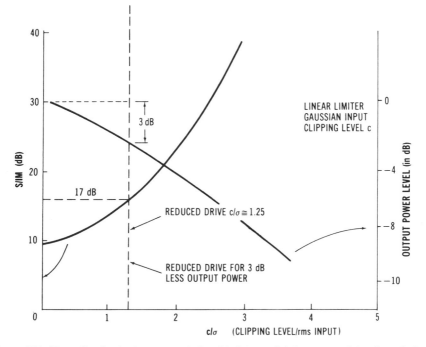

Figure 5.5. Normalized output power and signal-to-intermodulation spectral-density ratio for a linear limiter with Gaussian input. (From reference 5.)

By AM/PM Conversion Effects

Most amplifying devices also exhibit AM/PM conversion due to their nonlinear characteristics. A change in the envelope of a multiple sine-wave input causes a change in the output of each signal component. The following analysis demonstrates this effect:

A sinusoidal input with a small amount of AM amplitude can be expressed as

$$x(t) = A_i(t)\cos(w_0 t) \qquad (5.2.6)$$

Where $A_i(t)$ is the input envelope it can be expressed as

$$A_i(t) = A(1 + m \cos w_m t) \qquad (5.2.7)$$

and A is a small value. The output phase modulation $\theta_0(A) = \theta_0(A_i(t))$ is then approximately modeled proportional to the envelope squared for a small input level A_i.

$$\theta_0(A_i(t)) = KA_i^2(t) \tag{5.2.8}$$

where $\theta_0(A_i(t)) = \theta_0(t)$ is shown as the phase term of the output signal $y(t)$,

$$y(t) = KA_i(t)\cos[w_0 t + \theta_0(t)] \tag{5.2.9}$$

Substituting Eq. (5.2.7) into Eq. (5.2.8) yields

$$\theta_0(t) = KA^2(1 + 2m \cos w_m t + m^2 \cos^2 w_m t)$$

$$\simeq KA^2(1 + 2m \cos w_m t) \qquad m \ll 1 \tag{5.2.10}$$

Peak deviation from the mean phase is $\theta_p \simeq KA^2 \cdot 2m$ (in radian). This peak phase error can be expressed in radians/decibels of AM as

$$K_p = \frac{\theta_p}{20 \log_{10}(1 + m)} \qquad \text{(radian/dB)}$$

$$= \frac{KA^2 \cdot 2m}{8.69m} = 0.46KP_s \qquad \text{(radian/dB)}$$

$$= 26.38KP_s \qquad \text{(degree/dB)} \tag{5.2.11}$$

where $P_s \triangleq A^2/2$; thus K_p is linearly proportional to the input power P_s.

We express n multiple-input sinusoids as

$$x(t) = \sum_{i=1}^{N} A_i \cos[w_0 t + \phi_i(t)] \tag{5.2.12}$$

We could denote that the output phase modulation $\theta_0(A_i) = \theta_0(A_i(t))$ for each sinusoidal wave; then the output $y(t)$ after AM/PM conversion would contain a phase term $\theta(A(t))$:

$$y(t) = \sum_{i=1}^{N} A_i \cos[w_0 t + \phi_i(t) + \theta_0(A_i)]$$

$$\simeq \sum_{i=1}^{N} A_i \cos[w_0 t + \phi_i(t)] - \theta_0(A_i)$$

$$\cdot \sum_{i=1}^{N} A_i \sin[w_0 t + \phi_i(t)] \qquad \theta \ll 1$$

$$= \sum_{i=1}^{N} A_i \cos[w_0 t + \phi_i(t)] - \underbrace{\theta_0(A_i)A(t)\sin[w_0 t + \phi(t)]}_{\text{distortion term } \theta(A(t))} \tag{5.2.13}$$

Let

$$A(t) = \left[\sum_{i=1}^{N} A_i \cos \phi_i(t) \right]^2 + \left[\sum_{i=1}^{N} A_i \sin \phi_i(t) \right]^2 \qquad (5.2.14)$$

$$\phi(t) = \tan^{-1}\left(\frac{\sum_{i=1}^{N} A_i \sin \phi_i(t)}{\sum_{i=1}^{N} A_i \cos \phi_i(t)} \right) \qquad (5.2.15)$$

Then $\theta_0(A_i(t))$ can be found from Eq. (5.2.8). The distortion term becomes

$$\theta(A(t)) = \text{distortion term} = -K^2 A^2(t) A(t) \sin[w_0 t + \phi(t)]$$

$$= -K^2 A^3(t) \sin[w_0 t + \phi(t)] \qquad (5.2.16)$$

As seen in Eq. (5.2.16), the IM distortion products occur at the same frequency, have a different amplitude, and are shifted 90° in phase.

Combining AM/PM Conversion and Amplitude Nonlinearities Effects

The AM/PM conversion is obtained from Eq. (5.2.8), and the amplitude nonlinearity follows a simple cube law:

$$y = a_0 + a_3 x^3 \qquad (5.2.17)$$

where x is the input and y is the output. Now the input

$$x(t) = A(t)\cos(w_0 t + \phi) \qquad (5.2.18)$$

and the output

$$y(t) = F[A(t)]\cos[w_0 t + \theta_0(A) + \phi] \qquad (5.2.19)$$

where $F[A(t)]$ is the signal suppression and $\theta_0(A)$ is the AM/PM conversion. A model of these two effects is shown in Fig. 5.6. The intermodulation causes a distortion (unintelligible) when an FM signal is transmitted. If a filter precedes the power amplifier, it is possible to generate intelligible crosstalk. This is actually a two-step process, FM conversion to AM by the filter, followed by AM conversion to PM by the amplifier nonlinearity. The latter step causes modulation on the adjacent channel and produces intelligible crosstalk.

5.3 TIME-DIVISION MULTIPLEXING (TDM)

Time-division multiplexing can make power utilization 90% or more efficient compared to the 3- or 6-dB loss in power efficiency for FDM. In FDM, a

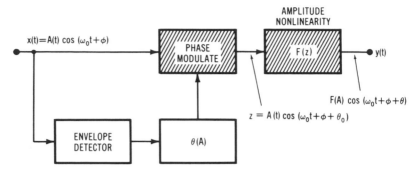

Figure 5.6. Model of AM/PM nonlinearities in a power amplifier.

3- to 6-dB power backoff is required in order to minimize intermodulation effects. TDM also can achieve better efficiencies in bandwidth utilization because no frequency guard band is required between channels.

TDM permits operating the output amplifier at full saturation, often resulting in a significant increase in useful power. Also IM product degradation is largely avoided by transmitting each signal with sufficient guard time between time slots to accommodate the following situations:

1. Timing inaccuracies due to clock instability
2. Delay spread
3. Transmission time delay due to propagation distance
4. "Tails" of the pulsed signal in TDM due to the transient response

5.3.1 TDM Buffers

Since the incoming bit streams continuously arrive at the TDM in real time, while the output of the TDM modulator is a periodic burst of RF energy, the TDM must contain a data buffer. This buffer stores the data bits received from one frame to the next. The total storage is M bits for N input bit streams at bit rate R_i and frame period τ_f

$$M = \sum_{i=1}^{N} (R_i \tau_f) \quad \text{bits} \qquad (5.3.1)$$

5.3.2 TDM Guard Time

Clock-timing terms are defined as follows:

t Universal or system time
$\tau(t)$ Clock pulse sequence at the mobile unit

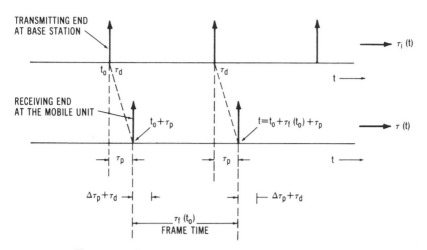

Figure 5.7. Timing at base station and at the mobile unit.

$\tau_i(t)$ Clock pulse sequence at the base station
$\tau_f(t)$ Frame-time interval at the transmitter
$\Delta\tau_f(t)$ Error in frame time due to the instability of the clock
$\tau_p(t)$ Propagation time related to the distance
$\Delta\tau_p(t)$ Delay spread at the receiver
$\tau_d(t)$ Die-out time of the tail of the pulsed signal
τ_g Guard time
$\Delta\tau_g$ Error tolerance in guard time

When the base station sends a pulse to the mobile unit, sufficient time is required for the mobile unit to receive this pulse. Two clock pulse sequences are shown in Fig. 5.7 with the die-out time τ_d and delay spread $\Delta\tau_p$—then the timing interval is called the *guard time*.

$$\tau_g = \tau_p + \Delta\tau_p + \Delta\tau_g + \tau_d \qquad (5.3.2)$$

where τ_p is not a constant, due to the deployment of mobile units in the field. Some are close to the base station while others are 10 miles away. The τ_p based on a 10-mile distance is

$$\tau_p = 10 \text{ miles} \times 5280 \text{ ft/miles} \times 1 \text{ ns/ft}$$

$$= 5.28 \times 20^{-5} \text{ sec}$$

The delay spread $\Delta\tau_p$ in the urban area is about 3 μs. Assume that the clock instability is 1 ppm. Then $\Delta\tau_g$ is the error in guard time due to the instability of the clock. Usually $\Delta\tau_g$ is negligible because 1 ppm of guard time is small. The die-out time τ_d for the tail of a pulsed signal depends on the amplitude and phase response of the filters.

Assume τ_d to be 1 μs. Then the guard time τ_g can be obtained from Eq. (5.3.2) as

$$\tau_g = (52.8 + 3 + 1) = 56.8 \ \mu s \tag{5.3.3}$$

5.3.3 The Bit Rate and the Frame Rate

The time between bits τ_b is

$$\begin{aligned}
\tau_b &= \Delta\tau_p + \tau_d + \Delta\tau_g \\
&\approx \Delta\tau_p + \tau_d \\
&= 3 \ \mu s + 1 \ \mu s = 4 \ \mu s
\end{aligned} \tag{5.3.4}$$

The bit rate has to be found by

$$R_b \leq \frac{1}{\tau_b} = 2.5 \times 10^5 \text{ bps} \tag{5.3.5}$$

The frame rate f_f is defined as frames per second:

$$f_f = \frac{1}{\tau_f} \tag{5.3.6}$$

If there are 10^4 bits per frame and $R_b = 2.5 \times 10^5$ bps, then $f_f = 25$ frames/sec.

5.3.4 TDM System Efficiency

The power efficiency of a mobile radio TDM system depends on (1) guard times between the transmission τ_{gi} of each terminal; (2) the preamble and postamble time (to provide addressing and carrier recovery), called the *addressing time* for each transmit/receiver terminal pair τ_{ai}; and (3) the frame-time duration $T_f(=N\tau_f + N\tau_g)$. The maximum efficiency for all terminals fully occupying the frame is

$$\eta_{\max} = \frac{T_f - \sum_{i}^{N} (\tau_{gi} + \tau_{ai})}{T_f} \tag{5.3.7}$$

where i is summed over all N terminals in the network (see Fig. 5.8). If guard

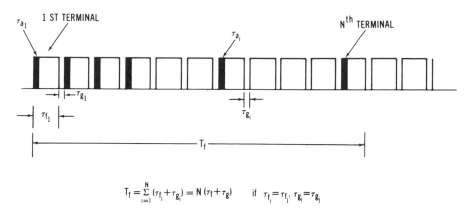

$$T_f = \sum_{i=1}^{N} (\tau_{f_i} + \tau_{g_i}) = N (\tau_f + \tau_g) \qquad \text{if } \tau_{f_i} = \tau_{f_j}, \ \tau_{g_i} = \tau_{g_j}$$

Figure 5.8. The frame-time duration T_f and the frame time interval of each terminal, τ_f.

times and address times of all terminals are identical and $\tau_f \gg \tau_g$, then the efficiency is

$$\eta_{\max} = \frac{T_f - [N(\tau_g + \tau_a)]}{T_f} = \frac{\tau_f - (\tau_g + \tau_a)}{\tau_f}$$

Increasing τ_f improves efficiency.

Example 5.1. Assume that $R_t = 250$ kbps, $\tau_g = 57$ μs (14 bits), and $\tau_a = 48$ bits (see Fig. 5.9). Also assume that the digital voice using LPC needs 2.4 kbps. The frame length can be computed as

$$2400 + 48 + 14 = 2462 \text{ bits}$$

the maximum efficiency $\eta_{\max}$

$$\eta_{\max} = \frac{2462 - 62}{2462} = 97\%$$

5.4 SPREAD SPECTRUM AND FREQUENCY HOPPING

It should be remembered that if the system is in a noise-limited environment, the SSB (3 to 4 kHz/channel) could be used to provide spectrum efficiency. But when the system is in an interference-limited environment, wideband techniques must first be used to suppress the interference. At the same time

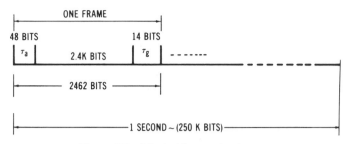

Figure 5.9. A typical frame structure.

more channels can be provided. In general there are two kinds of wideband techniques: spread spectrum and frequency hopping.

5.4.1   Spread Spectrum

Assume that a frequency of 10 MHz is commonly used by M users and the baseband is 10 kHz. Use a direct sequence to spread 10 kHz over a frequency band of 10 MHz so that every information bit is represented by 1000 bits. Now assume that

$$\frac{E_b}{\eta_0} = 15 \text{ dB} \sim 31.6$$

where E_b is the energy per information bit, and η is the noise power per Hz. At the IF or RF stage, the carrier-to-noise ratio of a single channel can be expressed as

$$\left(\frac{C}{N}\right) = \frac{E_b \times R_b}{\eta_0 \times B} = \frac{E_b}{\eta_0} \times \frac{10^4}{10^7}$$

$$= 31.6 \times \frac{1}{10^3} = 0.0316 \sim -15 \text{ dB} \qquad (5.4.1)$$

The C/N of -15 dB is required at a single-channel receiver. This result means that by using a spread spectrum transmission the carrier level can be 15 dB lower than the noise level. If the noise level is kTB $= -174 + 70 = -104$ dBm, the carrier level is -119 dBm. We assume that M users are in the field, and that each of M users has a (C/N) at each receiving terminal at -15 dB; then the value of M can be calculated based on the new carrier level x of each user is

$$\frac{C}{I + N} = \frac{10^{x/10}}{(M - 1) \cdot 10^{x/10} + 10^{-104/10}} = 10^{-15/10}$$

Then

$$M = 1 + 10^{1.5} - 10^{-10.4 - x/10}$$

If the new carrier level of each user is -104 dBm, the value of M is 31.6 or roughly 32. The total received level $(C + I + N)$ received at each receiver front end (rf) will be $-104 + 15$ which is -89 dBm. If the new carrier level of each user is much higher than -104 dBm, M is approaches to 32.6 or roughly 33.

Example 5.2. This example studies the near-end-to-far-end situation in a spread-spectrum system and assumes a (10-mile) far-end link and a 3.2 km (2-mile) near-end link. From Section 2.3.6, the suburban curve and the standard setup data are used. A 10-mile link ($h_1 = 100$ ft, $G_1 = 6$ dB/dipole, at 850 MHz) needs:

$$Tx_1 \text{ power} = 51 \text{ dBm results in a received signal of } -89 \text{ dBm}$$

A 2-mile link ($h_1 = 100$ ft, $G_1 = 6$ dB/dipole, at 850 MHz) with a same transmitted power needs:

$$Tx_2 \text{ power} = 51 \text{ dBm results in a received signal of } -74 \text{ dBm}$$

It is known that the base station always receives a strong signal from a mobile unit 3.2 km (2 miles) away and a weaker signal from 16 km (10 miles) away. If both signals are received simultaneously by the base station, the weaker signal will be interfered with by the stronger one. Therefore, in the spread-spectrum system, near-end-to-far-end ratio interference cannot be avoided, as shown in Fig. 5.10. By taking into account the same reception level at the base station (also shown in Fig. 5.10), the transmitting power of Tx_1 should be 15 dB higher than that of Tx_2.

Example 5.3. Suppose that the jamer's signal is 50 dB stronger than the desired signal received at the receiver. Assume that 1 Kbps data rate is sending from the desired transmitter how wide spread the bandwidth is needed in order to receive the 1 Kbps data rate with $E_b/\eta_0 = 10$ dB.
We may use Eq. 5.4.1 to solve this problem. Since we know that $C/I = -50$ dB $(=)$ 10^{-5},

$$R_b = 1 \text{ Kbps}$$

$$E_b/\eta_0 = 10 \text{ dB} (=) 10$$

Then

$$(C/N)_s = \left(\frac{E_b}{\eta_0}\right)\left(\frac{R_b}{B}\right)$$

and

$$B = 10^5 \times 10 \times 10^3 = 100 \text{ MHz}$$

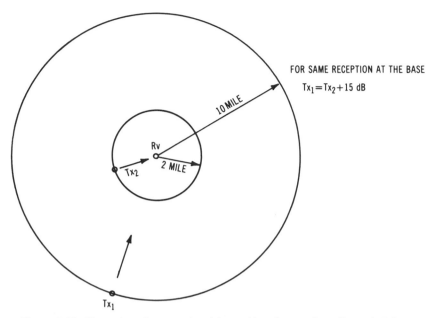

Figure 5.10. Illustration of near-end and far-end interference from Example 5.2.

In this case the bandwidth has to be 100 MHz wide in order to overcome the strong jamer. Under normal circumstances a data rate of 1 Kbps only could be transmitted using a bandwidth of 1 KHz. In this example the bandwidth under the jamer is 100 MHz. A system in which a bandwidth is spreaded from 1 KHz to 100 MHz is called a *spread spectrum system* which application is appeared in chapter 9.

5.4.2 Frequency-Hopped (FH) Systems

A frequency-hopping system that could serve a great number of users in the mobile radio environment was developed by Cooper and Nettleton.[6] It is a frequency-hopped differential phase-shift keying system (FH-DPSK). A frequency-hopped frequency-shift keying system (FH-FSK) has also been developed by Goldman.[7] These are both variants of the basic frequency-hopped system.

Reduction of Near-End-to-Far-End Interference
If all mobile units are equidistant from the base station, then there is no advantage to using an FH system, as shown in Fig. 5.11. However, in a real mobile radio environment the mobile units are scattered randomly. Near-end-to-far-end interference cannot be avoided. This interference can be reduced by having a frequency assignment plan or by using this frequency-hopping system. Assuming that a 12.5 MHz band is available and that each channel occupies 30 kHz, then there are 416 channels. If the frequencies are

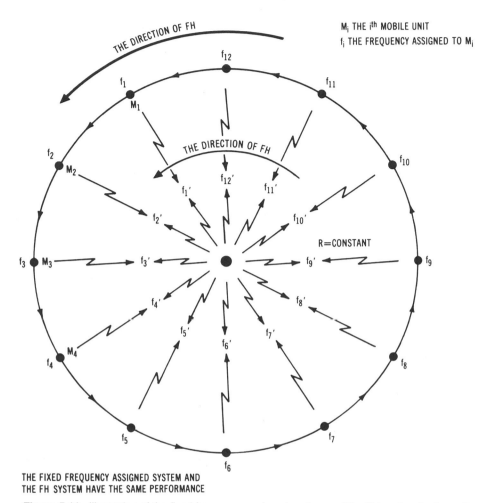

Figure 5.11. Illustration of the fixed frequency assigned system and the FH system for the cell radius R = constant.

hopped in different time slots, the FH system can serve 416 users. Consider a worst case, that of the mobile unit traveling near the rim of a 16-km (10-mile) radius from the base station. The probability that the mobile unit is in the region between 14 to 16 km (9 to 10 miles) is

$$p_1 = \frac{\pi(10^2 - 9^2)}{\pi(10^2)} = \frac{19}{100} = 0.19$$

The probability of another interfering mobile unit within a radius of 0.8 km

(0.5 mile) is

$$p_2 = \frac{\pi(0.5)^2}{\pi(10)^2} = 2.5 \times 10^{-3}$$

The total chance of having this situation is

$$p_t = p_1 \times p_2 = 4.75 \times 10^{-4}$$

In one case the separation is 20 channels, obtained from Eq. (4.4.5), to avoid the near-end-to-far-end interference. Assume that all channels are in use in the system. Then 40 of them (20 on either side of the desired channel) will interfere with the desired channel. The chance of having errors exceed a specified $P_e = 10^{-3}$ is

$$P_{e_t} = \left(1 - \frac{40}{416}\right)P_e + \frac{40}{416}(1 - P_e)$$

$$\simeq 0.09696 \tag{5.4.2}$$

Thus P_{e_t} is valid only in the event of $p_t = 4.8 \times 10^{-4}$. The result in P_{e_t} is very large even though the occurrence is small. We would like to show that the frequency-hopped system would reduce the near-end-to-far-end interference, which is shown in Fig. 5.12. The bit-error rate P_{e_t} for different frequency separations is shown in Fig. 5.13. However, in a FH system all 416 channels can be used in a cell. For instance, if an orthogonal arrangement is used, all the frequencies in use are different at any one instant. All 416 different frequencies can be used at the same time. Then one cell can serve 416 channels. The size of the cell increases; the number of channels remains the same. This means that if a 16-km (10-mile) cell serves 416 users, increasing the cell size will still only serve 416 users. If the FH system were to be reused in every adjacent cell, as shown in Fig. 5.14, then the interference calculation shown for a single cell would be changed because it is combined with both the near-end-to-far-end ratio, and the co-channel cell interference. The specified near-end-to-far-end ratio sets up a required isolation in dB. This isolation has to be achieved by using a filter with its characteristics, for example, 12 dB/oct, and determining a required frequency bandwidth separation according to the required isolation. Co-channel cell interference is reduced based upon a required carrier-to-noise ratio C/I.

Now consider co-channel interf rence C/I based on one interferer, but not the near-end-to-far-end ratio. A¹ ame the worst case: that the mobile unit is at 14.8 km (9.5 miles) away fr⌐ ı the base station and 16.8 km (10.5 miles)

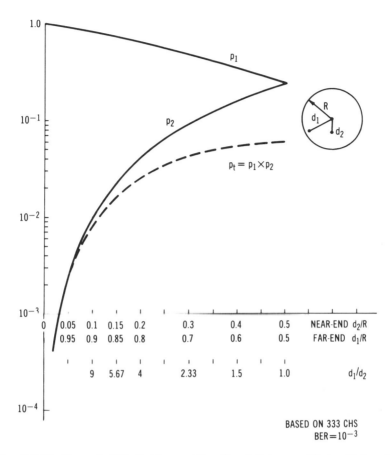

Figure 5.12. The probability that two mobile units are in two specified radii d_1 and d_2.

away from the co-channel interferers. Then the carrier-to-interference ratio (C/I) at the mobile unit is

$$\frac{C}{I} = \frac{P_s}{P_I} = \left(\frac{d_1}{d_3}\right)^{-4} = \left(\frac{9.5}{10.5}\right)^{-4} = (1.11)^4 = 1.49 \sim 1.74 \text{ dB}$$

For a six-interferer case the carrier-to-interference ratio becomes approximately

$$\frac{C}{I} = \frac{P_s}{P_I} = \frac{d_1}{6(d_3)} = \frac{1.49}{6} = 0.25$$

$$= -6 \text{ dB}$$

Then for a case of equal-power channels, the separation would be based on

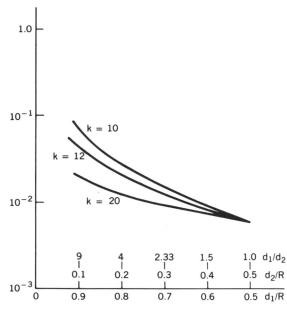

Figure 5.13. The bit-error rate at different specified ratios of d_1/d_2.

$$\frac{C}{I} \geq 18 \text{ dB}$$

which means that an additional isolation of 24 dB (18 + 6 dB) should be considered. The total isolation in this case (therefore the isolation for both co-channel interference and near-end-to-far-end interference) will be

$$24 + 52 = 76 \text{ dB}$$

Applying this to Eq. (4.4.3) yields

$$f_2 = f_1 \cdot 10^{(\text{Loss} \times 0.3)/K}$$

The separation is 40 channels or 1191 kHz. It is based upon a channel bandwidth of 30 kHz and a filter characteristic of 12 dB/oct. Any channel within the 40 channels on each side of the desired channel would cause interference. The chance of having errors exceed a specified $P_e = 10^{-3}$ in a total of 416 channels can be obtained by using an equation similar to Eq. (5.4.2) as follows:

$$P_{e_t} = \left(1 - \frac{80}{416}\right) P_e + \frac{80}{416} (1 - P_e) = 0.1929 \tag{5.4.3}$$

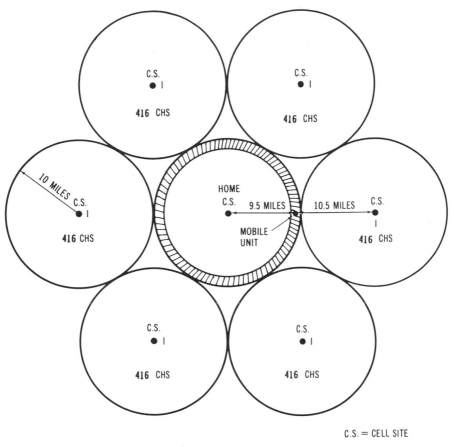

Figure 5.14. Near-end and far-end interference in a multi-FH system.

Comparing the result of Eq. (5.4.3) with Eq. (5.4.2), we find that the total error due to the reused frequency concept would increase twice and affect the FH system performance very seriously.

5.5 Cellular Concept[8,9]

It is a real challenge to try to serve a large number of subscribers at an affordable cost, with only limited frequency resources. There are several ways of achieving this: the narrowband or SSB approach, the spread-spectrum approach, and the cellular concept. The cellular concept will be covered since it has been used in commercial systems all over the world at 900 MHz.

5.5.1 Frequency Reuse and Cell Separation

A simple one-dimensional cellular system will do for an illustration. First f_1 will be used in an R-radius cell, and the same frequency will be used in a cell at a distance D away. Cell separation is an excellent way to avoid co-channel interference, since same-frequency users are in two different cells at the same time (see Section 4.2). The filter cannot isolate co-channel interference, and modulation schemes also cannot effectively reduce co-channel interference. Only a geographical separation can reduce the interference. We call the co-channel interference reduction factor (CIRF) a as

$$a = \frac{D}{R}$$

The value of D can be determined from a received signal-to-interference ratio, as demonstrated in Section 4.2. The value a is independent of the transmitted power. It means that as long as all the transmitted powers are the same in all the cells, increasing the transmitted power equally in each cell does not increase co-channel interference. If the separation D is reduced, a becomes small, and co-channel interference increases. The value of D is obtained from Eq. (4.2.3), assuming six co-channel interferers, as

$$\frac{C}{I} = \frac{C}{\sum\limits_{i=1}^{6} I_i} = \frac{R^{-4}}{6D^{-4}} = \frac{a^4}{6} = \frac{D^4}{6R^4} \qquad (5.5.1)$$

or

$$D = R\sqrt[4]{6\left(\frac{S}{I}\right)} \qquad (5.5.2)$$

where C is a received signal (carrier power) from the desired transmitter and I_i is the co-channel interference of the ith interferer among interferers (transmitters). The local noise at the receiver is assumed to be negligible. The interfering power I_i will be small when D increases.

Once the value D is determined, the many equal-size cells between the two co-channel cells must be filled with different frequencies in order to provide a continuity of frequency coverage in space that would facilitate the communication between the moving vehicles. The number of cells between the two co-channel cells is filled in a space such as shown in Fig. 5.15. In the figure we assume that the new cell frequency is F_2, then another co-channel cell of F_2 also has to meet the D/R ratio requirement.

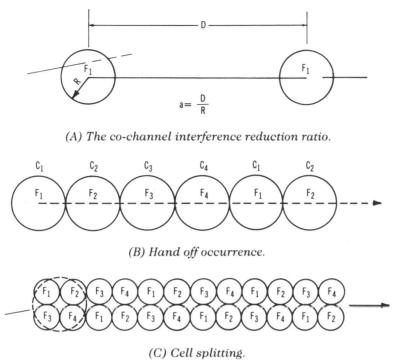

(A) The co-channel interference reduction ratio.

(B) Hand off occurrence.

(C) Cell splitting.

Figure 5.15. A cellular system illustrated in one dimension.

5.5.2 Hand-off (HO)

Suppose that a mobile unit is in the first cell F_1. The mobile unit travels along a path that passes through different cells. Each time the mobile unit enters a different cell associated with a different frequency, a hand-off (HO) occurs. The hand-off is an action controlled by the switching office. Some considerations of hand-off are described as follows:

1. The criterion of acquiring a hand-off will be based on a specified local-mean level, or phase-range information, or both. The phase-range information is not accurate in estimating the location of the mobile unit due to the obstructive built structure; depending on the height of the base-station antenna, errors of up to a half mile can occur. Usually for a large cell the hand-off algorithm needs only to be based on either local mean level or phase-range information, but not both. The latter is more costly than the former.

2. The system must allow enough time from the hand-off requisition to the actual hand-off. The switching office should minimize unnecessary handoff. Otherwise, the load on switching equipment increases and

performance is affected. The following information can help in making a decision about hand-off.

a. When the slope of the local-mean data in a given time interval is steep, the HO should be made quickly.

b. The mobile speed V can be found from a short-term fading signal; from Eq. (1.3.16), $V = f_d \lambda$. The fading frequency f_d can be measured from the fading signal; λ is the wavelength. Therefore the speed V is known. If V is high, the HO action should be taken very soon.

c. Phase-range information can be used to estimate the location of the mobile unit, and proper frequency assignment is also used accordingly to avoid co-channel interference.

An HO should be delayed in the following situations:

1. If there is a possibility that the local mean will increase so that no HO is needed
2. If the probability of making a hand-off into a right cell increases
3. If the mobile unit is traveling in a signal-strength hole within a cell
4. If no idle channel will be available during rush-hour traffic

Yet, if the delay is too long, by the time a hand-off is made, the control signal may be so weak that the mobile unit may not be able to receive or detect the information. The call will be lost.

5.5.3   Cell Splitting and Power Reducing

The concept of a cellular system is to serve an unlimited number of users. When the number of users reaches saturation in a startup cell and no more channels can be offered, the startup cell will be split. Usually a startup cell first splits into four smaller cells, and traffic increases by four. If after n splits, the traffic T_n will be

$$T_n = T_0(4)^n \qquad (5.5.3)$$

Then the power will be reduced

$$P_n = P_0 - n(12) \qquad \text{(in dB)} \qquad (5.5.4)$$

After two splits a cell splits into sixteen smaller cells; the power in each small cell should reduce 24 dB.

5.5.4   Reduction of Near-End-to-Far-End Ratio Interference

In the cellular system two schemes are used to reduce near-end-to-far-end ratio interference: One in the power control scheme which controls the mobile

New Frequency Managment (Full Spectrum)

Block A

1A	2A	3A	4A	5A	6A	7A	1B	2B	3B	4B	5B	6B	7B	1C	2C	3C	4C	5C	6C	7C
1	2	3	4	5	6	7	8	9	10	11	12	13	14	15	16	17	18	19	20	21
22	23	24	25	26	27	28	29	30	31	32	33	34	35	36	37	38	39	40	41	42
43	44	45	46	47	48	49	50	51	52	.53	54	55	56	57	58	59	60	61	62	63
64	65	66	67	68	69	70	71	72	73	74	75	76	77	78	79	80	81	82	83	84
85	86	87	88	89	90	91	92	93	94	95	96	97	98	99	100	101	102	103	104	105
106	107	108	109	110	111	112	113	114	115	116	117	118	119	120	121	122	123	124	125	126
127	128	129	130	131	132	133	134	135	136	137	138	139	140	141	142	143	144	145	146	147
148	149	150	151	152	153	154	155	156	157	158	159	160	161	162	163	164	165	166	167	168
169	170	171	172	173	174	175	176	177	178	179	180	181	182	183	184	185	186	187	188	189
190	191	192	193	194	195	196	197	198	199	200	201	202	203	204	205	206	207	208	209	210
211	212	213	214	215	216	217	218	219	220	221	222	223	224	225	226	227	228	229	230	231
232	233	234	235	236	237	238	239	240	241	242	243	244	245	246	247	248	249	250	251	252
253	254	255	256	257	258	259	260	261	262	263	264	265	266	267	268	269	270	271	272	273
274	275	276	277	278	279	280	281	282	283	284	285	286	287	288	289	290	291	292	293	294
295	296	297	298	299	300	301	302	303	304	305	306	307	308	309	310	311	312	667	668	669
670	671	672	673	674	675	676	677	678	679	680	681	682	683	684	685	686	687	688	689	690
691	692	693	694	695	696	697	698	699	700	701	702	703	704	705	706	707	708	709	710	711
712	713	714	715	716	X	X	X	991	992	993	994	995	996	997	998	999	1000	1001	1002	
1003	1004	1005	1006	1007	1008	1009	1010	1011	1012	1013	1014	1015	1016	1017	1018	1019	1020	1021	1022	1023
313*	**314**	**315**	**316**	**317**	**318**	**319**	**320**	**321**	**322**	**323**	**324**	**325**	**326**	**327**	**328**	**329**	**330**	**331**	**332**	**333**

New Frequency Managment (Full Spectrum)

Block B

1A	2A	3A	4A	5A	6A	7A	1B	2B	3B	4B	5B	6B	7B	1C	2C	3C	4C	5C	6C	7C
334*	**335**	**336**	**337**	**338**	**339**	**340**	**341**	**342**	**343**	**344**	**345**	**346**	**347**	**348**	**349**	**350**	**351**	**352**	**353**	**354**
355	356	357	358	359	360	361	362	363	364	365	366	367	368	369	370	371	372	373	374	375
376	377	378	379	380	381	382	383	384	385	386	387	388	389	390	391	392	393	394	395	396
397	398	399	400	401	402	403	404	405	406	407	408	409	410	411	412	413	414	415	416	417
418	419	420	421	422	423	424	425	426	427	428	429	430	431	432	433	434	435	436	437	438
439	440	441	442	443	444	445	446	447	448	449	450	451	452	453	454	455	456	457	458	459
460	461	462	463	464	465	466	467	468	469	470	471	472	473	474	475	476	477	478	479	480
481	482	483	484	485	486	487	488	489	490	491	492	493	494	495	496	497	498	499	500	501
502	503	504	505	506	507	508	509	510	511	512	513	514	515	516	517	518	519	520	521	522
523	524	525	526	527	528	529	530	531	532	533	534	535	536	537	538	539	540	541	542	543
544	545	546	547	548	549	550	551	552	553	554	555	556	557	558	559	560	561	562	563	564
565	566	567	568	569	570	571	572	573	574	575	576	577	578	579	580	581	582	583	584	585
586	587	588	589	590	591	592	593	594	595	596	597	598	599	600	601	602	603	604	605	606
607	608	609	610	611	612	613	614	615	616	617	618	619	620	621	622	623	624	625	626	627
628	629	630	631	632	633	634	635	636	637	638	639	640	641	642	643	644	645	646	647	648
649	650	651	652	653	654	655	656	657	658	659	660	661	662	663	664	665	666	X	X	X
X	X	X	X	X	717	718	719	720	721	722	723	724	725	726	727	728	729	730	731	732
733	734	735	736	737	738	739	740	741	742	743	744	745	746	747	748	749	750	751	752	753
754	755	756	757	758	759	760	761	762	763	764	765	766	767	768	769	770	771	772	773	774
775	776	777	778	779	780	781	782	783	784	785	786	787	788	789	790	791	792	793	794	795
796	797	798	799																	

Bold face numbers indicate 21 control channels for Block A and Block B respectively

Figure 5.16. Frequency management (full spectrum).

transmit power from the cell site. The mobile transmit power of each mobile unit is controlled so that the power received at the cell site is the same for every mobile unit. Another scheme is to draw a frequency management plan. The plan used in the present cellular system is shown in Fig. 5.16. In each column of the chart, a frequency set is designated. Two frequency channels are never nearer than seven channels apart. This arrangement circumvents both the IM generated at the antenna combiner and near-end-to-far-end ratio interference.

Comparison of FH System and Cellular System

A seven-cell cellular system has 21 sectors as shown in Fig. 5.16. If there are a total of 416 channels among them, 21 channels are setup channels and 395 channels are voice channels. Then each sector can have 19 voice channels. The locations and the sizes of the sectors are planned according to traffic conditions. The seven-cell system should be planned well ahead of need. Suppose that in a given sector, more channels are needed to meet some special traffic conditions. Without splitting a cell or channel sharing (as in Chapter 8) a cellular system cannot serve more than 19 users in a sector at one time. An FH system can serve up to 395 users, but its performance—with 395 users per sector—is poorer than that of 19 users per sector. This is because severe near-end-to-far-end ratio interference always occurs in the FH system. A reasonable compromise is to have the same 395 channels in both systems. Then a seven-cell cellular system needs seven cells, while an FH system only needs one. Moreover the reuse of an FH system, as shown in Fig. 5.11, has much interference (see Section 5.4.2).

5.6 SPECTRAL EFFICIENCY AND CELLULAR SCHEMES[10]

Spectral efficiency is different from channel efficiency. *Channel efficiency* is defined as the maximum number of channels that can be provided over a given overall spectral bandwidth. *Spectral efficiency* is defined as the maximum number of calls that can be served in a given area. In most systems channel efficiency is directly related to spectral efficiency. But in cellular systems where channels are reused over and over again, spectral efficiency is not equal to channel efficiency. So spectral efficiency, not channel efficiency, is the parameter we hope to maximize in a cellular system.

Narrowing the channel bandwidths or increasing the number of channels does not necessarily increase spectral efficiency.[11] In a cellular system spectral efficiency is based on the number of channels per cell. In cellular analog systems[12,13] roughly 60 channels per cell are estimated for an allocated bandwidth of 25 MHz regardless of each individual channel bandwidth, which can be 30, 15, or 5 kHz.*

Underlay/overlay[14] and diversity schemes[15] have proven to have better spectral efficiency. In this section we present two new schemes that improve spectral efficiency by more than 60 channels per cell. The two schemes are multiple channel bandwidth systems and one-third channel offset systems. We also suggest a hybrid scheme for integrating all four schemes—underlay/overlay, diversity, multiple-channel bandwidth, and one-third channel offset—to achieve maximum spectral efficiency.

---

*The total bandwidth for full duplexed channels is 25 MHz.

5.6.1 Multiple-Channel Bandwidth Systems

Two or three different channel bandwidths can be implemented to improve spectral efficiency. Each cell can have two or three rings. The 30-kHz channels are assigned to the outer ring. Values of 15 or 7.5 kHz are assigned to the middle and/or inner rings. The area of each ring for a two-ring system is calculated so that the areas of two rings are approximately equal. The area of each ring in a three-ring system can be engineered to the required traffic conditions.

The notion of improving spectral efficiency using multiple channel bandwidth systems rests on the fact that to achieve the same voice quality, 30-kHz channel bandwidth systems require a lower carrier-to-interference (C/I) ratio than a 15-kHz channel bandwidth systems. Suppose that the transmitted power at a cell site always remains constant. The 30-kHz channel bandwidth then can serve a large cell, but the 15-kHz channel bandwidth can only serve a relatively small cell. Since a 30-kHz channel bandwidth requires a smaller C/I ratio at the cell boundary, it can tolerate a higher level of interference and therefore results in a smaller D/R ratio (co-channel cell separation to cell radius ratio).

The 15-kHz channel bandwidth requires a relatively higher C/I ratio and therefore tolerates a lower level of interference. The 15-kHz channel bandwidth system requires a larger D/R ratio. The larger the D/R ratio, the higher will be the number of frequency-reuse pattern cells. Hence a 15-kHz channel system requires a higher number of smaller diameter cells to cover a given area than that of a 30-kHz channel system.

Carrier-to-Interference Ratio Requirement

From the subjective voice quality tests of the present cellular analog system, we find that the C/I requirement is

$$\frac{C}{I} \geq 18 \text{ dB} \quad \text{(30 kHz channel bandwidth)}$$

$$\frac{C}{I} \geq 24 \text{ dB} \quad \text{(15 kHz channel bandwidth)}[13,16]$$

$$\frac{C}{I} \geq 30 \text{ dB} \quad \text{(7.5 kHz channel bandwidth}$$
$$\text{within 1-mile radius of the cell).*} \qquad (5.6.1)$$

Areas of Assigning Different Channel Bandwidths

For two-channel bandwidth systems (Fig. 5.17A), assume that the cell radius is R_0, the area of outer ring is serviced by 30-kHz channels, and the inner

*In this 7.5-kHz bandwidth, the value of $C/I \geq 30$ dB needs to be verified from a subjective test under a Rician fading environment instead of Rayleigh fading environment.

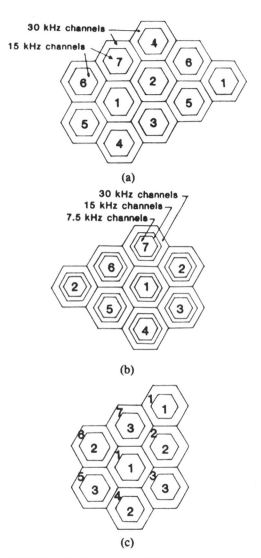

Figure 5.17. Multiple-channel bandwidth systems. (a) Two-channel system. (b) Three-channel system. (c) Hybrid system combining two-channel scheme with overlay/underlay scheme.

ring is serviced by 15-kHz channels. To keep the C/I requirement as shown in Eq. (5.6.1), the size of the inner ring can be found by

$$-24 + 18 = 40 \log\left(\frac{R_1}{R_0}\right) \tag{5.6.2}$$

in the case of a 40-dB/decade path-loss slope. From Eq. (5.6.2), R_1 can be determined as

$$R_1 = 0.70R_0 \tag{5.6.3}$$

The area of the inner ring A_1 is obtained as a portion of the total area of a cell A_0:

$$A_1 = \pi R_1^2 = 0.49A_0 \tag{5.6.4}$$

For three-channel bandwidth systems (Fig. 5.17B), three rings in a cell are created. The outer one is serviced by 30-kHz channels, the middle ring by 15-kHz channels, and the inner ring by 7.5-kHz channels. The relation between the cell radius R_0 and the radius R_2 of the inner ring based on the requirement of C/I shown in Eq. (5.6.1) and can be expressed as

$$-30 + 18 = 40 \log\left(\frac{R_2}{R_0}\right). \tag{5.6.5}$$

Equation (5.6.5) can be solved as

$$R_2 = 0.5R_0 \tag{5.6.6}$$

and the area of inner ring is

$$A_2 = \pi R_2^2 = 0.25A_0 \tag{5.6.7}$$

Channel Distribution

For two-channel bandwidth systems, assume that the number of channels in each ring is kept approximately equal:

$$\frac{2}{3} \times \frac{12.5 \text{ MHz}}{30 \text{ kHz}} + \frac{1}{3} \times \frac{12.5 \text{ MHz}}{15 \text{ kHz}} = 277 + 277$$

30-kHz channels   15-kHz channels

Total number of channels = 555

Based on the seven-cell reuse pattern ($K = 7$), the number of channels per ring in each cell is

39 channels/cell   (30-kHz channels in the outer ring)*

39 channels/cell   (15-kHz channels in the inner ring)

*Some additional 30-kHz channels should be assigned to serve the present 30 kHz only mobile units.

For three-channel bandwidth systems in which the inner ring has a higher traffic density than the outer ring, a three-channel bandwidth system can be used. One method of distributing channels can be expressed as

$$\frac{1}{3} \times \frac{12.5 \text{ MHz}}{30 \text{ kHz}} + \frac{1}{3} \times \frac{12.5 \text{ MHz}}{15 \text{ kHz}} + \frac{1}{3} \times \frac{12.5 \text{ MHz}}{7.5 \text{ kHz}}$$

$$= 138 \quad + \quad 227 \quad + \quad 555$$

$$(30 \text{ kHz}) \qquad (15 \text{ kHz}) \qquad (7.5 \text{ kHz}) \tag{5.6.9}$$

The number of channels per ring in each cell is

      20 channels/cell   (30-kHz channels in the outer ring)

      39 channels/cell   (15-kHz channels in the middle ring)

      79 channels/cell   (7.5-kHz channels in the center ring)†   (5.6.10)

Hybrid System (Integrating the Overlay/Underlay Scheme with the Multichannel Bandwidth System)

In an underlay/overlay ($K = 3/K = 7$) scheme we need to reduce the power of the underlay (inner ring) region to maintain the D_1/R_1 ratio (where D_1 is the separation of two co-channel rings in different cells and R_1 is the radius of the inner ring) at 4.6. However, the D_1/R_0 ratio (R_0 is the radius of the cell) is reduced to only three. In this overlay/underlay system, the power of inner ring must be reduced by 6 dB (or one-quarter) below that of the outer ring. Therefore the channel-reuse pattern K becomes three ($K = 3$) in the inner ring and $K = 7$ remains the same for the outer ring.

The radius of the inner ring R_1 can be found from the following two conditions:

$$\frac{D_1}{R_1} = 4.6$$

$$\frac{D_1}{R_0} = 3$$

or

$$K = \frac{(D_1/R_0)^2}{3} = 3 \qquad \text{for inner ring}$$

Then

$$R_1 = 0.65 \, R_0 \tag{5.6.11}$$

---

†Use of 7.5-kHz channel bandwidth for voice channels may have to lower the signaling rate of function control from 10 kbit/sec.

By comparing Eq. (5.6.11) with Eq. (5.6.3), we see that the area served by both the two-channel bandwidth system and the overlay/underlay system is nearly identical. This result encourages the use of overlay/underlay schemes on top of the two-channel bandwidth system (see Fig. 5.17C)

From Eq. (5.6.8), we see that the channels per cell for the outer ring remain the same:

$$277 \text{ channels/7 cells} = 39 \text{ channels/cell} \qquad (5.6.12)$$

The numbers 1–7 indicate seven different channel sets used in the outer ring as shown in Fig. 5.17C.

From Eq. (5.6.8) we see that the channels per cell for the inner ring are

$$277 \text{ channels/3 cells} = 92 \text{ channels/cell} \qquad (5.6.13)$$

The numbers 1–3 indicate three different channel sets used in the inner ring as shown in Fig. 5.17C.

Evaluation

The channels per cell of the present cellular system ($K = 7$) is 57 channels per cell. We use this number to compare this system with other systems. The calculation of traffic capacity is based on the following assumptions:

1. Total frequency band for full duplex channels = 25 MHz
2. Blocking probability = 1%–10%
3. Holding time = 100 seconds

Figure 5.18 compares the traffic capacity (based on the Erlang B model) for the following four schemes:

1. Today's $K = 7$ scheme
2. Two-channel bandwidth scheme
3. Three-channel bandwidth scheme
4. Hybrid system (two-channel bandwidth with underlay/overlay).

The hybrid system shows a great deal of improvement in spectrum among all but the three-channel bandwidth scheme. The three-channel bandwidth scheme needs to use two different signaling rates and formats, and this complicates the system. After all, the three channel bandwidth scheme is more or less an idealized scheme.

5.6.2 One-third Channel Offset Scheme

By offsetting each channel by one-third of its channel bandwidth, different offset channels are assigned to neighboring co-channel cells, as shown in

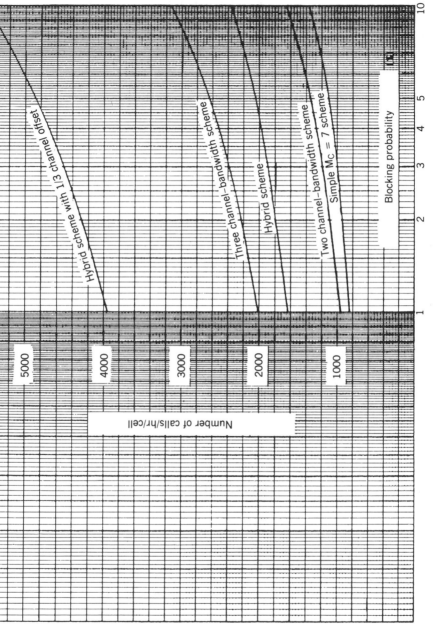

Figure 5.18. Comparison of spectrum efficiencies with different schemes.

192

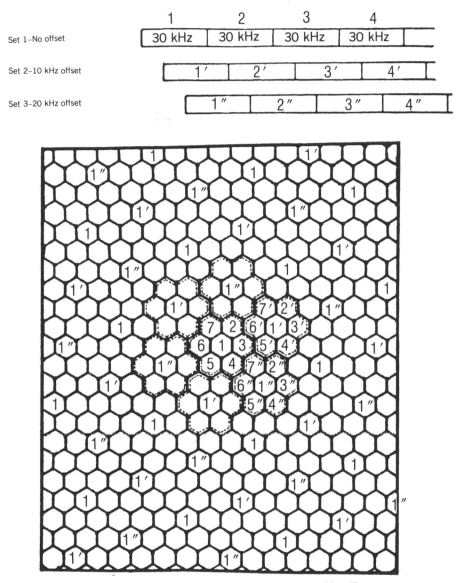

Figure 5.19. One-third channel offset system ($K = 7$).

Fig. 5.19. As a result the co-channel cells surrounding the center cell have the different offset channels. The co-channel interference of this scheme is reduced for two reasons:

1. The energy of co-channel interference reduces in each channel.
2. The intelligibility of speech from the co-channel cells (crosstalk) is drastically reduced.

The one-third channel offset scheme can be used in today's $K = 7$ system, and it can further reduce the number of K to less than seven. A method of reducing K from seven to four by using this scheme is demonstrated as follows.

Estimate of Separation between Co-channel Cells

The general formula for deriving a $K = 7$ system[10] is

$$\frac{c}{I} = \frac{c}{\sum_{i=1}^{3} (I'_i + I''_i)} \geq 18 \text{ dB} \tag{5.6.14}$$

where I'_i and I''_i are interference terms from the other two offset channels, respectively. The relationship between the offset channel interference I'_i or I''_i and the co-channel interference I_i is

$$I'_i = \alpha'_i I_i$$
$$I''_i = \alpha''_i I_i \tag{5.6.15}$$

where α'_i and α''_i are the offset channel advantages that reduce interference. Substituting Eq. (5.6.15) into Eq. (5.6.14) yields

$$\frac{c}{I} = \frac{c}{\sum_{i=1}^{3} (\alpha'_i + \alpha''_i) I_i} \geq 18 \text{ dB} \tag{5.6.16}$$

From Fig. 5.19 we found that the interference comes from two groups of offset channels:

$I'_i = 0.333 I_i$ (interference for channels
                            offset 20 kHz from desired channels)

$I''_i = 0.666 I_i$ (interence for channels offset
                            10 kHz from desired channels)

Then Eq. (5.6.16) can be derived as

$$\frac{c}{3(0.333 + 0.666) I_i} \geq 18 \text{ dB} \tag{5.6.17}$$

or

$$\frac{C}{3 I_i} \geq 63$$

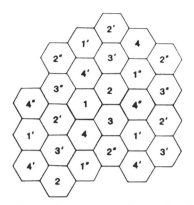

Figure 5.20. One-third channel offset system ($K = 4$).

Then

$$\left(\frac{D}{R}\right)^4 = 189$$

or

$$D = 3.71R \qquad (5.6.18)$$

Configuration Based on the One-third Channel Offset Scheme
Since

$$\frac{D}{R} = \sqrt{3K}$$

then

$$K = 4.58 \qquad (5.6.19)$$

The channel-reuse pattern is reduced from $K = 7$ to $K \approx 4$. It assumes that three sectors are implemented in each cell so that $K = 4$ can be used without generating noticeable interference. The channel assignment is shown in Fig. 5.20.

5.6.3 An Application of a Hybrid System

The two major schemes introduced in this section can be integrated in a hybrid system composed of three schemes: two-channel bandwidth, overlay/underlay, and one-third channel offset. The performance is shown as follows:

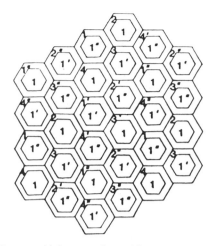

Figure 5.21. Hybrid system combining two-channel bandwidth scheme with underlay/overlay scheme and one-third channel offset scheme.

In the outer ring region, the same number of 30-kHz channels are used as shown in Section 5.6.3. However, $K = 4$ is obtained by using the one-third offset channel scheme. The number of channels per cell is

$$\frac{277 \text{ channels (30 kHz channels)}}{4 \text{ cells}} = 68 \text{ channels/cell}$$

In the inner ring region, based on the overlay/underlay and one-third offset channel schemes, the number of channels per cell is

$$\frac{277 \text{ channels (15 kHz channels)}}{2 \text{ cells}} = 137 \text{ channels/cell}$$

The increased spectrum efficiency of using a hybrid system (shown in Fig. 5.21) is demonstrated in Fig. 5.18. By implementing multiple-branch diversity reception at the cell site, we can further improve the spectral efficiency.

REFERENCES

1. Ehrlich, N., R. E. Fisher, and T. K. Wingard, "Cell Hardware," *Bell Sys. Tech. J.* 58 (Jan. 1979): 153–200.
2. Lee, W. C. Y., "Elements of Mobile Cellular System," *IEEE Trans. Veh. Tech.* (May 1986):
3. Spilker, J. J., Jr., *Digital Communicating by Satellite* (Prentice Hall, 1977): 214.
4. Ibid., 226.

5. Ibid., 230, 240.

6. Cooper, G. R., and R. W. Nettleton, "A Spread Spectrum Technique for High Capacity Mobile Communications," *IEEE Trans. Veh. Tech.* VT-27 (Nov. 1978): 264–275.

7. Goodman, D. J., P. S. Henry, and V. K. Prabhu, "Frequency Hopped Multilevel FSK for Mobile Ratio," *Bell Sys. Tech. J.* (Sept. 1980): 1257–1275.

8. Lee, W. C. Y., *Mobile Cellular Telecommunication Systems* (McGraw-Hill, 1989): Ch. 2.

9. MacDonald, V. H., "The Cellular Concept," *Bell Sys. Tech. J.* 58 (Jan. 1979): 15–42.

10. Lee, W. C. Y., "New Cellular Schemes for Spectral Efficiency," *IEEE Trans. Veh. Tech.* VT-36 (Nov. 1987): 188–192.

11. Lee, W. C. Y., "Spectrum Efficiency: A Comparison between FM and SSB in Cellular Mobile Systems," presented to the FCC (Washington, DC, Aug. 2, 1985; a condensed version appeared in *Telephony,* pp. 82–92, Nov. 11, 1985).

12. Swerup, J., and J. Uddenfeldt, "Digital cellular," *Personal Commun. Tech.* (May 1986): 6–12.

13. Lee, W. C. Y., "Narrowbanding in Cellular Mobile Systems," *Telephony,* pp. 44–46, (Dec. 1986).

14. J. F. Whitehead, "Cellular Spectrum Efficiency via Reuse Planning," *35th IEEE Vehicular Technology Conference Record* (Boulder, CO, 1985): 16–20.

15. Yeh, Y. S., and D. O. Reudink, "Efficiency Spectrum Utilization for Mobile Radio System Using Space Diversity," *IEEE Trans. Commun.* COM-30 (Mar. 1982): 447–455.

PROBLEMS

5.1   Find the probability that a mobile is within a circular band region between 9.6 and 11 km (6 and 7 miles) of an 11-km cell, and an interference is within 4 km from the cell site. What are the chances of this situation occurring?

5.2   If a mobile unit is 10 km away from the base station and 15 km away from the co-channel interference, find the signal-to-interference ratio at the mobile unit.

5.3   If a signal-to-interference ratio has to be 20 dB, and the mobile is 10 km away from its own cell site, how far away should the co-channel site be located?

5.4   After cell splitting N times, the required power can be reduced by

$$P_n = P_0 - N(12) \qquad \text{(in dB)}$$

Derive this equation.

5.5 Derive the traffic and power relation

$$\frac{P_0}{P_n} = \left(\frac{T_n}{T_0}\right)^2$$

5.6 Let a required C/I = 18 dB in a system. What is the required D/R ratio for designing a cellular system?

5.7 Consider the case of two adjacent channels, one on each side of a desired channel. The desired channel is 8 km away, and the two adjacent channels are 16 km away. Assuming that the channel isolation is 15 dB between the two adjacent channels, what would be the carrier-to-interference ratio received on the desired frequency channel?

5.8 A signal is received at 850 MHz with a mobile speed at 90 kmph. What is the fading frequency?

5.9 Suppose that a frequency-hopping system is hopping 10 frequencies and that nine frequencies have a C/I of 17 dB while one has a C/I of 9 dB. What is the resultant C/I after the frequency hopping?

5.10 Given E_b/N_0 = 10 dB and C/N = $10^{-5}(-50$ dB$)$. Use the equation

$$\frac{C}{N} = \frac{E_b}{N_0} \times \frac{R_b}{B}$$

to find out the processing gain (B/R_b).

DESIGN PARAMETERS AT THE BASE STATION

6.1 ANTENNA LOCATIONS

It is difficult to select an optimum antenna location for a base station. First the signal-strength coverage at a distance such as 13 km (8 miles) from a base-station antenna does not exhibit a uniform pattern. This irregular pattern is due to the irregular terrain configuration. Another important aspect is avoiding interference. Therefore a plan for a base-station antenna location should consider both its coverage range and its interference with other antennas. In a large system all the potential base-station location should be considered at the same time. This is because all base-station locations chosen to satisfy these two requirements are closely related among themselves. If one base station is moved to a different position, then all the other station locations are affected.

There are several steps in choosing a base-station location:

1. First, decide on a reception level at the cell boundary. This is based on the features of the mobile transreceiver and the system performance required. Assume that a level of -100 dBm* is at the coverage boundary of a cell. Then according to a given power, antenna height, antenna gain, and terrain configuration of the area, a cell size can be determined. Suppose we use the conditions listed in Section 2.3.6 for a suburban area. The radius of the cell will be 16 km (10 miles) for a reception of -100 dBm.

2. Choose a location where land is usually available for a first-choice base station.

3. Follow the new path-loss prediction model to make a point-to-point prediction. An equal-strength contour can be drawn on the map, based on the instructions appearing in Section 2.4.

---

*0 dBm means 0 dB with respect to 1 milliwatt. Therefore 0 dBm means 1 mw; -10 dBm means 0.1 mw; -20 dBm means 0.01 mw, 30 dBm means 1 watt, etc.

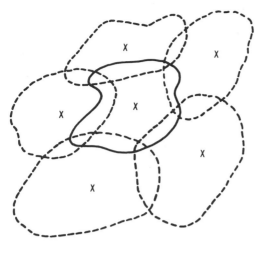

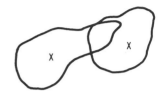

(A) Ideal design.   (B) Undesired design.

Figure 6.1. Cell site location design.

4. Choose other locations and draw -100-dBm equal-strength contours of them. The equal-strength contours of all locations should have roughly the same portion overlapping as shown in Fig. 6.1A.
5. Avoid the equal-strength contour condition shown in Fig. 6.1B.

Two guidelines for choosing a location are

1. Do not select a high spot (see Fig. 6.2A). This is to (a) avoid interference with other cells and (b) avoid weakening signal strength in its own cell.
2. Select a low spot but increase the base-station antenna height (see Fig. 6.2B).

6.2 ANTENNA SPACING AND ANTENNA HEIGHTS

In the real mobile radio environment there are two effects that make mobile radio communication very difficult: multipath fading and excessive path loss. At the base station diversity schemes can be used to reduce fading, and antenna heights can be raised to increase signal reception level. As we mentioned in Section 3.4.2, there are many diversity schemes that can reduce fading. Among them, space diversity has a great advantage over the others. The space diversity scheme does not use more frequency spectrum than just channel bandwidth. If a two-branch diversity is incorporated, each of two received signal strengths is 3 dB higher than that of either the polarization or the frequency diversity. The only drawback of using the space diversity

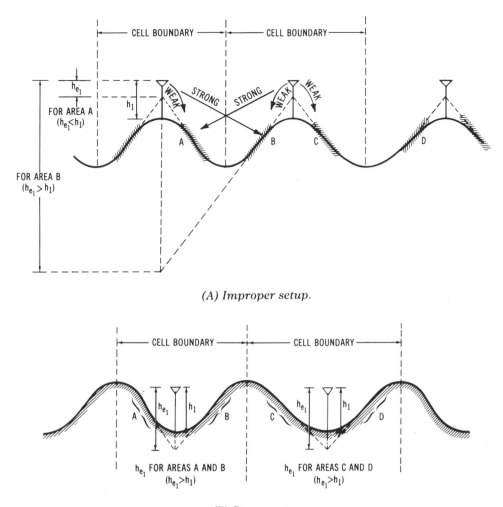

(A) Improper setup.

(B) Proper setup.

Figure 6.2. The rule of setting up base-station antennas in hilly areas.

scheme is that it needs a sufficient space separation. The space separation is determined by the correlation coefficients of two fading signals received by two base-station receiving antennas. The correlation coefficients must come from two different fading envelopes. The wider the separation, the lower will be the correlation coefficients (i.e., the variations of two fading signals will be less alike). Combining two fading signals with lower correlation coefficients reduces the detrimental effects of fading. The mechanism for choosing this base-station separation is shown in Fig. 6.3. The effective scatterer radius around the mobile unit shown in the figure is about 100λ (as derived in Section 4.7). The scattered waves reaching the base-station antennas come from that

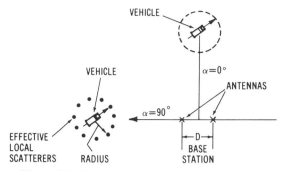

Figure 6.3. Antenna orientation at the base station.

active scatterer region. An active scatterer region is formed wherever the mobile unit is located.

6.2.1 Antenna Orientation Dependency

Waves transmitted from a mobile unit miles away, at an angle α (as shown in Fig. 6.3), and received by two base-station antennas would propagate through different scatterers in the medium. The difference in correlation coefficients of two fading signals depends on the separation of the two antennas and the direction of the angle α. Intuitive logic reasons a lower value of correlation coefficient would be obtained from a broadside case than from an inline case at a given antenna separation. This is because the two received fading signals will tend to be the same in the inline case. The fading signals received at the two base-station antennas in the inline case come up to the first antenna from the same propagation path. The only cause in lowering correlation coefficients of these two signals obtained from two base-station antennas is that the signal propagates an additional distance arriving at the second antenna due to the separation.

6.2.2 Antenna Height/Separation Dependency

The antenna separation required is determined by the correlation coefficient data. The correlation coefficients of two fading signals are obtained by experimenting with different antenna heights and different antenna spacing. A new parameter η is proposed as

$$\eta = \frac{\text{antenna height}}{\text{antenna separation}} = \frac{h}{d} \qquad (6.2.1)$$

Plot the experimental correlation coefficients in a suburban area on a new parameter η as shown in Fig. 6.4 at 850 MHz with different orientation angles.

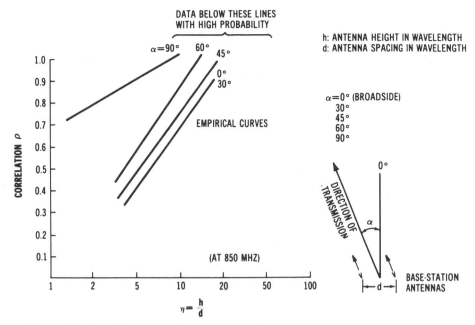

Figure 6.4. Correlation versus the parameter η for two antennas in different orientations.

All the measured data are below an empirical curve with high probability for each orientation angle case. For a given η the correlation coefficient values are always smaller in the broadside case ($\alpha = 0°$) than in any other case. The highest values of correlation coefficients are in the inline case ($\alpha = 90°$). Lowering the correlation coefficient ρ lowers the value of η, as shown in Fig. 6.4. The experimental correlation coefficients in an urban area with respect to the parameter η would be much smaller than those in the suburban area. This is because more scatterers exist along the path between the mobile unit and the base station in the urban area. The correlation coefficients of two signals received at urban base-station antennas have a tendency to be reduced. Reducing a fading signal in a Rayleigh fading environment can be achieved by using a diversity scheme. The correlation coefficient of up to 0.7 between two maximal-ratio diversity branches† gains a large reduction of signal fading. The performance of $\rho = 0.7$ compared to other values of ρ is shown in Fig. 6.5. The percentage shown in Fig. 6.5 means the percentage of the signal below its corresponding dB level. Some percentage figures will appear in this section. At a -10-dB level with respect to the rms value, the fading reduction is from 9.5% at $\rho = 1.0$ (no diversity condition) to 1.3% at $\rho = 0.7$ and then to 0.5% at $\rho = 0$. This observation encourages us to use

†Maximal-ratio combining technique is described in Section 3.5.

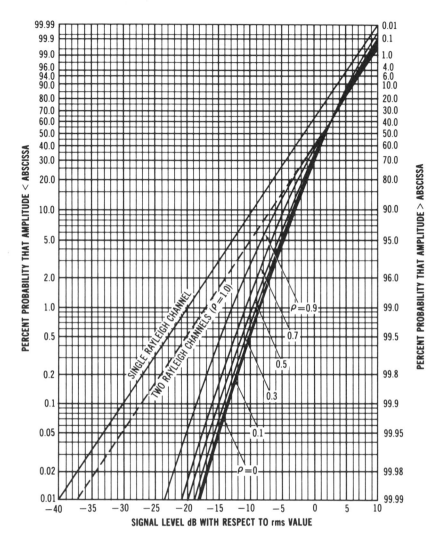

Figure 6.5. Performance of two-branch maximal ratio combiner with different correlation coefficients between branches.

$\rho = 0.7$. Since great improvement in performance is shown from $\rho = 1.0$ to $\rho = 0.7$, then a relatively slight improvement is shown from $\rho = 0.7$ to $\rho = 0$. This $\rho = 0.7^*$ is chosen for its cost effectiveness in realizing physical antenna spacing, as is shown in the following example: Given $\rho = 0.7$, then $\eta = 11$; given $\rho = 0.125$, then $\eta = 2$ for the broadside case. With a given antenna

---

*The $\rho = 0.7$ is not a desirable correlation coefficient for normal or log-normal fading.

height of 30 m (100 ft), but different values of η, the antenna spacing can be determined

$$d = \frac{h}{\eta} = \frac{100 \text{ ft}}{11} = 9 \text{ ft} \qquad (\rho = 0.7)$$

$$= \frac{100 \text{ ft}}{2} = 50 \text{ ft} \qquad (\rho = 0.125)$$

From this example it is clear that an antenna spacing of 3 m (9 ft) for $\rho = 0.7$ and 15 m (50 ft) for $\rho = 0.125$. Therefore lowering the value of η widens the antenna spacing d. From Fig. 6.5, the percentage of a signal below -10 dB is 1.3% at $\rho = 0.7$, and about 0.52% at $\rho = 0.2$. Combining the information observed from Figs. 6.4 and 6.5, we found that at a given $h = 30$ m (100 ft), a 1.3% signal will be below a -10-dB level if an antenna spacing is 9 ft, and 0.52% of total signal below a -10-dB level if antenna spacing is 50 ft. It is obvious that while making an antenna separation from 9 to 50 ft is a big effort, the improvement is not significant. Therefore a separation of 9 ft at an antenna height of 100 ft is suggested. We also can determine the antenna separation for different antenna heights, but we maintain the same value of η, $\eta = 11$:

$$d = \frac{h}{\eta} = \frac{150 \text{ ft}}{11} = 13.6 \text{ ft} \qquad (\text{for } h_1 = 150 \text{ ft})$$

$$= \frac{100 \text{ ft}}{11} = 9 \text{ ft} \qquad (\text{for } h_1 = 100 \text{ ft})$$

The example indicates that the higher the antenna height, the wider will be the separation. Usually the antenna height h used in Eq. (6.2.1) is the effective antenna height. This means that even though the physical antenna height is 30 m (100 ft), the instant effective antenna height can be taller or shorter than the physical one due to the terrain contour between the base-station antenna and the mobile unit in real time. The instant effective antenna height is illustrated in Section 2.4.

For an actual antenna height of 30 m (100 ft), the separation required between two receiving antennas for the broadside case is 3 m (9 ft) based on $\eta = 11$. Since the degree of reducing fading by using diversity is based on the two effective antenna heights and their separation, not their actual antenna height. It is found that if the effective antenna height is low, perhaps 15 m (50 ft), and the separation is kept the same (3 m or 9 ft), then $\eta = 5.5$, which is equivalent to the correlation coefficient ρ. From the curve of Fig. 6.5, $\rho = 0.4$. Therefore the smaller the correlation coefficient, the better will be the diversity performance.

As a matter of fact, when an effective antenna height is lower, the signal reception level drops. For instance, the drop in reception level for an effective antenna height of 15 m (50 ft) compared with the actual height of 30 m (100 ft) can be found from Eq. (2.3.21) as

$$\text{Antenna height gain} = 20 \log_{10} \frac{50}{100} = -6 \text{ dB} \qquad \text{(loss)}$$

In this case the signal drops 6 dB, but the diversity advantage increases as ρ decreases from 0.7 to 0.4 (see Fig. 6.5). On the other hand, if the effective antenna height increases to 60 m (200 ft), then an antenna height gain results:[2]

$$\text{Antenna height gain} = 20 \log_{10} \frac{200}{100} = 6 \text{ dB} \qquad \text{(gain)}$$

Here the value of η becomes

$$\eta = \frac{200}{9} = 22$$

The correlation coefficient ρ for $\eta = 22$ can be obtained from Fig. 6.4 as $\rho = 0.9$, which diminishes the diversity improvement. Therefore when the effective antenna height increases, the antenna height gain increases, but so does the correlation coefficient (i.e., the diversity advantage is reduced). The effective antenna height gain and space diversity advantage actually help each other.

For the inline case ($\alpha = 90°$), the value of η for a correlation coefficient of 0.7 is one (see Fig. 6.4). This means that the required separation of two receiving antennas is the same as the antenna height. If the antenna height is 30 m (100 ft), then the separation also is 100 ft. In reality a separation of 100 ft between two 100-ft receiving antennas is impossible. A practical antenna configuration is described in Section 6.3.

6.2.3 Frequency Dependency

In designing and calculating antenna spacing for space diversity in a mobile radio environment, the same experimental curves used for the frequency of 850 MHz shown in Fig. 6.4 can be used to compute the antenna separation for other frequencies.

$$d' = d\left(\frac{850}{f'}\right) \qquad (6.2.2)$$

where f' is the frequency in megahertz. The formula is valid for $f \geq 30$ MHz. If $f' = 85$ MHz, the required antenna separation d' is 10 times larger than the required antenna separation d at 850 MHz. If $d = 3$ m (9 ft), then $d' = 30$ m (90 ft), which is obviously impractical. Therefore space diversity is not recommended at lower frequencies, especially at the base station because the physical separation between the two antennas has become impractically large. At the mobile receiver the required antenna spacing is roughly half a wavelength; this case is explained in Chapter 7.

6.3 ANTENNA CONFIGURATIONS

Transmitting and receiving antennas will be physically separated to provide an additional isolation, though the transmitted band and received band are widely apart. This section will concentrate on the directional antenna, the tilting antenna configuration, and the diversity antenna configuration.

6.3.1 Directional Antennas

Besides depending on frequency band isolation, and physical space separation to reduce signal interference, the directional antenna could be positioned to eliminate unnecessary contamination of radiation in certain areas. In the cellular system the sectionized cells shown in Fig. 5.16 are used to reduce co-channel interference. Each sectionized cell can be formed by using three directional antennas at each cell site. The co-channel interference can be reduced by more than one-half because strong co-channel interference comes only from the back cells, not from the front cells.

In most cases omnidirectional antennas are used. The coverage of the signal strength from an omnidirectional antenna should take the shape of a smooth circle if the ground is perfectly flat. However, in reality the area of coverage always has a distorted circular shape because the ground is not ever perfectly level. In a hilly terrain the coverage of an omnidirectional antenna may be an irregular starfish shape. In some directions the transmitted power might be so intensified as to cause severe interference. In these circumstances the directional antenna can be used to eliminate the transmitted power in some directions and enhance the transmitted power in other directions. Although mostly omnidirectional antennas are used (one for each cell site) in omni-directional-antenna cell systems, we can still use directional antennas in certain cells to control the coverage patterns of those cells.

6.3.2 Tilting Antenna Configuration[3]

An effective way to confine the signal in its own coverage cell and reduce its interference in the other co-channel cells is to tilt a directional antenna beam pattern downward at a certain angle. A typical antenna vertical beam pattern is shown in Fig. 6.6. When the beam pattern is tilted downward, the field strength received by a distant mobile unit diminishes. When the antenna

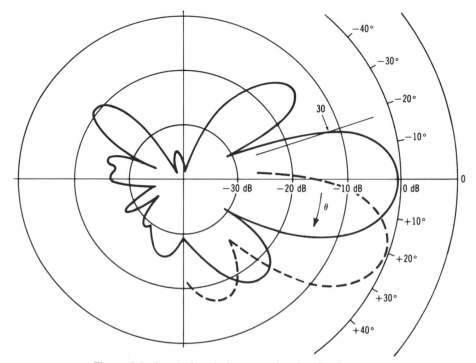

Figure 6.6. A typical vertical pattern of a directional antenna.

vertical beam pattern is tilted to a certain angle, a notch at the center of the horizontal beam pattern is produced. The antenna horizontal beam pattern is shown in Fig. 6.7. The notch becomes larger when the tilt angle increases. This notch can be used to effectively reduce interference in the co-channel cells as shown in Fig. 6.8.

The average signal-to-interference ratio in an interfered cell region interfered by a serving cell antenna site is improved by tilting the antenna gain pattern downward some predetermined amount. A limited loss of antenna gain is accepted in some parts of the serving cell to achieve an increase of average signal-to-interference ratio in the interfered cell area. Moreover the antenna-gain pattern of a directional antenna is tilted downward enough to create a notch in the horizontal pattern. Such a notch is characterized by a sector of reduced field intensity, as shown in Fig. 6.8. This sector spans a nearby co-channel interfering cell, reducing reception from that cell while also reducing serving cell interfering transmission to that co-channel cell.

It must be understood that the signal-to-interference ratio in any serving cell will increase if antenna beam patterns are tilted downward from other cell sites. This means that the degree of tilting of the antenna beam pattern at each cell site has to be especially chosen to reduce further co-channel interference in the system other than using directional antennas alone. Be

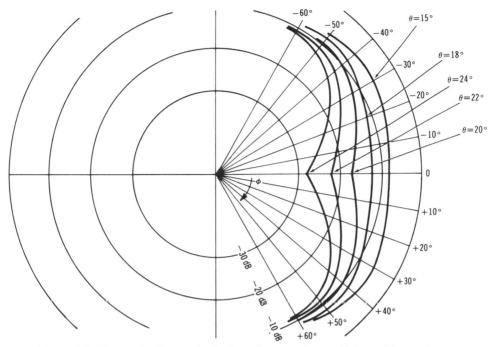

Figure 6.7. The notch effect on the horizontal pattern shown with large tilting angles.

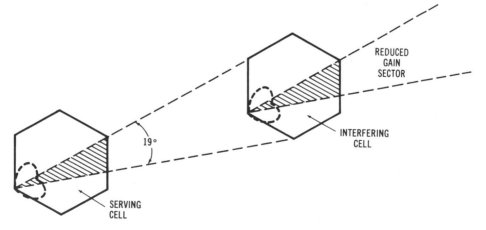

Figure 6.8. Reduction of co-channel interference by creating a notch in the pattern.

aware that if the vertical beam of Fig. 6.6 is very narrow, the tilting scheme may not work.

6.3.3   Diversity Antenna Configuration

Diversity antenna configuration is very important in systems design. As mentioned in Section 6.2, to achieve the same degree of diversity advantage, the separation requirement for $\alpha = 0°$ is much less than that for $\alpha = 90°$. However, the mobile units travel around the base station with no restriction in any direction α (i.e., $0° \leq \alpha \leq 90°$). At $\alpha = 90°$ the required separation is the same as the antenna height for a correlation coefficient of 0.7. A triangular configuration for three omnidirectional antennas is shown in Fig. 6.9A. This is the arrangement that will overcome the difficulty of $\alpha = 90°$. The combiner can always choose two out of the three received signals for diversity, or it can form a three-branch diversity. With this arrangement the $\alpha = 90°$ case will never occur. The worst case would be $\alpha = 60°$. Figure 6.4 shows that the $\alpha = 60°$ curve is much closer to the $\alpha = 0°$ case than to the $\alpha = 90°$ case. The $\alpha = 0°$ case may be used for design purposes. If a three-sector cell is used, the antenna configuration is as shown in Fig. 6.9B. The worse case in this configuration is also $\alpha = 60°$. Therefore Fig. 6.9A and B show configurations that avoid the case of $\alpha = 90°$.

Two Omnidirectional Antenna Configurations
Effective antenna height conditions were reviewed in Section 2.4. If there is a certain direction at the base station whose antenna height shows a higher effective antenna height gain, then diversity gain will not be needed to receive such a strong signal from the mobile transmitter in that direction. Two received base-station antennas can be aligned in this direction with a separation according to the $\alpha = 0°$ case. An illustration appears in Fig. 6.9C. For example, if the operating frequency is 850 MHz and the antenna height is 30 m (100 ft), the separation would be 9 ft as in the $\alpha = 0°$ case. The orientation of the two antennas is based on the terrain configuration as shown in Fig. 6.9C. One kind of terrain configuration does not need diversity antennas, but the other kind of terrain configuration does. If the terrain configuration surrounding the base station is such that $\alpha = 90°$ is impossible to avoid, then the three-antenna configuration of Fig. 6.9B should be used.

6.3.4   Comments on Vertical Separation

The notion of vertically separating two diversity antennas is an appealing one. First, it is easier to separate two antennas vertically than horizontally on an antenna mast. Second, vertical separation avoids the complications of horizontal separation mentioned earlier for different angles α, the location of the mobile unit. But, unfortunately, vertical separation's diversity performance is poor because, as the distance increases the correlation between two received

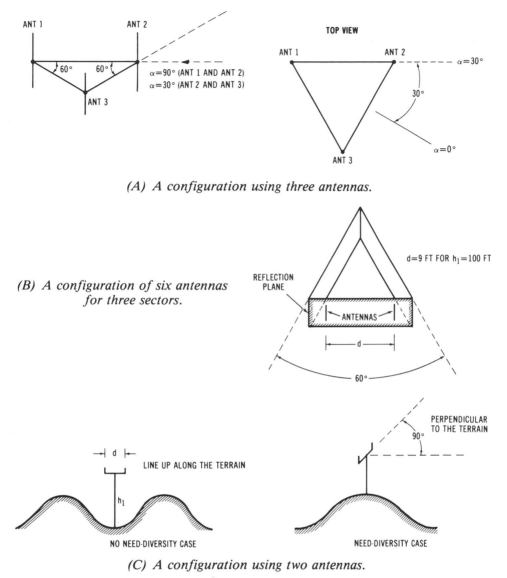

(A) A configuration using three antennas.

(B) A configuration of six antennas for three sectors.

(C) A configuration using two antennas.

Figure 6.9. Arrangement of base-station antennas for space-diversity schemes.

base-station antennas separated by a fixed space increases at a much faster rate with horizontal spacing than with vertical. For this reason vertical separation must be discouraged. In the horizontal plane the signal arrives at the base station over an angle $\Delta\theta$ based on (see Fig. 6.10A).

$$\Delta\theta = \frac{2r}{D} \qquad (6.3.1)$$

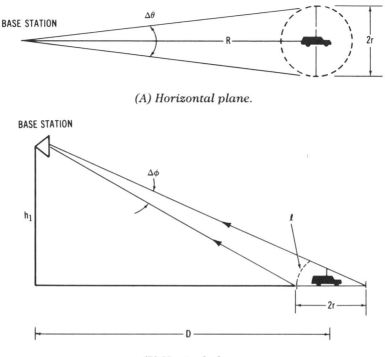

(A) Horizontal plane.

(B) Vertical plane.

Figure 6.10. The vertical and horizontal angles of a scattered area observed from the base station.

where D is the distance from the base station to the mobile unit, and r is the radius of active scatterers surrounding the mobile units. Equation (6.3.1) was shown in Section 4.7. However, the signal arrival at the base-station antenna (h_1) from a mobile transmitter at a vertical angle $\Delta\phi$ is limited to (see Fig. 6.10B):

$$\Delta\phi = \frac{2r(h_1/D)}{\sqrt{D^2 + h_1^2}} \simeq \frac{2rh_1}{D^2} \qquad (6.3.2)$$

Since r is much smaller than D, a comparison of Eq. (6.3.1) with Eq. (6.3.2) yields

$$\Delta\theta \gg \Delta\phi \qquad (6.3.3)$$

To gain the same correlation coefficient of two received branches, the vertical separation between two base-station antennas should be much larger than a horizontal separation. As the effective antenna height at the base station becomes much smaller than its actual height, the received signal at the base becomes weaker. When this situation occurs, diversity schemes need to be used. To find the difference in reception gain ΔG between the two antennas, we write for two vertically spaced antennas erected at the base station,

$$\Delta G = 20 \log_{10} \frac{h'_{e_1}}{h_{e_1}} \simeq 20 \log_{10} \frac{(h_1 + s) - \Delta h}{h_1 - \Delta h}$$

$$= 20 \log_{10}\left[1 + \frac{s}{h - \Delta h} \right] \qquad (6.3.4)$$

where h_{e_1} and h'_{e_1} are the effective antenna heights of two vertical antennas. The vertical separation is s, and Δh is the difference in heights between the actual antenna height and the effective antenna height. The difference in heights Δh can be a positive value or a negative value. When Δh is a positive value, the effective antenna height is shorter than the actual antenna height. If the separation and the effective antenna height $h_{e_1} = h - \Delta h$ of the lower antenna are the same, then the difference in gain between the upper and lower antennas is 6 dB. Usually when the difference between two signal levels is more than 4 dB, the advantage of diversity in reducing fading diminishes. The advantage of using diversity with different received signal levels is seen in Fig. 6.11. The biggest advantage to using diversity comes when two received signal levels are the same. The diversity advantages (gain) decrease as the difference between two received signal levels increases. Little advantage would be found in the case of $\Gamma' = 10\Gamma$ where Γ and Γ' are two signal-to-noise ratios of two respective branches.[4] When the difference in heights Δh increases, the effective antenna height is much shorter than the actual antenna height. The reception at the base station is weak and needs more diversity gain to enhance it. Nevertheless, Eq. (6.3.4) shows that as Δh increases, ΔG becomes larger, and there is less diversity gain. Therefore in a vertically spaced antenna arrangement when signal reception dictates a large diversity gain, such diversity gain cannot always be provided as we have planned.

The previous observations point out the disadvantages of vertically spaced antennas with large separations. When the diversity advantage is insignificant, and there is no concern of interference, a single high-tower antenna can be used for reception.

6.3.5 Physical Considerations in Horizontal Separation

Antenna separation can be determined based upon the parameter $\eta = 11$, as mentioned in Section 6.2.2.

$$\eta = \frac{h}{d} = 11 \qquad (6.3.5)$$

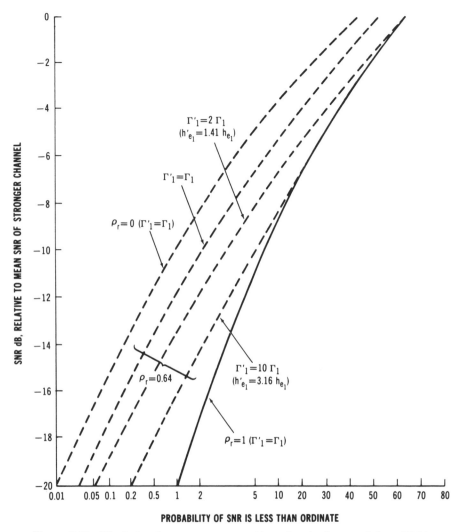

Figure 6.11. Effect of performance on unequal branches for the correlation of 0.64.

where h is the antenna height and d is the horizontal separation. If $h = 100$ ft, then $d = 9$ ft; if $h = 50$ ft, then $d = 4.5$ ft. However, the physical separation of either 4.5 ft or 9 ft may cause a severe ripple effect on the two base-station antenna patterns, depending on its antenna length.

For a 9-dB gain (with respect to a dipole) antenna its physical length is four wavelengths. At 850 MHz, four wavelengths equal approximately 4 ft. Then the far field distance of this antenna is

$$D \approx \frac{2L^2}{\lambda} = \frac{2 \times (4 \text{ ft})^2}{1 \text{ ft}} = 32 \text{ ft} \tag{6.3.6}$$

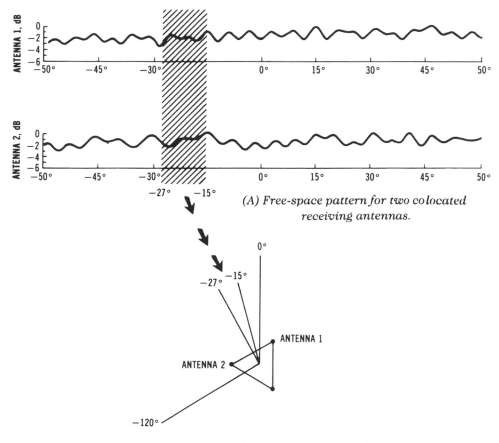

(A) Free-space pattern for two colocated
receiving antennas.

(B) Physical orientation of antennas in a multiple-array
configuration.

Figure 6.12. Free-space patterns for two receiving antennas in a multiple-array antenna configuration.

Now if we separated two 4-ft antennas only 9 ft apart, then one antenna is in the other's near field and causes the ripple effect.

The two antenna patterns altered by the ripple effect are shown in Fig. 6.12[5]. The ripple effect is caused by having three 4-ft antennas in a triangular configuration with a separation of 9 ft while operating at 850 MHz. The variation of the ripples is ±1 dB. The difference between two rippled patterns at any angle in azimuth can be ±2 dB when comparing two antenna patterns. Since the two diversity signals are received by a difference of less than 2 dB (most of the time), we can still have diversity gain, as shown in Fig. 6.11. If the antenna separation is 4.5 ft, the severe ripple effect will occur on the two antenna patterns. The advantage of using diversity then diminishes.

6.4   NOISE ENVIRONMENT

The human-made noise at the base station provides the noise level for the signal-to-noise ratio at base-station reception. Two kinds of spurious noise are considered: automotive noise and industrial plant noise. In the mobile radio environment, the dominant noise is automotive noise; the secondary noise source is power-generating facilities; and the third-ranked noise source is industrial equipment. Other noises such as from consumer products, lighting systems, medical equipment, electrical trains, and buses are low noise and can be ignored.

6.4.1   Automotive Noise

The ignition system of a gasoline engine is a source of high voltage and electrical currents that function continuously during vehicle operation. Radiation produced by high-pulse currents and voltages in cabling, and at points of ignition-circuit discontinuity is the primary source of automotive radio noise.

At the base station (see Fig. 6.13), the antenna is usually 36 m (120 ft) or more from the mobile-ignition noise source, which makes the noise level received by the base-station receiver lower than that at street level as received by a mobile-unit receiver. The ignition noise levels from the ignition system of the mobile unit itself and from another mobile unit are about the same. (An explanation of this appears in Section 7.6.) Assume that the ignition noise level at a mobile unit is based on the number M of the ignition systems (including its own and surrounding mobile units) and on an average distance d_1 from it to the surrounding mobile units. Assume that the ignition noise level at the base-station antenna is based on the number N of the mobile ignition systems surrounding it, and an average distance d_2 from it to the surrounding mobile units. Since the base-station antenna height is 33 m (100 ft) or more, it receives an ignition noise level from a large number of mobile ignition systems because of its height. The number of N is larger than the number M, say

$$N = 10M$$

However, the distance d_2 is also greater than the distance d_1 (see Fig. 6.13). The noise level will be proportionally lower to the square of their distances. The difference in noise level ΔN between base-station reception and mobile-unit reception is

$$\Delta N = 10 \log_{10}\left[\frac{N}{M}\left(\frac{d_1}{d_2}\right)^2\right] = 10 \log_{10}\left[10\left(\frac{30 \text{ ft}}{120 \text{ ft}}\right)^2\right] = -2 \text{ dB} \qquad (6.4.1)$$

Once the average ignition noise level at a mobile-unit receiver in a particular

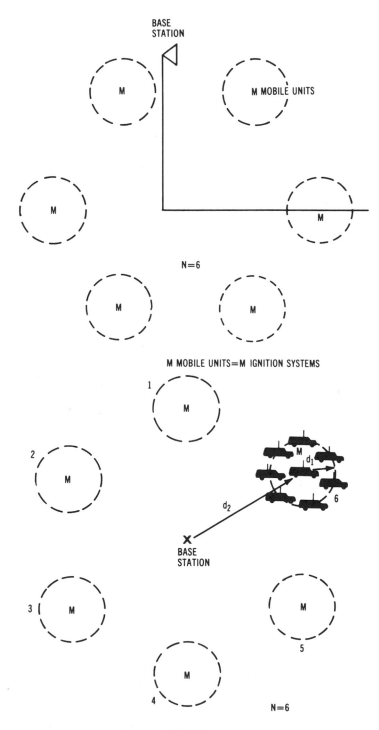

TOP VIEW
Figure 6.13. A model of estimating the base-station noise.

area is known (see Section 7.4), the base-station receiver noise level is about 1 to 2 dB lower because its antenna is located away from the ignition noise sources.

6.4.2 Power-Line Noise and Industrial Noise[6]

Radio noise arises within the spectral range from the fundamental generation frequency either 50 Hz or 60 Hz into the UHF range. Power-line noise dominates the low frequencies as shown in Fig. 6.14. The electric field strength in dB relative to 1 μv/m/MHz is shown in the y-axis. The industrial noise level is found as follows: Suppose that at a frequency of 100 MHz the high-level noise of a power line (345 kv) at 65 m (200 ft) away is 90 dBμv/m/MHz or 110 dBμv/m at 100 MHz. Changing this to a power scale with a 50-ohm resistance, the poynting vector is expressed

$$\rho = \frac{E^2}{R} \text{ watts/m}^2 \tag{6.4.2}$$

where E is expressed in μv/m. A field strength E of 0 dBμv/m becomes

$$0 \text{ dB}\mu\text{v/m} = 20 \log \frac{1 \ \mu\text{v/m}}{1 \ \mu\text{v/m}}$$
$$\Rightarrow 10 \log \frac{10^{-10.7} \text{ mw/m}^2}{1 \text{ mw/m}^2} = -107 \text{ dBm/m}^2 \tag{6.4.3}$$

Then the field strength 110 dBμv/m is equivalent to the poynting vector 3 dBm/m^2. The received power can be found by

$$P = \rho \times A_{e_r} \tag{6.4.4}$$

where A_e is the effective aperture. For a dipole receiving antenna, the aperture is $0.13\lambda^2$. At 1 GHz, one wavelength equals 0.3 m (1 ft). The industrial noise level then becomes

$$P = 3 \text{ dBm} + 10 \log_{10}[0.13 \times (0.3)^2]$$
$$= 3 \text{ dBm} - 19 \text{ dB} = -16 \text{ dBm}$$

This is a very high noise level—higher than ignition noise. When selecting a base-station site, the surroundings must be considered. As in Fig. 6.14, the arc-welder noise dominates at the frequency band from 100 MHz to 16 GHz, or even higher, measured from 30 m (100 ft) away. It too would cause a noise level of 90 dBμv/m/MHz—the same as the power-line noise at 100 MHz.

Information on automotive ignition noise can be gathered statistically by counting the number of vehicles passing in an area. (See Section 7.6.) However, power-line noise and industrial noise will only be a problem if the base

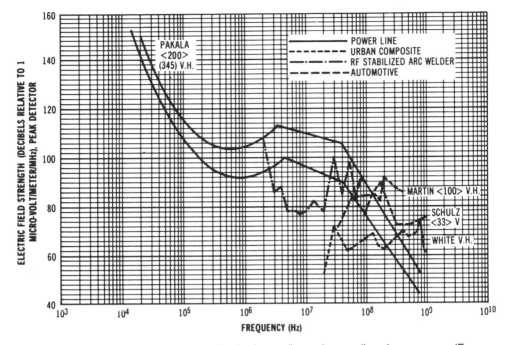

Figure 6.14. Electric-field strength (peak detected) spurious radio-noise sources. (From reference 6.)

station is located in an area with factories and power lines. The way to avoid all noise is to increase antenna height. This of course creates other considerations such as cost and zoning approval. It is crucial to investigate the noise level of any site before deciding to install a base station.

6.5 POWER AND FIELD STRENGTH CONVERSIONS

The signal strength at the reception is usually called *carrier strength* and is measured in dBm or dBμ. The carrier strength needs to be as high as possible with a limited transmitted effective radiation power (ERP). ERP is power transmitted from an omnidirectional antenna. Good design of a transmitting antenna at the cell site can ensure good reception. For example, space diversity antennas provided at the cell site can reduce multipath fading and improve reception, as mentioned in Section 6.2.

The noise level is based only on the environment and the receiver at the receiving end. In the mobile radio environment the spurious environmental noise level usually dominates. The noise level (measured in dBm or dBμ) can be found from the equations in Section 6.4.

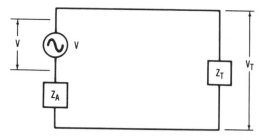

$$Z_A = R_A + j X_A$$
$$Z_T = R_T + j X_T$$

Figure 6.15. A circuit used for calculating power delivery.

The carrier-to-noise ratio (C/N) then can be obtained by

$$\frac{C}{N} \text{ (dB)} = \text{carrier strength level (dBm)} - \text{noise level (dBm)}$$

$$= \text{carrier strength level (dB}\mu) - \text{noise level (dB}\mu) \qquad (6.5.1)$$

6.5.1   Conversion between dBμ and dBm in Power Delivery

Sometimes, the carrier strength (signal strength) or the noise level is measured in dB with respect to 1 mwatt as dBm and sometimes in dB with respect to 1 μv as dBμ. If the carrier strength is measured in dBm and the noise level in dBμ, we have to convert the measured units before applying Eq. (6.5.1). We may use the circuit diagram shown in Fig. 6.15 for the conversion calculation.

In the figure Z_T is the load impedance, $Z_T = R_T + jX_T$, in ohms. Z_A is the antenna impedance, $Z_A = R_A + jX_A$, in ohms. Two cases are considered:

CASE 1. The maximum power received from an emf (electromotive force) induced by a passing wave can be expressed as

$$P_m = \frac{V^2}{4R_A} \qquad (6.5.2)$$

provided that

$$R_A = R_T$$
$$Z_T = Z_A$$

Equation (6.5.2) can be found from circuit textbooks. Assume that $V = 1$

μv (i.e., 0 dBμ) and R_A = 50 ohms. Substituting these figures into Eq. (6.5.2), we obtain the maximum received power:

$$P_m = \frac{V^2}{4R_A} = \frac{(10^{-6})^2}{4 \times 50} = 5 \times 10^{-15} \text{ watts}$$

or

$$P_m \text{ (in dBm)} = 10 \log (5 \times 10^{-15}) = -143 \text{ dBw} = -113 \text{ dBm}$$

The relation is

$$1 \ \mu v \leftrightarrow 5 \times 10^{-12} \text{ mwatt}$$

or

$$0 \text{ dB}\mu \leftrightarrow -113 \text{ dBm}$$

CASE 2. The power density delivery to the load is

$$P_r = \frac{V_T^2}{R_T} \tag{6.5.3}$$

If V_T = 1 μv is measured and R_t = 50 ohm, then the power density delivery to the load is

$$P_r = \frac{V_T^2}{R_T} = \frac{(10^{-6})^2}{50} = 2 \times 10^{-14} \text{ watt}$$

or

$$P_r \text{(in dBm)} = 10 \log (2 \times 10^{-14}) = -137 \text{ dBw}$$
$$= 10 \log (2 \times 10^{-11}) = -107 \text{ dBm}$$

The relation is

$$1 \ \mu v \leftrightarrow 2 \times 10^{-11} \text{ mwatt}$$
$$0 \text{ dB}\mu \leftrightarrow -107 \text{ dBm}$$

This relationship is commonly used for conversion.

Example 6.1. What is the dBm value of a received level of 65 dBμ?

$$65 \text{ dB}\mu - 107 \text{ dBm} = -42 \text{ dBm}$$

$$65 \text{ dB}\mu \leftrightarrow -42 \text{ dBm}$$

6.5.2   Relationship between Field Strength and Received Power

The field strength ρ is measured either in microvolts per meter or in milliwatts per square meter. The power received at the receiving antenna can be expressed as

$$P = \rho \cdot A \tag{6.5.4}$$

where A is the receiving antenna aperture expressed as

$$A = \frac{G\lambda^2}{4\pi} \tag{6.5.5}$$

where G is the gain of the receiving antenna and λ is the wavelength. Then the relationship between the field strength and the received power is as expressed in Eq. (6.5.4) or as follows:

$$P = \rho \cdot \frac{G \cdot \lambda^2}{4\pi} \tag{6.5.6}$$

The received power is a function of field strength and aperture as shown in Eq. (6.5.4) or a function of field strength and wavelength as shown in Eq. (6.5.6).

6.5.3   A Simple Conversion Formula

To convert a field strength to a received power, we assume the frequency, the type of the probe to collect the field strength, and the optimum terminal impedance. In general, we use a half dipole whose effective length is λ/π and a 50-ohm optimum terminal whose equivalent circuit is shown in Fig. 6.15.

The notation dBμ commonly used by the industry is confusing. It should be written as dB μv/m or dB to refer to microvolts per meter. However, today dBu can mean either a voltage or a field strength.

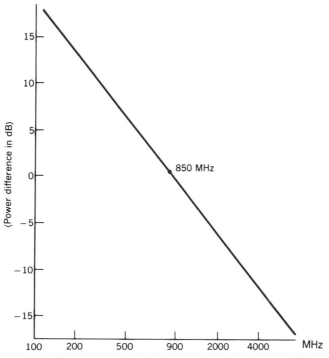

Figure 6.16. Power conversion for frequencies other than 850 MHz.

The relationship between the field strength in dBμ and the power in dBm can be expressed as follows:

$$0 \text{ dB}\mu \ (=) \ 10 \log\left[\frac{(10^{-6})^2(\lambda/\pi)^2}{4 \times R_A} \times 1000\right] \text{dBm}$$

$$(=) \ 10 \log\left[22.8 \times 10^{-7} \frac{1}{R_A \cdot f^2}\right] \text{dBm}$$

At 850 MHz, $R_A = 50$ ohms

$$0 \text{ dB}\mu = -132 \text{ dBm}$$
$$39 \text{ dB}\mu = -93 \text{ dBm}$$
$$32 \text{ dB}\mu = -100 \text{ dBm}$$

At another frequency or using another kind of probe: ($R_A = 50$ ohms)

$$0 \text{ dB}\mu = -132 \text{ dBm} - 20 \log\left(\frac{f}{850}\right) + G$$

where f is the other frequency in MHz and G is the gain of the probe over a dipole in dB. This equation is plotted in Fig. 6.16 with different frequencies where $G = 0$ dB.

Some Cautions on dBm Calculation

The dBm units cannot be added. dBm + dBm means "power squared," which has no physical meaning.

10 dBm + 10 dBm $\neq$ 20 dBm   (does not follow the arithmetic rule nor have physical meaning)

$$10 \text{ dBm} + 10 \text{ dB} = 20 \text{ dBm}$$

$$30 \text{ dBm} - 10 \text{dB} = 20 \text{ dBm}$$

$$30 \text{ dB} + 10 \text{ dB} = 40 \text{ dB}$$

$$30 \text{ dB} - 10 \text{ dB} = 20 \text{ dB}$$

$$30 \text{ dBm} - 10 \text{ dBm} = 20 \text{ dB}$$

REFERENCES

1. Lee, W. C. Y., "Mobile Radio Signal Correlation versus Antenna Height and Spacing," *IEEE Trans. Veh. Tech.* 25 (Aug. 1977): 290–292.
2. Lee, W. C. Y., "Studies of Base-Station Antenna Height Effects on Mobile Radio," *IEEE Trans. Veh. Tech.* 29 (May 1980): 252–260.
3. Lee, W. C. Y., "Cellular Mobile Radiotelephone System Using Tilted Antenna Radiation Patterns" (U.S. Patent 4,249,181, Feb. 3, 1981).
4. Schwartz, M., W. R. Bennet, and Seymour Stein, *Communication Systems and Techniques* (McGraw-Hill 1966): 473.
5. Lee, W. C. Y., *Mobile Communications Engineering* (McGraw-Hill, 1982): 154.
6. Skomal, E. N., "Man-made Radio Noise" (Van Nostrand Reinhold, 1978): 9.
7. Lee, W. C. Y., "The Decibel: A Confusing Issue" Cellular Business, March 1992, p. 52.

PROBLEMS

6.1   Design a diversity antenna separation using at the base the parameter

$$\eta = \frac{h}{d}$$

Why must the antenna separation be found by the correlation coefficient of 0.7? How about $\rho = 0.2$?

6.2   Why is the antenna separation at the base station wider when the antenna height is higher? Give the physical reason.

6.3   Why is the antenna separation at the base station wider when the orientation of the two antennas is in line to the direction of the mobile unit?

6.4   When the mobile unit is traveling in the field, the effective base-station antenna height is measured based on the location of the mobile unit. In what kind of terrain contour is the effective antenna height greater than or less than the actual antenna height?

6.5   Assume that the antenna height is 100 ft and that the co-channel cell is $4.6R$ away, where the cell radius R is 4 miles. What would be the power difference between the signal received at the cell boundary and that at the co-channel cell if the downtilting of the antenna in Fig. 6.6 is 20°?

6.6   The vertical separation of two diversity antennas is not desired. Explain why.

6.7   Explain the nature of the limitation shown in Fig. 6.10. Why does the vertical separation need more spacing than the horizontal separation for the same correlation coefficient?

6.8   Two signals are received from a two-branch diversity receiver. Their signal correlation coefficient is 0.64. If the two signals are equal in strength

$$\Gamma'_1 = \Gamma_1$$

or unequal in strength

$$\Gamma' = 10\Gamma$$

what is the probability of CNR being less than -10 dB (which is a level of the stronger channel below its average power).

6.9   The 30 dBμ signal delivers at the load terminal with a 75-ohm load impedance. What is the equivalent dBm?

6.10   Given an antenna whose height is 100 ft and space-diversity separation is 9 ft at the base. The effective antenna height measured at the base station based on the present mobile unit location is 40 ft. Is the diversity gain increasing? How much? If the effective antenna height were 200 ft, would the diversity gain be increasing? Why?

7

DESIGN PARAMETERS AT THE MOBILE UNIT

7.1 ANTENNA SPACING AND ANTENNA HEIGHTS

At the mobile unit the antenna height is always supposed to be lower than its surroundings. This is an assumption of the model in Fig. 1.1. If the mobile antenna is higher than its surroundings, it cannot be called a mobile antenna; this case was considered in Chapter 6. For the purpose of design considerations, examine a worst case scenario of reception at the mobile-unit site. A worst case would be a direct wave between base-station transmitter and mobile receiver that is blocked by buildings or houses in between.* The scattered waves thus arrive from 360° around the mobile unit and two signal fadings can be less correlated with a small antenna spacing at the two mobile antennas. Therefore the two spaced diversity antennas at the mobile unit are located as close as 0.5 wavelength (see Section 3.1.4). This closed spacing is based on the phenomenon that the correlation coefficient ρ of two signal fadings is less than 0.2 at that spacing and the antenna height is roughly 3 m (10 ft) above the ground.[1] The 0.5-wavelength spacing applied to a frequency of 900 MHz is about 18 cm (6 in.), which is very practical for mounting on the roof of a mobile unit. However, at 90 MHz the 0.5λ spacing is 1.5 m (5 ft), an impractical length.

If the same argument that uses the parameter $\eta = h/d$ shown in Eq. (6.2.1) is followed, then when the antenna height h is lower, the spacing d is smaller. This is also true at the mobile antenna site.

The computation used to obtain the spacing requirement of the base-station antenna separation shown in Fig. 6.4 cannot be applied here. The required separation for a roof-mounted mobile-unit antenna to achieve a diversity gain is always about 0.5λ to 0.8λ if the antenna height is 3 m (10 ft) or less. The correlation coefficient in this case is less than 0.2 (see Fig. 7.1), which is a very

---

*If a direct wave is blocked by the terrain contour, no reflected waves will occur. This situation is dealt with by the obstruction condition in Section 2.4.2.

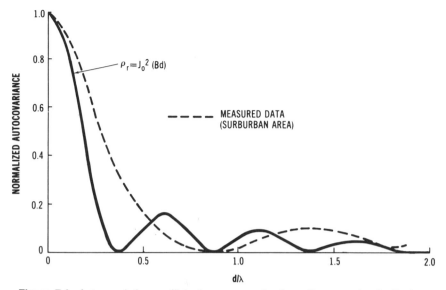

Figure 7.1. Autocorrelation coefficient versus spacing for uniform angular distribution.

satisfactory value for diversity application. At 850 MHz a wavelength 0.5λ is 0.5 ft long. Therefore the value of η for applying diversity at the mobile unit is

$$\eta = \frac{h}{d} = \frac{10 \text{ ft}}{0.5 \text{ ft}} = 20 \qquad \text{at 850 MHz}$$

which is based on the correlation coefficient of 0.2. Clearly Fig. 6.4 cannot be applied here.

Since the physical height of the mobile antenna is always in the range from 1.5 to 3 m (or 5 to 10 ft) and the antenna separation is always 0.5λ, which is measured in wavelengths, lowering the frequency reduces the value of η. The correlation coefficient is independent of the individual signal gain.[2] If two signals are s_1 and s_2, the correlation coefficient $\rho(\tau)$ of these two signals $s_1(t)$ and $s_2(t + \tau)$ is

$$\rho = \frac{\overline{s_1 s_2} - \overline{s_1}\,\overline{s_2}}{\sqrt{\overline{s_1^2} - \overline{s_1}^2}\,\sqrt{\overline{s_2^2} - \overline{s_2}^2}} \tag{7.1.1}$$

assuming that

$$s_1(t) = Ae_1(t) \tag{7.1.2}$$

and

$$s_2(t + \tau) = Be_2(t + \tau) \tag{7.1.3}$$

where $\overline{e_1^2(t)} = \overline{e_2^2(t)} = 1$, and A^2 and B^2 are the values of gain of two signals respectively. Substituting Eqs. (7.1.2) and (7.1.3) into Eq. (7.1.1), we obtain a correlation coefficient

$$\rho(\tau) = \frac{\overline{e_1 e_2} - \overline{e_1}\,\overline{e_2}}{\sqrt{\overline{e_1^2} - \overline{e_1}^2}\,\sqrt{\overline{e_2^2} - \overline{e_2}^2}} \qquad (7.1.4)$$

that is independent of the values of gain. Equation (7.1.4) indicates that any values of gain result in the same correlation coefficient. Therefore in considering an adequate diversity performance, not only the correlation coefficient but also the difference between two signal gains must be taken into account. If the correlation coefficient is satisfied but either two signal levels are the same but weak, or two signal levels are different by more than 4 dB (see Fig. 6.11), then no diversity advantages can be seen. The experimental data show that when both h_2 and h_2' are less than 5 m (15 ft), the 3-dB/oct antenna height-gain rule is applied (see Section 2.3.7)

$$\text{Gain (or loss)} = 10 \log_{10} \frac{h_2'}{h_2} \qquad (7.1.5)$$

At a mobile-unit site only the actual antenna height counts. There is no effective antenna height for the mobile antenna. To predict the path loss, the mobile antenna height does not need to be adjusted, based on the terrain configuration. When 5 m $< h_2' <$ 10 m, Eq. (2.3.29) is used:

$$\text{Gain (or loss)} = 2\, h_2' \log_{10}\left(\frac{h_2'}{3\text{ m}}\right) \qquad (7.1.6)$$

When $h_2' > 10$ m (30 ft), h_2' will be considered as the base-station antenna height h_1 and the gain-height relation will be applied for the base-station antennas, as shown in Eq. (2.5.2). When the mobile antenna heights, h_2' is within a range from 5 to 10 m (15 to 30 ft), and h_2'' is any height below 5 m, then

$$\text{Gain (or loss)} \simeq 2\, h_2' \log_{10} h_2' - 2 h_2'' \log_{10} h_2'' - 0.954\, (h_2' - h_2'') \qquad (7.1.7)$$

Assume that two mobile antenna heights $h_2' = 8$ m (26 ft) and $h_2'' = 4$ m (13 ft), then the difference in gain is 5.82 dB, which is not exactly 6 dB, as shown in Eq. (2.5.2). For the case of $h_2' = 6$ m (20 ft) and $h_2'' = 3$ m (10 ft), the difference in gain is 3.62 dB. Equation (7.1.6) is an empirical formula.

7.2 MOBILE UNIT STANDING STILL AND IN MOTION

When the mobile unit is standing still, it may be situated in a signal fade. The probability of this situation can be calculated as shown in Section 3.7.

In this case the signal will never be received. If the mobile unit is transmitting, the mobile user may realize the situation since there will not be an acknowledgment from the base station. The solution is to move the vehicle more than half a wavelength and try to call again. If the call still cannot go through, then the vehicle may be in a field-strength hole, where the average field strength is below the threshold. In this situation there is nothing the user can do. However, in a well-designed system, the probability of running into a field-strength hole is very slight.[3]

When the mobile unit is standing still in a fade, the called party may never realize that an incoming call is being missed. To avoid this situation, two antennas separated by a required separation (0.5λ) can be mounted on the roof of the vehicle. This is a space-diversity arrangement. A proper separation for the two signals received from two separated antennas to combat signal fading can be determined by the correlation[2] of two fading signals,

$$\rho_r(d) = J_0^2(\beta d) \tag{7.2.1}$$

where $\rho_r(d)$ is the correlation coefficient of two signal-fading envelopes, and a function of antenna separation d. The range of $\rho_r(d)$ is

$$0 \le \rho_r(d) \le 1 \tag{7.2.2}$$

Thus $\rho_r(d)$ should be as small a value as possible within a practical physical separation between two antennas. Equation (7.2.1) is plotted with the solid curve shown in Fig. 7.1. The first null of the correlation is roughly $d = 0.5\lambda$. For a separation d greater than 0.5λ, ρ_r is less than 0.2. Any value of ρ_r less than 0.2 can be considered as uncorrelated between two fading signal envelopes. Therefore

$$d = 0.5\lambda \tag{7.2.3}$$

is the required antenna separation against signal fading. The data measured in the suburban area are also shown in Fig. 7.1. The first null of the measured $\rho_r(d)$[4] is at $d = 0.8\lambda$. Sometimes a separation of 0.8λ may be used if there is enough space. At 900 MHz the separation of 0.8λ is equivalent to 24 cm (9.6 in)—still no problem to mount on top of the roof of the car.

7.3   INDEPENDENT SAMPLES AND SAMPLING RATE

On many occasions, the raw signal-fading envelope must be sampled and an uncorrelation assumed between two adjacent samples. Then the sampling interval d should be in the range

$$0.5\lambda \le d \le 0.8\lambda$$

If d is less than 0.5λ, the adjacent samples are correlated. If d is greater than 0.8λ, overkill occurs.

The sampling rate should not be based on the length T of a received signal measured in the time domain, but on the velocity information and the required spacing between adjacent bits, as

$$\text{Number of samples } (n) = \frac{VT}{0.5\lambda} \tag{7.3.1}$$

$$\text{Sampling rate } (R_s) = \frac{n}{T} = \frac{V}{0.5\lambda} \tag{7.3.2}$$

Example 7.1. A piece of one-minute data is received while the vehicle moves at 20 m/s. The wavelength is 0.3 m (1 ft). Then the total number of independent samples is

$$n = \frac{(60 \text{ s})(20 \text{ m/s})}{(0.5\lambda)(0.3 \text{ m/}\lambda)} = 8000 \text{ samples}$$

The sampling rate is

$$R_s = \frac{n}{T} = \frac{8000 \text{ samples}}{60 \text{ s}} = 133.33 \text{ samples per second}$$

Example 7.1 shows that if the signaling transmission rate is 133 samples per second or less under the given conditions, each bit will fade independently of the next bit. Of course this slow rate of transmission is unacceptable. Then, if a new transmission rate increases 10 times to 1330 samples per second in the same environment described in Example 7.1, a bit will be uncorrelated with another bit 10 bits away. Within the 10 bits, the bits will be correlated among themselves. The adjacent bits yield the strongest correlation. This means that if a bit falls in the fade, the chance that the adjacent bit will do likewise is very great.

7.4 DIRECTIONAL ANTENNAS AND DIVERSITY SCHEMES

7.4.1 Directional Antennas[5]

Directional antennas with an N-element array mounted on the roof of the mobile unit can only reduce the fading frequency compared to a single antenna, but cannot increase the received gain due to N elements, compared to the single antenna. A physical explanation can be found by reviewing Fig. 7.2. Since the multipath signals come from all directions around the mobile unit, a directional antenna only received a sector $\Delta\theta$ of the incoming multipath

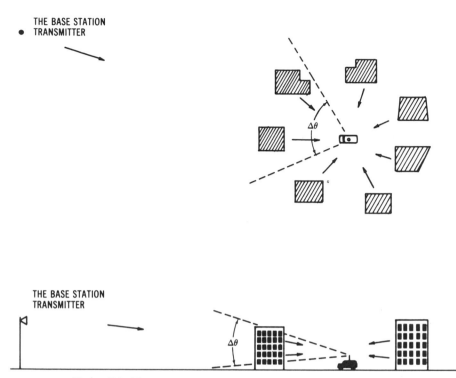

Figure 7.2. The reception of a directional antenna at the mobile unit.

signal. Sector $\Delta\theta$ is smaller; the fading is less. It can be seen from the fading frequency of Eq. (1.3.14) as

$$f_d = \frac{V}{2\lambda} (1 - \cos \theta) \qquad (7.4.1)$$

and the maximum fading frequency is

$$f_{max} = \frac{V}{\lambda} \qquad (7.4.2)$$

If the signal comes from all directions, then the maximum fading frequency is obtained from Eq. (7.4.1). If in the equation $\theta = 180°$, then $f_d = f_{max}$. When sector $\Delta\theta$ becomes 45°, the fading reduces to 0.15 f_{max} according to $\theta = 45°$. When sector $\Delta\theta$ becomes 30°, the fading reduces to 0.067 f_{max} according to $\theta = 30°$. However, sector $\Delta\theta$ is related both to the gain of the directional antenna and to the loss associated with the mobile radio environment. Here the free-space formula is used to show the relationship between the antenna directivity gain and the loss of the received signal due to the limited sector $\Delta\theta$. In a conventional line-of-sight communication link, the direction of an incoming signal is usually known. Then the received signal power P_r can be predicted as

$$P_r = \frac{P_t g_t g_r}{(4\pi r_1/\lambda)^2} \tag{7.4.3}$$

where P_t is transmitted power, g_t is gain of transmitting antenna, r_1 is distance between the antennas, λ is wavelength, and g_r is gain of receiving antenna.

In a mobile radio environment the signal can come from any direction with equal probability, so no particular direction is preferred. The loss is actually due to the limited sector $\Delta\theta$ of the directional antenna receiving a portion of total incoming waves. Equation (7.4.3) should be modified for the mobile radio environment as

$$P_r = \frac{P_t g_t}{(4\pi r_1/\lambda)^2} (\alpha g_r) \tag{7.4.4}$$

where α is the attenuation factor associated with the limited beam width of the directional antenna in the mobile radio environment. The two parameters α and g_r can be expressed as

$$\alpha = \frac{(\Delta\theta)(\Delta\phi)}{2\pi^2} \tag{7.4.5}$$

$$g_r = \frac{k}{(\Delta\theta)(\Delta\phi)} \tag{7.4.6}$$

where k is a constant, $\Delta\theta$ and $\Delta\phi$ are the directional antenna beamwidths in two planes. Then the product of αg_r becomes

$$\alpha g_r = \frac{k}{2\pi^2} = k_2$$

This shows that the received signal power in a mobile radio environment is a constant, independent of the directivity of the directional antenna.

$$P_r = k_2 \left[\frac{P_t g_t}{(4\pi r_1/\lambda)^2} \right] \tag{7.4.7}$$

7.4.2 A Diversity Scheme for Mobile Units

At the mobile unit, the received signal has suffered a severe fade that affects both voice and data transmission. The traffic noise level on the highways or

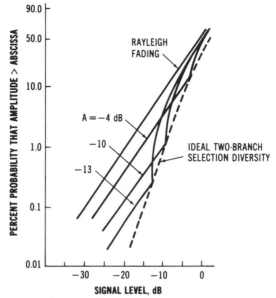

Figure 7.3. Performance of a two-branch switched-combined signal with various threshold levels.

in a business area is also very high compared with the thermal noise level. For this reason, the signal-to-noise ratio in a mobile receiver can be improved by reducing signal fading, but not by increasing receiver sensitivity, because of the high environmental noise level. A switch-combined, space-diversity, low-cost receiver may be recommended, since it has only one RF front end. The received signal is based on a threshold level, which controls the switches.[6] When the signal of one antenna falls below the threshold level, the unit can switch to the other antenna. The level of performance is shown in Fig. 7.3. The best level of performance of the switch-combined scheme is at the threshold level, where performance approaches that of a selection-combined scheme (see Section 3.5). This indicates that if a variable threshold level scheme can be implemented, based on past and present received information, then the switch-combined scheme will be an effective combining technique. However the switch-combined scheme in reality is hard to achieve an expected performance. Therefore sometimes to make a choice based on the cost may not be a wise policy.

Two vertical separated space-diversity antennas with a separation of 1.5λ mounted on the roof of a vehicle can provide a low correlation between two received signals. The difference in two signals strength is only 0.5 dB stated in Section 6.3.4. Therefore a vertical-separated diversity scheme can be applied at the mobile unit.

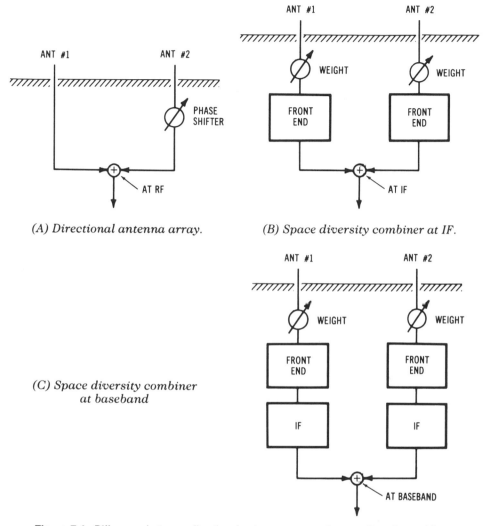

(A) *Directional antenna array.*

(B) *Space diversity combiner at IF.*

(C) *Space diversity combiner at baseband*

Figure 7.4. Difference between directional antenna array and space-diversity combiner.

7.4.3 Difference Between Directional Antenna Arrays and Space-Diversity Schemes

Directional antenna array elements are combined at the RF. With a proper phase relationship among the elements in space and the lengths of connecting wires or RF phase shifters among the elements, its antenna pattern can be directed to a broadside or an endfire, as shown in Fig. 7.4A. The space-diversity scheme is usually combined either at IF or at the baseband as shown in Figs. 7.5B and C. The combining techniques used can be either maximal-ratio or equal-gain techniques. The maximal-ratio technique will achieve a maximal signal-to-noise ratio at the IF output by weighing the input of each individual branch at RF frequency. Equal-gain combining will co-phase the random phase of each individual branch at IF. Adding the power outputs at the baseband is equivalent to maximal-ratio combining. The selection-combined technique selects the strongest signal among two antenna inputs. All three combining techniques require two RF front ends. The switch-combined technique uses only one RF at a time for the desired signal output. All these diversity schemes do not affect the individual antenna reception patterns after combining all the branches. This is the main difference between the directional antenna array and the space-diversity schemes. At the mobile site the directional antenna array reduces fading without increasing the signal reception level. On the other hand, diversity schemes reduce fading and increase the reception level.

7.5   FREQUENCY DEPENDENCY AND INDEPENDENCY

7.5.1   Operating Frequency Dependency on Space Diversity

Whether or not the space-diversity scheme can be used is based on the operating frequencies and the amount of physical space available. The space-diversity scheme is relatively simple. It does not need additional spectrum for diversity as the frequency-diversity scheme does. It does not suffer a 3-dB loss for transmitted power as the polarization-diversity scheme does. Space diversity should be a first choice when considering diversity scheme. At the mobile site the required separation of antennas for a space-diversity scheme is a half-wavelength. At an operating frequency of 1 GHz, a half-wavelength is only about 15 cm (6 in). Two antennas separated by 6 in. can easily be mounted on the roof or bumpers of any type of vehicle. When operating at a lower frequency such as 100 MHz, the half-wavelength is about 2 m (6 ft), which is an impractical length for separating two antennas on the roof of a mobile unit. Therefore in low frequencies another scheme, called a *field-component diversity* (see Section 3.4 and Section 7.8), can be considered. This diversity scheme has all the advantages of space diversity, without a required spatial separation.[7]

7.5.2   Operating Frequency Independence of Frequency Diversity

In some circumstances frequency diversity can be used effectively against multipath fading. The same signal information carried by two frequencies with enough separation in the frequency spectrum will have two uncorrelated fading characteristics at the receiving end. Combining these two received fading signals, the fading is reduced and the signal information is less likely to be affected by the fading. Another scheme such as transmitting the same information in two frequencies in two consecutive time intervals is recommended. Both schemes can overcome severe fading of the signal transmission. The former scheme sacrifices spectrum while the latter reduces throughput. For a 40-bit word, if the repetition is 5 times at the same frequency or different frequencies, in order to ensure correct reception, the throughput is 1/5. If transmitting the same word in five different frequencies at one time, the spectrum bandwidth increases five times.

The frequency separation for the application of frequency diversity is based on the coherence bandwidth mentioned in Section 3.3:

$$B_c = \frac{1}{2\pi\Delta} \qquad (7.5.1)$$

where Δ is the delay spread. In a suburban area $\Delta = 0.5 \ \mu s$, so

$$B_c \simeq 300 \ \text{kHz} \qquad \text{(in suburban area)}$$

In an urban area $\Delta = 3 \ \mu s$, so

$$B_c \simeq 50 \ \text{kHz} \qquad \text{(in urban area)}$$

In an open area $\Delta \leq 0.2 \ \mu s$, so

$$B_c \simeq 800 \ \text{kHz} \qquad \text{(in open area)}$$

From the preceding data we choose a frequency separation ΔF:

$$\Delta F > 300 \ \text{kHz}$$

Frequency diversity is usually designed to combat severe fading in suburban or urban areas. Therefore in an open area the signal reception should be strong enough so that no frequency diversity is needed. In an urban area the frequency separation ΔF only needs to be greater than 50 kHz. The design rule is to always cover the fading reduction in a suburban area first; that is, $\Delta F > 300$ kHz. This frequency separation also meets the frequency diversity requirement in the urban area.

Two Important Facts about Frequency Diversity

1. The delay spread Δ is not a function of operating frequencies as long as the sizes of the scatterers surrounding the mobile unit are much larger than one wavelength of the operating frequency. In that case the frequency range of a mobile radio environment with Δ as a constant should be greater than 30 MHz. Therefore above 30 MHz, Δ is independent

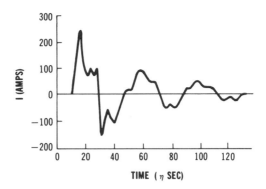

Figure 7.5. Ignition circuit current waveforms.

of operating frequency and only dependent on human-made structures in open, suburban, or urban areas.

2. Frequency diversity can be used on both base-station and mobile-unit sites, with the same frequency separation. However, the space separations used for space diversity on two sites are different. In a space-diversity scheme the antenna separation at the mobile site is 0.5λ; at the base station the antenna separation is quite large in terms of multiple wavelengths, depending on the height of the base-station antennas. This is the basic difference between the two kinds of diversities.

7.6 NOISE ENVIRONMENT

Human-made noise is overshadowed by the ignition noise of the vehicles. Because all vehicles travel on the road, very close to each other, the signal reception at each vehicle is affected by its own ignition noise and that of the surrounding vehicles. The best information on the mobile noise environment is a measurement of traffic flow. The human-made noise level can be extrapolated from the traffic flow information.

To estimate the ignition spike width and the number of spikes in one second, a simple calculation is as follows: One sharp spike of ignition noise exceeding 200 amperes typically lasts 1 to 5 ns.[8] This high portion of the spike will get into a frequency range from 200 mHz to 1 GHz. The spike width below 100 amperes (see Fig. 7.5) is 20 ns. Assume that an engine has eight cylinders and each of them has a speed of 3000 rpm. Since at any instant only half of them are fired, then

$$4 \times 3000 \text{ rpm} = 12,000 \text{ rpm} = \frac{12,000 \text{ rpm}}{60 \text{ s/m}} = 200 \text{ spikes/s}$$

If there are many vehicles on the road, the number of spikes is multiplied

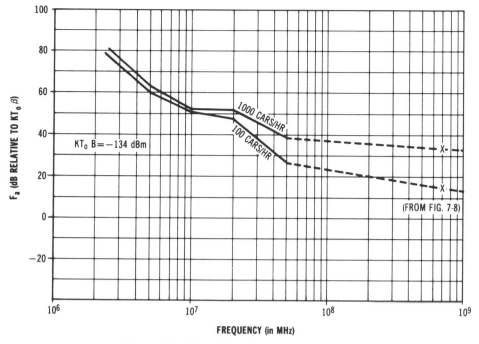

Figure 7.6. Automotive noise versus frequency.

by the number of vehicles. This is why spurious noise is dependent on the traffic. The low portion of the spike (below 100 smperes) will affect the low frequency. For a spike lasting more than 20 ns, the effect on the frequency will be 50 MHz or less. The average automotive traffic-noise power expressed as a noise figure F_a is shown in Fig. 7.6 as a function of frequency. The noise bandwidth at the detector is 10 kHz. The thermal noise kT_0B at room temperature (290°K) and a filter bandwidth of 10 kHz is about -134 dBm. Two traffic densities are shown in Fig. 7.6: The lower the frequency, the higher is the ignition noise. Traffic density also affects the noise level more at the higher frequency as the difference in two noise levels along the frequency scale with two different traffic densities shows. There are noise figures of 10 dB received from low-traffic density (100 cars/hr) and 34 dB received from a relatively high-traffic density (1000 cars/hr) at the 700 to 1000 MHz range. The distribution of the automotive noise follows log-normal distributions, as shown in Fig. 7.7. The data were measured from a vertical dipole at a frequency of 700 MHz. The received filter bandwidth is 10 kHz, and the distance to the nearest vehicle is 10 m (30 ft).

The noise level is always higher with higher-traffic density. The average automotive noise level is at 50%. The average automotive noise versus the number of cars/hr is plotted in Fig. 7.8. Visualize that the noise level increases linearly with traffic density in a logarithmic scale up to 1000 cars/hr, and then

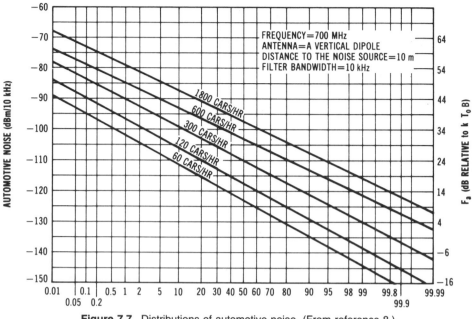

Figure 7.7. Distributions of automotive noise. (From reference 8.)

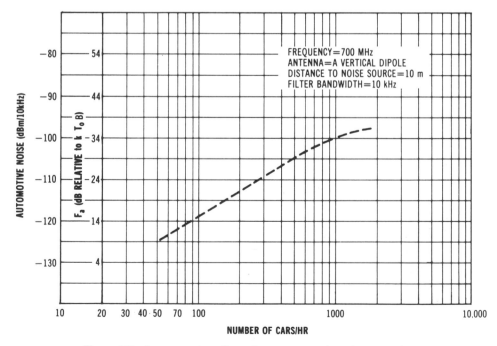

Figure 7.8. Average automotive noise versus number of cars per hour.

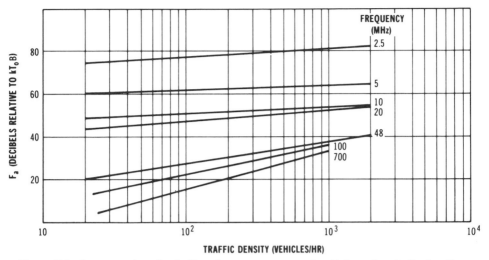

Figure 7.9. Average automotive traffic-noise power expressed as F_a for various traffic densities and frequencies. Detector noise bandwidth, 10 kHz. (From reference 8.)

tapers off as the traffic density increases beyond that. Using the information of Fig. 7.8, an extrapolation is made on two traffic-density curves extending to 1 GHz, as shown in Fig. 7.6 by two dotted lines. In Fig. 7.6 with the two dotted lines added, we see that when the traffic density is high, the automotive noise drops slowly and the frequency increases from 100 MHz to 1 GHz. Also the traffic densities are more effective at high frequencies than at the low frequencies. Figure 7.9 shows this phenomenon whereby the slope becomes steeper along the traffic density scale as the frequency increases. The difference in the two noise levels at the mobile unit and the base station is about 1 or 2 dB. An explanation using a simple model appears in Section 6.4.

7.7 ANTENNA CONNECTIONS AND LOCATIONS ON THE MOBILE UNIT

To get a good reception at the mobile unit, one should place the antenna as high as possible. However, there are physical limitations: The higher the antenna is placed, the greater is the chance of damage. Of course, most users also do not want a hole drilled in the center of the car roof and are willing to suffer 3-dB reception loss by using glass mounting. Some users may choose to mount the antenna on the top of the car trunk so that the antenna can be easily detached while not in use. Others may mount the antenna on the car bumper.

7.7.1 The Impedance Matching at the Antenna Connection

The antenna load impedance and the wire impedance to the mobile transceiver have to be matched to eliminate the reflected wave at the antenna load. Reflected waves occur if manufacturers do not carefully design their products to have good impedance matches; then either transmitted power or received power is reduced. Mobile phone installers usually do not bother to reduce this loss.

We measure the loss using the standing wave ratio (SWR) (see Fig. 7.10A)

$$\text{SWR} = \frac{V_0 + V_1}{V_0 - V_1} = \frac{1 + (V_1/V_0)}{1 - (V_1/V_0)} = \frac{1 + |\rho_v|}{1 - |\rho_v|} \tag{7.7.1}$$

where the voltages V_0 and V_1 of two standing waves are as shown in Fig. 7.10A and ρ_v is the voltage reflection coefficient which can be expressed by

$$\rho_v = \frac{Z_L - Z_0}{Z_L + Z_0} = \frac{(R_L - R_0) + j(X_L - X_0)}{(R_L + R_0) + j(X_L + X_0)} \tag{7.7.2}$$

where Z_L is a load impedance equal to $R_L + jX_L$, Z_x is an impedance at distance x looking toward a load equal to $R_x + jX_x$, and Z_0 is the characteristic impedance of the line equal to $R_0 + jX_0$. For a good match, $V_1 = 0$. In Eq. (7.7.1) SWR $= 1$.

Example 7.1. What are the losses for the two cases $Z_0 = 0.7Z_L$ and $Z_0 = 0.5Z_L$?

CASE 1. We know that $R_0 = 0.7R_L$, $X_0 = 0.7X_L$. Then if we substitute these values into Eq. (7.7.2), we get

$$|\rho_v| = \frac{0.3(R_L + jX_L)}{1.7(R_L + jX_L)} = 0.18$$

$$\text{Loss} = 20 \log(1 - 0.18) = 1.7 \text{ dB}$$

$$\text{SWR} = \frac{1.18}{0.823} = 1.43$$

CASE 2. Now substituting the values $R_0 = 0.5 R_L$ and $X_0 = 0.05 X_L$ into Eq. (7.7.2) yields

$$|\rho_v| = \frac{0.5}{2} = 0.25$$

$$\text{Loss} = 20 \log(1 - 0.25) = 2.5 \text{ dB}$$

$$\text{SWR} = \frac{1.25}{0.75} = 1.67$$

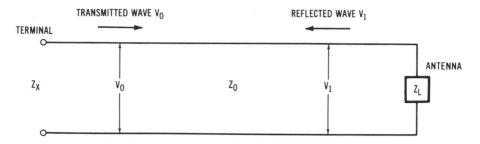

(A) A circuit diagram to illustrate the standing waves.

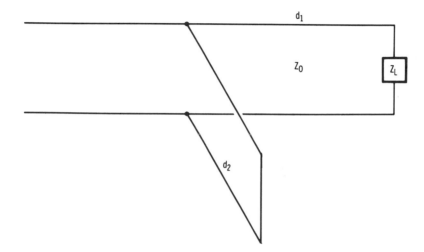

(B) A single-stub tuner.

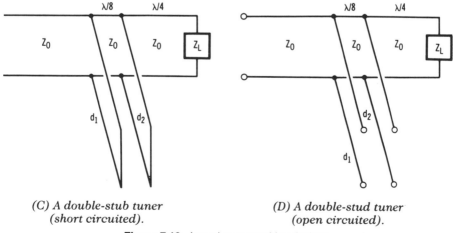

(C) A double-stub tuner
(short circuited).

(D) A double-stud tuner
(open circuited).

Figure 7.10. Impedance matching devices.

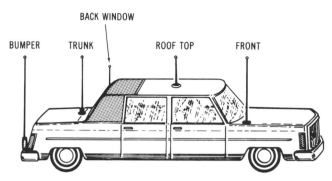

Figure 7.11. Mobile antenna mounting.

When the difference between Z_0 and Z_L is great, the losses are also great. To reduce the losses, we could use a single-stub tuner, as shown in Fig. 7.10B. Another tuner is a double-stub tuner with short-circuited stub or open-circuited stub, as shown in Figs. 7.10C and 7.10D. In using these tuners, we need to adjust two valuable lengths, d_1 and d_2, as shown in either Fig. 7.10C or 7.10D.

7.7.2   Antenna Location on the Car Body

There are five locations—front, roof top, back window, trunk, and bumper—where the antenna can be placed on a car, as shown in Fig. 7.11. But no matter where it is located, the antenna has to be higher than the roof of the car for it to provide a good reception of waves arriving from all directions. The advantage of using a glass-mounted antenna is that it does not require that a hole be drilled. The disadvantage is that there is a loss of signal coupling through the glass. This varies with the thickness of the glass and the frequencies, usually about 3 dB.

7.7.3   Vertical Mounting

In the mobile radio environment the energy of the vertical polarized wave can be coupled to the horizontal polarized wave.[9] This crosscoupling does not produce a strong effect; nevertheless, to have a good reception, the antenna on the mobile unit must be properly mounted in the vertical direction.

7.8   FIELD COMPONENT DIVERSITY ANTENNAS[10,11]

The advantage of using field component diversity antenna is discussed in Section 3.4.2. A feasibility study of the field component diversity antenna is shown in Fig. 7.12. There are two configurations. One is called an *energy density antenna,* and the other is called an *uncorrelated signal antenna.*

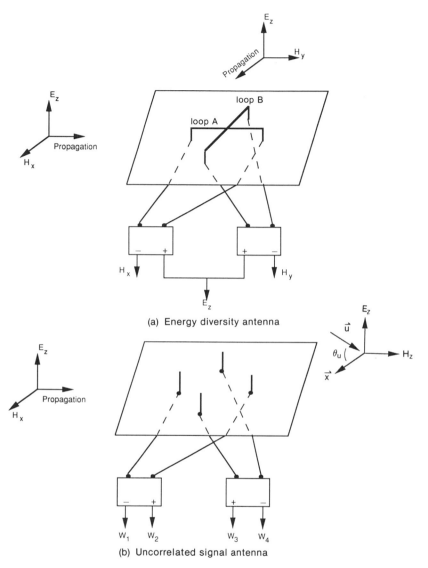

(a) Energy diversity antenna

(b) Uncorrelated signal antenna

Figure 7.12. Field component diversity antenna.

7.8.1 The Energy Density Antenna

As the top center of Fig. 7.12(a) shows, the extremities of the two cross semiloops of the energy density antenna are each connected to a 180° hybrid. One output of the hybrid is a summing port, and the other is a differencing port. The E_z wave propagates in the y-direction where it is received as the sum of the two inputs from loop A, which is proportional to the voltage. The H_x is received as the difference between the two inputs of loop A, which is proportional to the current. The same relationship takes place with the two components E_z and H_y from loop B. As seen in Fig. 1.5C, the field E reflected from a scatterer will create a standing wave. The same is true for the H field. Two standing waves will have a 90° phase shift. For example, let one standing wave represent two E waves at the summing port and the other represent two H waves at the differencing port:

$$E \text{ field} \sim e^{+j\beta x} + e^{-j\beta x} \sim \cos \beta x \qquad (7.8.1)$$

$$H \text{ field} \sim e^{+j\beta x} - e^{-j\beta x} \sim \sin \beta x \qquad (7.8.2)$$

where β is the wave number as $\beta = 2\pi/\lambda$, and x is the propagating distance. Now by squaring Eq. (7.8.1) and Eq. (7.8.2), respectively, and adding the result, we have

$$\cos^2 \beta x + \sin^2 \beta x = 1 \qquad (7.8.3)$$

which shows that the fading signal is totally eliminated and that the envelope becomes a constant. Equation (7.8.3) can only work in a standing wave environment. In a general fading environment a multiple number of standing waves can occur, so the result will not be a constant:

$$s(x) = \left(\sum_1^N A_i \cos \beta x \right)^2 + \left(\sum_1^N B_i \sin \beta x \right)^2 \neq 1 \qquad (7.8.4)$$

where A_i and B_i are the amplitudes of the standing waves. They are random in nature, and E_z, H_x, and H_y are always uncorrelated. (See Chapter 3, Refs. 1, 9.) The field component diversity antenna is represented by

$$s(x) = \sum_1^N A_i \cos \beta x + \sum_1^N B_i \sin \beta x \qquad (7.8.5)$$

Both the energy density diversity of Eq. (7.8.4) and the field component diversity of Eq. (7.8.5) are clearly working. Both energy density diversity and field component diversity do not require any antenna spacing but the space diversity does require antenna spacing in terms of the wavelength of the carrier frequency. At low frequencies, space diversity is hard to achieve due to the requirement of physical antenna separation.

7.8.2 Uncorrelated Signal Diversity Antenna

The uncorrelated signal diversity antenna follows the same principle of the field component diversity antenna. Assume that there are two electric fields, E_{z_1} and E_{z_2} expressed as

$$E_{z_1} = \sum_{u=1}^{N} A_u \exp(-j\beta\mathbf{u} \cdot \mathbf{x}_1) = X_1 + jY_1 \qquad (7.8.6)$$

and

$$E_{z_2} = \sum_{u=1}^{N} A_u \exp(-j\beta\mathbf{u} \cdot \mathbf{x}_2) = X_2 + jY_2 \qquad (7.8.7)$$

where A_u is a complex amplitude of an electric wave propagating at a direction $\mathbf{u}$, and $\mathbf{u}$ is a unit vector related to an angle θ_u between $\mathbf{u}$ and $\mathbf{x}$ as shown in Fig. 7.12(b). β is the wave number and N is the number of wave arrivals. E_{z_1} and E_{z_2} can also be expressed in real and imaginary parts, as shown in Eqs. (7.8.6) and (7.8.7), respectively.

The two outputs W_1 and W_2 of a field component diversity antenna as shown can be expressed as:

$$W_1 = E_{z_1} + E_{z_2} = (X_1 + X_2) + j(Y_1 + Y_2) \qquad (7.8.8)$$

$$W_2 = E_{z_1} - E_{z_2} = (X_1 - X_2) + j(Y_1 - Y_2) \qquad (7.8.9)$$

The correlation of the two signals W_1 and W_2 is

$$\begin{aligned}
\overline{W_1 W_2^*} &= \overline{(X_1^2 - X_2^2)} + \overline{(Y_1^2 - Y_2^2)} + 2j\overline{(X_1 Y_2 - Y_1 X_2)} \\
&= 2j\overline{(X_1 Y_2 - Y_1 X_2)} \\
&= 0 \qquad\qquad\qquad (7.8.10)
\end{aligned}$$

since

$$\overline{X_1 Y_2} = \overline{Y_1 X_2} = 0$$

It can be proved by taking Eqs. (7.8.6) and (7.8.7) and averaging the product terms as indicated in Eq. (7.8.10) that the two outputs W_1 and W_2 are uncorrelated.

Second, Eqs. (7.8.8) and (7.8.9) will be examined. If E_{z_1} and E_{z_2} are highly correlated, then the mean value of W_2, W_2 is much smaller than W_1. There is no advantage in combining them even when W_1 and W_2 are uncorrelated.

To keep W_1 and W_2 the same, it may be necessary to let the envelope correlation of E_{z_1} and E_{z_2} be 0.5 as follows:

$$\rho_{|E_{z1} - E_{z2}|} = J_0^2[\beta(x_1 - x_2)] = 0.5 \qquad (7.8.10)$$

where $J_0(.)$ is the Bessel function of the first kind of zero order. From Eq. (7.8.10)

$$J_0[\beta(x_1 - x_2)] = 0.707$$

$$\beta(x_1 - x_2) = 1.1 \qquad (7.8.11)$$

$$x_1 - x_2 = \frac{1.1}{2\pi}\lambda = 0.175\lambda$$

Hence, from the theoretical analysis, the spacing between the two elemental antenna elements is 0.175λ, but from the experimental results, the spacing between the elemental antenna elements can be as small as 0.125λ, which is smaller than the theoretical value.

REFERENCES

1. Lee, W. C. Y., "A Study of the Antenna Array Configuration of an M-Branch Diversity Combining Mobile Radio Receiver," *IEEE Trans. Veh. Tech.* VT-20 (Nov. 1971): 93–104.
2. Lee, W. C. Y., "An Extended Correlation Function of Two Random Variables Applied to Mobile Radio Transmission," *Bell Sys. Tech. J.* 48 (Dec. 1969): 3423–3440.
3. Lee, W. C. Y., "Introduction to Mobile Cellular Concepts," *Microwave Sys. News Commun. Tech.* (June 1985).
4. Lee, W. C. Y., "Antenna Spacing Requirement for a Mobile Radio Base-Station Diversity," *Bell Sys. Tech. J.* 50 (July–Aug. 1971): 1859–1874.
5. Lee, W. C. Y., "Preliminary Investigation of Mobile Signal Fading Using Directional Antennas on the Mobile Unit," *IEEE Trans. Veh. Tech.* 15 (Oct. 1966): 8–15.
6. Rustako, A. J., Y. S. Yeh, and R. R. Murray, "Performance of Feedback and Switch Space Diversity 900 MHz, FM Mobile Radio System with Rayleigh Fading," *IEEE Trans. Commun.* 21 (Nov. 1973): 1257–1268.
7. Lee, W. C. Y., "Close-Spaced Diversity Antenna at HF," *IEEE Milcom 85* Boston, MA (Oct. 1985): 21–23.
8. Skomal, E. N., "Automotive Noise," *Man-made Radio Noise* (Van Nostrand Reinhold, 1978): ch. 2.
9. Lee, W. C. Y., and Y. S. Yeh, "Polarization Diversity System for Mobile Radio," *IEEE Trans. Commun.* Com-20 (Oct. 1972): 912–923.

10. Lee, W. C. Y., "Statistical Analysis of the Level Crossings and Duratio of Fades of the Signal from an Energy Density Mobile Radio Antenna," *Bell Sys. Tech. J.*, Vol. 46, No. 2, Feb. 1967, pp. 417–448.

11. Lee, W. C. Y., "An Energy Density Antenna for Independent Measurement of the Electric and Magnetic Field," *Bell Sys. Tech. J.*, Vol. 46, No. 7, September 1967, pp. 1587–1599.

PROBLEMS

7.1 Explain why physically the space-diversity scheme requires larger antenna separation at the base station but smaller antenna separation at the mobile unit?

7.2 Two waves arrive at the mobile unit. One incoming wave is always perpendicular to the motion of the vehicle. Derive a simple equation that expresses this condition.

7.3 Does the required frequency separation for frequency-diversity scheme depend on the carrier frequency?

7.4 The signal received by the roof-mounted antenna has been found to be 3 dB stronger than the signal received by the glass-mounted antenna. What would the effect be on the roof-mounted antenna in a system designed according to D/R, where R is the cell radius based on the glass-mounted antenna coverage?

7.5 A car's body is a conductor that can alter the mobile antenna's wave patterns. Also a glass-mounted antenna may receive a weak signal in the front of the car and a strong signal at the rear of the car. Can irregular antenna wave patterns cause mobile reception problems or system performance problems?

7.6 Prove that the average signal strength received from a directional antenna on a mobile unit is constant regardless of the beam width of the directional antenna.

7.7 Find the fading frequency for four signals coming from the four directions 0°, 90°, 180°, and 270°.

7.8 Why is it that the power delivered through the wireline has to be shared by the total number of traffic channels in the wireline, but the power delivered in the space does not need to be shared? Each channel receives the same amount of power regardless of whether other channels have received any power.

7.9 Why is wire loss measured in dB/ft or dB/km and radio loss measured in dB/oct or dB/dec?

7.10 If the beam width of a directional antenna is very small, say, 0.3°, can we use dB/ft or dB/km to express the power loss?

8

SIGNALING AND CHANNEL ACCESS

8.1 CRITERIA OF SIGNALING DESIGN

Signaling conveys information used to set up, control, and terminate calls. Signaling is usually represented by a digital form that consists of a number of bits called a *word*. A signal usually only needs one word. The criteria of signaling design will be based on two kinds of performance: *false-alarm rate* and *word-error rate*.

8.2 False-Alarm Rate

The false-alarm rate has to do with the probability of false recognition among all the codes being assigned to different functions or addresses. It is very annoying to the customer to have an address number answered by a wrong party, or an assigned operational function such as "request a handoff" taken by the receiving unit and turns out to be interpreted as an unexpected "turn out" in operation. This situation usually occurs due to noise contamination from the medium. If the medium is very smooth, a stream of bits will be received with a high rate of correctness; the false-alarm rate is very low. If the medium is rough, many bits in a stream of bits will be in error at the receiving end due to the medium; the false-alarm rate is very high. The false-alarm rate is calculated based on the Hamming distance. A Hamming distance of d bits means that the difference between two designated code words with the same word length L is the difference in their bits in d places over L bits. The false-alarm rate is used as a criterion for designing a set of code words based on the roughness of the medium and the importance of the functions. The Hamming distance of d bits out of a word length of L bits is chosen based on the individual bit-error rate due to the medium. The bit-error rates are discussed in Section 3.6.

A proper design ensures that all d bits in a, or d bits in a length of L bits, are false-alarm free in the given transmission medium. A proper design is based on a properly chosen number of d. If the medium is rough, the value of d should be increased. The false-alarm rate is set for each communication system and can be achieved at a given condition. The false-alarm rate is expressed as

$$P_f = \text{false-alarm rate} = P_e^d (1 - P_e)^{L-d} \qquad (8.2.1)$$

where P_e is the bit-error rate of each bit. As an example, observe two code words of 9 bits. Let $L = 9$:

$$
\begin{array}{ccccccccc}
1 & 0 & 1 & 0 & 1 & 1 & 0 & 1 & 0 \\
1 & 1 & 0 & 1 & 0 & 1 & 0 & 0 & 0
\end{array}
$$

There are five places where the bits differ, so $d = 5$. Assume that the bit-error rate of each bit is 10^{-2}, the $P_e = 10^{-2}$. Then substituting $L = 9$, $d = 5$, and $P_e = 10^{-2}$ into Eq. (8.2.1), P_f becomes

$$P_f = (0.01)^5 (1 - 0.01)^{9-5} \simeq 10^{-7}$$

The chance of a false alarm in this case is one in ten million. If ten thousand users call the same area at the same time, then the probability of having a false alarm is 10^{-3}. It means one out of every thousand users will receive a false alarm. This of course is an undesirable and very rare situation. After code words have been generated, they must be received free of errors. If there are more than five erroneous bits in a code word of nine bits, then the code detector, instead of recognizing the errors in a word, interprets it as a correct word for some other user's identification or for cuing another totally different operation or function. This too is undesirable and should be eliminated whenever possible. The control of word-error rate is described in the following section.

8.3 WORD-ERROR RATE

The word-error rate is the probability of sending an erroneous word or code word. It differs from the false-alarm rate in that an error in a word occurs relatively often when compared with the rate of misinterpreted words. The word-error rate is always higher than the false-alarm rate. In a typical communication system the word-error rate is 10^{-2} to 10^{-3}, but the false-alarm rate is 10^{-6} to 10^{-7}. In this section we examine the word-error rate in different environments.

8.3.1 In a Gaussian Environment

In a Gaussian environment, once the bit-error rate P_e (see Section 3.6) is known, the word-error rate of a message-word with length L in bits can be obtained as

$$P_{ew} = \text{word-error} = 1 - (1 - P_e)^L \qquad (8.3.1)$$

for no error correction. The code-word-error rate of a code word consisting of N bits, with an error-correction capability of correcting t errors or less, can be expressed as

$$P_{ew} = 1 - \sum_{k=0}^{t} C_k^N P_e^k (1 - P_e)^{N-k} \qquad (8.3.2)$$

where

$$C_k^N = \frac{N!}{(N-k)!k!} \qquad (8.3.3)$$

The price of having a t-error-correcting code on an L-digit message is to add N-L parity check bits for a binary cyclic code or add N-L redundant bits for a linear block code. Usually the number of N-L bits is much greater than t bits, and the throughput is

$$\text{Throughput} = \frac{L}{N} \qquad (8.3.4)$$

A code with a Hamming distance d generally has an error-correcting capability

$$t = \frac{d-1}{2} \qquad (8.3.5)$$

Example 8.1. The Golay (23,12) code has 12 message bits, and 11 redundant bits. The Hamming distance is seven, and is capable of correcting any combination of three or fewer random errors in a block of 23 bits:

$$\text{Throughput} = \frac{12}{23} = 0.52$$

What is the trade-off between the throughput and the error reduction? Use a noncode word with $L = 12$ and a Golay code world with $N = 23$ to illustrate the differences in throughputs and error reductions by applying Eqs. (8.3.1) and (8.3.2), as shown in Fig. 8.1. At a word-error rate of 10^{-3} (which is usually used for communication), the required S/N level is 5 dB for a noncode word and 9.5 dB for a Golay code word. At a word-error rate of 10^{-7}, which may be accepted by computer, the required S/N level is 7.7 dB for a noncode word and 12.5 dB for a Golay code word. With a given word-error rate, the S/N level of a noncode word transmission is always 4.5 dB higher than the

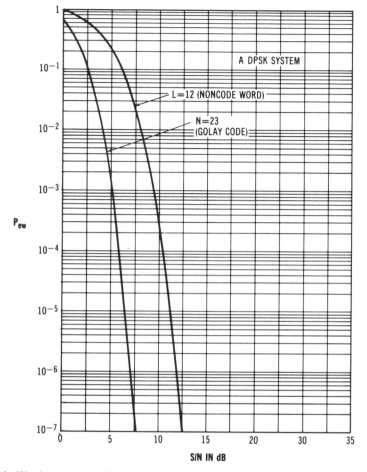

Figure 8.1. Word-error rates of a noncode word and a Golay code in a Gaussian noise environment.

level of a Golay code word. It is quite possible to be willing to pay the price of increasing its S/N level 4.5 dB for a noncode word in order to keep the same word-error rate of the Golay code-word while doubling the throughput.

Comparison of False-Alarm Rate and Word-Error Rate

The false-alarm rate of Eq. (8.2.1) and the word-error rate of Eq. (8.3.1) can be compared by combining their expressions as follows:

$$\frac{P_{ew}}{P_f} = \frac{1 - (1 - P_e)^L}{P_e^d(1 - P_e)^{L-d}} \simeq \left(\frac{1}{P_e}\right)^d \left[\left(\frac{1}{1 - P_e}\right)^L - 1\right] \qquad (8.3.6)$$

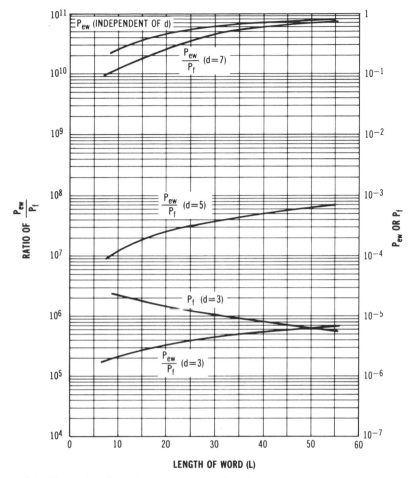

Figure 8.2. The ratio of word-error rate to false-alarm rate versus hamming distance ($P_e = 0.03$).

Since $P_e \ll 1$, the approximation can be found from the Taylor expansion as

$$(1 - P_e)^{-L} \simeq 1 + LP_e$$

Therefore Eq. (8.3.6) becomes

$$\frac{P_{ew}}{P_f} = \left(\frac{1}{P_e}\right)^d LP_e = LP_e^{1-d} \gg 1 \qquad (8.3.7)$$

provided that $P_e^{-1} \gg L$. Equation (8.3.7) is plotted with $P_e = 0.03$ in Fig. 8.2 which clearly indicates that the word-error rate is much higher than the false-alarm rate under normal conditions. Note that P_{ew} increases as the length L increases, but P_f decreases as the length L increases.

8.3.2 In a Rayleigh Environment

In a Rayleigh fading environment, the transmitted signals encounter a burst error because of the duration of fades. The duration of fade is related to the velocity of the mobile unit. When the velocity is high, the duration of the fade is short. When the velocity is low, the duration of the fade is long. The word-error rate, or code-word-error rate, is dependent on the speed of the vehicle. However, it is very hard to find an analytic expression for the word-error rate in terms of the speed of vehicles. Therefore we examine two extreme cases, a fast-fading case and a slow-fading case.[1,2]

8.3.3 A Fast-Fading Case in a Rayleigh Fading Environment

The fast-fading case is defined as if the mobile speed approaches infinity. Of course this is an unrealistic idea. Yet it serves to obtain an upper bound value for the word-error rate. When the speed of a vehicle goes to infinity, the distribution of amplitude fading remains unchanged, but the duration of fades approaches zero. When a stream of bits is being transmitted at a constant bit rate with $V = \infty$, there can be no correlation between adjacent bits in terms of the fading characteristics of each bit.

Each bit then has to be treated independently. The average bit-error rate $\langle P_e \rangle$ of each digital system through a Raleigh fading environment can be found in Section 3.6 [from Eqs. (3.6.8) to (3.6.11)]. The code-word-error rate of a code word (N, k) with t-error-correction capability is the same form as Eq. (8.3.2):

$$\langle P_{cw} \rangle = 1 - \sum_{k=0}^{t} C_k^N \langle P_e \rangle^k \, (1 - \langle P_e \rangle)^{N-k} \qquad (8.3.8)$$

The notation C_k^N is shown in Eq. (8.3.3).

A DPSK system is used to illustrate the difference in $\langle P_{cw} \rangle$ between this case and a slow-fading case (discussed later in this section) because of its simple expression [see Eq. (3.6.8)] in a Rayleigh environment. The following analyses are for different signaling transmission strategies.

Plain Transmission (t = 0)

The word-error rate with no error correction capability can be found from Eq. (8.3.8). Let $t = 0$. Then

$$\langle P_{cw} \rangle = 1 - (1 - \langle P_e \rangle)^N \qquad (8.3.9)$$

Equation (8.3.9) is plotted in Fig. 8.3 for $N = 22$ bits.

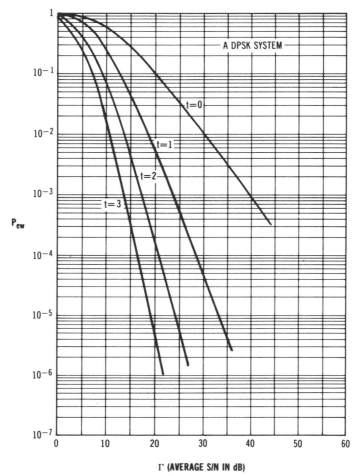

Figure 8.3. A code-word-error rate of 22 bits in a Rayleigh fading environment.

Error-Correction Code Transmission ($t \neq 0$)

The word-error rate with an error-correction capability for one error ($t = 1$) can be obtained from Eq. (8.3.8):

$$\langle P_{ew} \rangle = 1 - (1 - \langle P_e \rangle)^N - N(1 - \langle P_e \rangle)^{N-1}\langle P_e \rangle \qquad (8.3.10)$$

Equation (8.3.10) is also plotted in Fig. 8.3 for $N = 22$ bits.

Repetition Transmission and Majority-Voting Process

The word-repetition format for sending words repeatedly requires a majority-voting process for detecting the repeated words at the time of reception. Assuming that each word is repeated J times during transmission, the J repeats of the received message-bit stream must be aligned bit by bit, as shown in

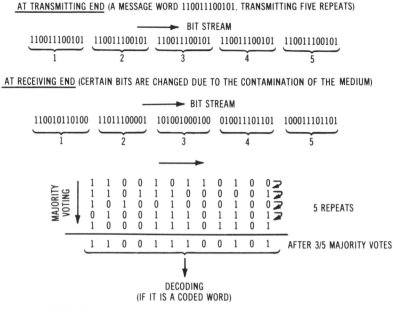

AT TRANSMITTING END (A MESSAGE WORD 11001100101, TRANSMITTING FIVE REPEATS)

Figure 8.4. The illustration of repetition transmission and a majority voting process.

Fig. 8.4. Usually J is an odd number. For every message bit, there are J repeats. Among them if $(J + 1)/2$ repeats or more are 1s, then the received bit is 1. This majority-voting process is used to determine each valid message bit. The resulting majority-voted message words then constitute the improved message stream. To illustrate this word-error reduction strategy, assume that a single-bit error-correction code is used at the transmission end. Then at the receiving end, after the improved message stream is formed by applying the majority-voting process, the one-bit error-correction capability further improves the chances for obtaining an error-free message stream. Under fast-fading conditions, and on the assumption that no correlation exists between any two repeated bits among J repeats, the improved bit-error rate $\langle P'_e \rangle$ for J repeats with a majority-voting process can be expressed as

$$\langle P'_e \rangle = \sum_{k=(J+1)/2}^{J} C_k^J \langle P_e \rangle^k (1 - \langle P_e \rangle)^{J-k} \qquad (8.3.11)$$

Equation (8.3.11) is plotted in Fig. 8.5 for a two-out-of-three and three-out-of-five majority-voting process using a DPSK system. The improved bit-error rates of a repetition transmission are lower than those of the non-repetition transmission, since in a fast-fading case all the message bits are uncorrelated. A comparison of the improved bit-error rate in two kinds of environment, Rayleigh and Gaussian, is shown in Fig. 8.5. The bit-error rate in Rayleigh environment after the majority-voting process still cannot be lower

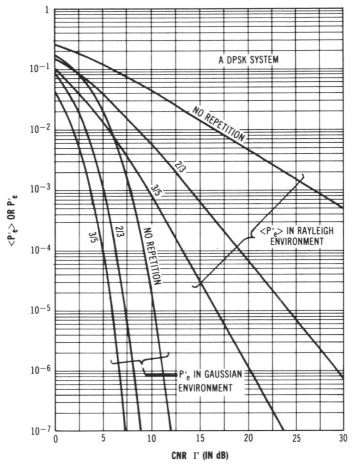

Figure 8.5. Comparison of improved bit-error rates in Rayleigh environment and in Gaussian environment (fast-fading case).

than that in a Gaussian environment. Given a word length of N uncorrelated message bits, the bit-error rate that 1 or more bits will be in error can be expressed as

$$\langle P'_{cw} \rangle = 1 - (1 - \langle P'_e \rangle)^N \quad \text{(no error correction)} \quad (8.3.12)$$

Note the similarity to Eq. (8.3.8). Given a word length of N uncorrected message bits, the bit-error rate that more than t bits are in error can be obtained is

$$\langle P'_{ew} \rangle = 1 - \sum_{k=0}^{t} C_k^N \langle P'_e \rangle^k (1 - \langle P'_e \rangle)^{N-k} \quad (t \text{ error correction}) \quad (8.3.13)$$

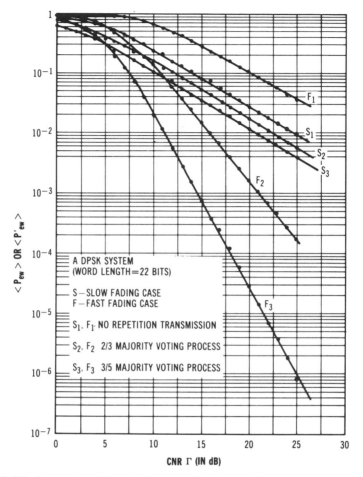

Figure 8.6. Word-error rates with and without repetition transmission—no-error-correction code.

Equations (8.3.12) and (8.3.13) are plotted in Figs. 8.6 and 8.7 where the performance of repetition transmissions is compared with that of nonrepetition transmissions. In contrast to the case of no-coding and nonrepetition transmission, the coding and repetition transmissions provide considerable performance improvement as shown in these two figures.

8.3.4   A Slow-Fading Case in a Rayleigh Fading Environment

A slow-fading case occurs when the mobile unit is moving very slowly, yet not stopping. Then all adjacent bits are correlated among themselves; that is, if one bit is in a fade, then the probability that the adjacent bits are also in the fade is very high. Assume that all of the bits in a word are under the same fading conditions, that is, all of them are either above the fade or in

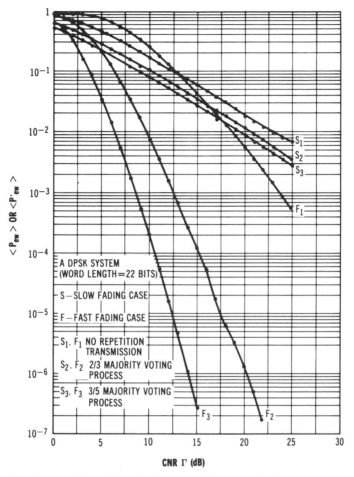

Figure 8.7. Word-error rates with and without repetition transmission—one-error-correction code.

the same fade. A whole word in this slow-fading case is treated like a single bit in a fast-fading case. The result is the average word-error rate for a slow-fading condition in a Rayleigh fading environment.

Plain Transmission ($t = 0$)

The word-error rate P_{ew} of a signal transmitting in a Gaussian environment is shown in Eq. (8.3.1). Since the whole word is treated like a single bit and sent over a Rayleigh fading environment, the average word-error rate under this condition becomes

$$\langle P_{ew} \rangle = \int_0^\infty P_{ew} p(\gamma) \, d\gamma \qquad (8.3.14)$$

where $p(\gamma)$ is the pdf of Rayleigh fading shown in Eq. (3.6.7). Equation (8.3.14) is plotted in Fig. 8.6.

Error-Correction Code Transmission ($t \neq 0$)

The word-error rate with an error-correction capability for one error ($t = 1$) in this slow-fading case ($V \to 0$) can be obtained by averaging the whole code word as a single bit in a Rayleigh fading environment. It means that by substituting Eq. (8.3.2) into Eq. (8.3.14), we can find the word-error rate for the code word shown in Fig. 8.7.

Repetition Transmission and Majority-Voting Process

After a majority-voting process over J-repetition transmission, the improved bit-error rate P_e is given by

$$P'_e = \sum_{k=(J+1)/2}^{J} C_k^J P_e^k (1 - P_e)^{J-k} \tag{8.3.15}$$

where P_e is the bit-error rate in the Gaussian environment, as in Eqs. (3.6.1) through (3.6.4). A DPSK system will be used to illustrate performance. Substituting Eq. (3.6.1) into Eq. (8.3.14), the improved bit-error rate P'_e is shown in Fig. 8.5 as in a Gaussian environment. The P'_e of a fast-fading case is higher than that of a slow-fading case in both strategies of the majority-voting processes: two out of three and three out of five. The average word-error rate to come out of this repetition transmission is

$$\langle P'_{ew} \rangle = \int P'_{ew} p(\gamma) \, d\gamma \tag{8.3.16}$$

where

$$P'_{ew} = 1 - \sum_{k=0}^{t} C_k^N (1 - P'_e)^{N-k} P'^k_e \quad \text{(general form)}$$

$$= 1 - (1 - P'_e)^N \quad (t = 0, \text{no-error-correction case})$$

$$= 1 - (1 - P'_e)^N - N(1 - P'_e)^{N-1} P'_e$$

$$(t = 1, \text{one-error-correction case}) \tag{8.3.17}$$

The no-error-correction case and one-error-correction case of Eq. (8.3.16) are shown in Figs. 8.6 and 8.7, respectively. The majority-voting process with a larger number of repetitions and higher number of error-corrections always improves the performance. However, the trade-off is the inefficiency of its throughput.

8.3.5 A Comparison between a Slow-Fading Case and a Fast-Fading Case

In a plain transmission in which no error correction nor repetition transmission occurs, the performance of word-error rate in the fast-fading case is worse than that in the flow-fading case. When a repetition transmission is implemented, then immediately the performance of word-error rate in a fast-fading case supercedes that in a slow-fading case, as shown in Figs. 8.6 and 8.7. Furthermore the performance becomes more effective with coding and repetition transmission in the fast-fading case than in the slow-fading case. Because the actual word-error rate lies between the two cases, there is an advantage to using a repetition transmission.

8.4 CHANNEL ASSIGNMENT

In designing a mobile communication system with finite resources of frequency allocation, a channel-assignment scheme must be considered. Many strategies in channel assignment are described in this section, such as co-channel assignment, channel assignment within a cell, channel sharing, and channel borrowing.

8.4.1 Co-channel Assignment

As mentioned previously, because of limited channel resources channels have to be reused in different geographical locations. If a channel-reuse plan is not well designed, there will be co-channel interference in the system. This affects the performance of the whole system. Co-channel interference therefore has to be eliminated. The minimum distance at which co-channel interference can be ignored is determined by first specifying the required carrier-to-interference ratio (C/I) at the signal reception, then relating to the C/I to the propagation path loss, which is in terms of the propagation distance as described in Section 4.2. A parameter, a so-called co-channel reduction factor, is used to separate two co-channels in different areas, as noted in Section 4.2. In an ideal flat terrain with a required $C/I = 18$ dB, the factor a obtained is

$$a = \frac{D}{R} = 4.6 \qquad \left(\text{for } \frac{C}{I} \geq 18 \text{ dB}\right) \qquad (8.4.1)$$

where R is the radius of a cell and D is the distance between two co-channel cells. Therefore any co-channel has to be assigned at a distance D that is 4.6 times the radius of the cell:

$$D = 4.6R \qquad (8.4.2)$$

Yet in a real environment, $D = 6R$ is used for a system with omnidirectional antenna cells. In hilly terrain the number of a will be generally larger. There are two different strategies that can be used for co-channel assignment as described in the next section.

Constant Minimum C/I for All Channels

Under this strategy the threshold level of mobile receivers remains the same based on a minimum C/I. All the channels follow a required D/R ratio for setting up their co-channel separation. If the cell size is smaller, the separation is less. The advantage of using this strategy is that the system operation is simpler, since no control is necessary at the base station (or so-called cell site) to adjust the threshold levels of the mobile receiver.

Different Minimum C/I among All Channels[3]

A group of channels is assigned to each cell. Some of the channels are based on the received level $C/I = 18$ dB at the mobile unit; some of them are based on the received level higher than $C/I = 18$ dB, perhaps 24 dB. The co-channel reduction factor will no longer be 4.6. It is calculated by replacing $C/I = 24$ dB in Eq. (4.2.3) based on six interferers, as

$$\frac{C}{I} = \frac{a^4}{6} \geq 251 \tag{8.4.3}$$

then

$$a \geq 6.23 \quad \text{(for } C/I \geq 24 \text{ dB)} \tag{8.4.4}$$

A system plan can be based on this strategy, as shown in Fig. 8.8. If a 10-mile cell is based on $C/I = 18$ dB, then the size of the cell will be smaller based on $C/I = 24$ dB. The size of the reduced cell can be found from either Eq. (4.2.3) or Eq. (8.4.3) because these two equations are valid for any size cell. Therefore we use the mobile radio propagation path loss rule of 40 dB/dec to calculate the reduced cell size. The difference in cell radii is related to the difference in the received carrier power, which is 6 dB. From Eq. (2.3.20),

$$6 \text{ dB} = 40 \log \frac{10}{x} \tag{8.4.5}$$

The new radius x is 7 miles, as shown in Fig. 8.8. The co-channel separation distances for different conditions are found as

$$D_1 = 10 \times 4.6 = 46 \text{ miles} \quad \text{(for a 10-mile cell, and } C/I \geq 18 \text{ dB)}$$

$$D_2 = 7 \times 6.23 = 43.6 \text{ miles} \quad \text{(for a 7-mile cell and } C/I \geq 24 \text{ dB)}$$

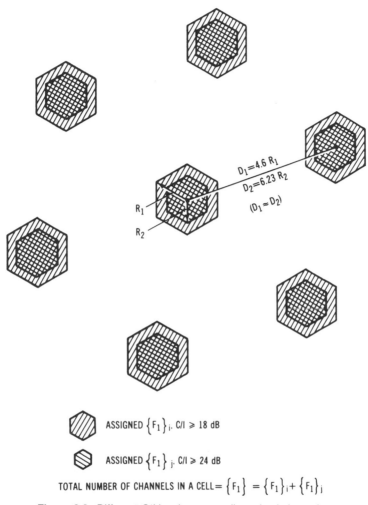

$$D_1 = 4.6 \, R_1$$
$$D_2 = 6.23 \, R_2$$
$$(D_1 \approx D_2)$$

◇ ASSIGNED $\{F_1\}_i$. $C/I \geqslant 18$ dB

◆ ASSIGNED $\{F_1\}_j$. $C/I \geqslant 24$ dB

TOTAL NUMBER OF CHANNELS IN A CELL= $\{F_1\}$ = $\{F_1\}_i + \{F_1\}_j$

Figure 8.8. Different C/I levels among all received channels.

Since two separation distances are almost the same, no new cell sites would be added. In applying this strategy to a system design, the total frequency resources are divided into two sets of channels. One set is used to serve general traffic conditions in a 16-km (10-mile) cell, and another set aims for areas that need more channels to handle local traffic. This strategy does not increase the total traffic capacity but does improve system performance in certain heavy traffic spots in a cell.

8.4.2 Channel Assignment within a Cell

Within a cell the near-end-to-far-end ratio interference (mentioned in Section 4.4) determines the channel assignment. The channel must be assigned in such a way that the near-end-to-far-end ratio interference and the filter characteristics of the mobile receivers are suppressed. Referring to Figure 4.2, note that if the near-end-to-far-end distance ratio is 10 and the filter characteristic is 10 dB/oct, then the two assigned frequency channels in a cell should be separated by 16 natural channels. In section 5.5, the cellular concept was introduced, and Fig. 5.16 showed the 21 subsets. Among them the closest two channels in a set are 7 channels apart. Therefore we have to carefully assign one or more sets among 21 sets in a cell in order to reduce the near-end-to-fare-end ratio interference. From Fig. 5.12 we know the probability that two mobile units are in two specified radii, r_1 and r_2, in a cell. The bit-error rates caused by the near-end-to-far-end ratio interference at different specified ratios of d_1/d_2 are shown in Fig. 5.13. These two figures also help to demonstrate a proper frequency assignment strategy.

8.4.3   Channel Sharing

Depending on the local traffic conditions, a group of frequency channels may have to be shared between two cell sites if omnidirectional antennas are used or shared between two faces in a cell site if directional antenna cells are used. Therefore in frequency assignment one must consider that the offered load capacity (defined as the serving capacity with a given number of channels) always increases with channel-sharing schemes.

Channel Sharing in an Omnidirectional Antenna Cell
Assume that 45 channels are assigned to each cell site. If among 45 channels at each site, 15 channels are assigned to share with another cell, then the highest number of available channels is 60 and the lowest is 30, as shown in Fig. 8.9A. The traffic density of this channel-sharing scheme can be compared with that of the no-channel-sharing scheme. The Erlang B model is used to illustrate the difference.[4] The Erlang B model is based upon a model of serving without queuing; that is, all the blocked calls cleared. This is called a *loss system* because the cell site loses the customers who arrive when all channels are busy. The Erlang C model is formed under the condition of blocked calls held. Both Erlang B and Erlang C models appear in Tables 8.1 and 8.2, respectively. Given a same number of servers N and a same value of block probability P(B), Erlang B model offers more Erlangs than Erlang C. From the limit data, the cellular traffic model may be formed between two Erlang models. In this section, we only use Erlang B table to calculate the traffic capacity.

The following assumptions are made:

N is the number of channels per cell.
$\bar{t}$ is average calling time = 1.76 min.
B is blocking probability = 2%.
$A(N, B)$ is the offered load (a function of N and B).

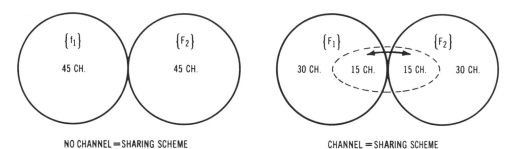

NO CHANNEL＝SHARING SCHEME   CHANNEL ＝SHARING SCHEME

(A) Illustration of the difference between two schemes in an omnidirectional antenna system.

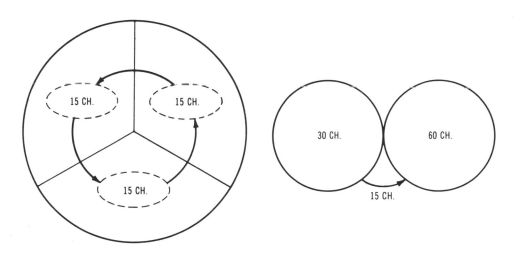

(B) Channel sharing scheme in a   *(C) Channel borrowing scheme in an*
directional antenna system.   *omnidirectional antenna system.*
Figure 8.9. Different channel assignment schemes.

CASE 1. No-Channel-Sharing Case

N is 45 channels/cell (no sharing).

A (45, 0.02) is 35.6 Erlang (from Table 8.1).

M is number of mobile units that can be served or number of users.

$$M = \frac{A \times 60 \text{ min/hr}}{\bar{t}}$$

$$= \frac{35.6 \times 60}{1.76} = 1214 \text{ users} \qquad (8.4.6)$$

TABLE 8.1 Erlang B Model—Blocked Calls Cleared

(Offered Load)						A in Erl						
N					B	(Blocking Probability)						
	0.01%	0.02%	0.03%	0.05%	0.1%	0.2%	0.3%	0.4%	0.5%	0.6%	0.7%	0.8%	0.9%
1	0.0001	0.0002	0.0003	0.0005	0.0010	0.0020	0.0030	0.0040	0.0050	0.0060	0.0070	0.0081	0.0091
2	0.0142	0.0202	0.0248	0.0321	0.0458	0.0653	0.0806	0.0937	0.105	0.116	0.126	0.135	0.144
3	0.0868	0.110	0.127	0.152	0.194	0.249	0.289	0.321	0.349	0.374	0.397	0.418	0.437
4	0.235	0.282	0.315	0.362	0.439	0.535	0.602	0.656	0.701	0.741	0.777	0.810	0.841
5	0.452	0.527	0.577	0.649	0.762	0.900	0.994	1.07	1.13	1.19	1.24	1.28	1.32
6	0.728	0.832	0.900	0.996	1.15	1.33	1.45	1.54	1.62	1.69	1.75	1.81	1.86
7	1.05	1.19	1.27	1.39	1.58	1.80	1.95	2.06	2.16	2.24	2.31	2.38	2.44
8	1.42	1.58	1.69	1.83	2.05	2.31	2.48	2.62	2.73	2.83	2.91	2.99	3.06
9	1.83	2.01	2.13	2.30	2.56	2.85	3.05	3.21	3.33	3.44	3.54	3.63	3.71
10	2.26	2.47	2.61	2.80	3.09	3.43	3.65	3.82	3.96	4.08	4.19	4.29	4.38
11	2.72	2.96	3.12	3.33	3.65	4.02	4.27	4.45	4.61	4.74	4.86	4.97	5.07
12	3.21	3.47	3.65	3.88	4.23	4.64	4.90	5.11	5.28	5.43	5.55	5.67	5.78
13	3.71	4.01	4.19	4.45	4.83	5.27	5.56	5.78	5.96	6.12	6.26	6.39	6.50
14	4.24	4.56	4.76	5.03	5.45	5.92	6.23	6.47	6.66	6.83	6.98	7.12	7.24
15	4.78	5.12	5.34	5.63	6.08	6.58	6.91	7.17	7.38	7.56	7.71	7.86	7.99
16	5.34	5.70	5.94	6.25	6.72	7.26	7.61	7.88	8.10	8.29	8.46	8.61	8.75
17	5.91	6.30	6.55	6.88	7.38	7.95	8.32	8.60	8.83	9.03	9.21	9.37	9.52
18	6.50	6.91	7.17	7.52	8.05	8.64	9.03	9.33	9.58	9.79	9.98	10.1	10.3
19	7.09	7.53	7.80	8.17	8.72	9.35	9.76	10.1	10.3	10.6	10.7	10.9	11.1
20	7.70	8.16	8.44	8.83	9.41	10.1	10.5	10.8	11.1	11.3	11.5	11.7	11.9
21	8.32	8.79	9.10	9.50	10.1	10.8	11.2	11.6	11.9	12.1	12.3	12.5	12.7
22	8.95	9.44	9.76	10.2	10.8	11.5	12.0	12.3	12.6	12.9	13.1	13.3	13.5
23	9.58	10.1	10.4	10.9	11.5	12.3	12.7	13.1	13.4	13.7	13.9	14.1	14.3
24	10.2	10.8	11.1	11.6	12.2	13.0	13.5	13.9	14.2	14.5	14.7	14.9	15.1
25	10.9	11.4	11.8	12.3	13.0	13.8	14.3	14.7	15.0	15.3	15.5	15.7	15.9
26	11.5	12.1	12.5	13.0	13.7	14.5	15.1	15.5	15.8	16.1	16.3	16.6	16.8
27	12.2	12.8	13.2	13.7	14.4	15.3	15.8	16.3	16.6	16.9	17.2	17.4	17.6
28	12.9	13.5	13.9	14.4	15.2	16.1	16.6	17.1	17.4	17.7	18.0	18.2	18.4
29	13.6	14.2	14.6	15.1	15.9	16.8	17.4	17.9	18.2	18.5	18.8	19.1	19.3
30	14.2	14.9	15.3	15.9	16.7	17.6	18.2	18.7	19.0	19.4	19.6	19.9	20.1
31	14.9	15.6	16.0	16.6	17.4	18.4	19.0	19.5	19.9	20.2	20.5	20.7	21.0
32	15.6	16.3	16.8	17.3	18.2	19.2	19.8	20.3	20.7	21.0	21.3	21.6	21.8
33	16.3	17.0	17.5	18.1	19.0	20.0	20.6	21.1	21.5	21.9	22.2	22.4	22.7
34	17.0	17.8	18.2	18.8	19.7	20.8	21.4	21.9	22.3	22.7	23.0	23.3	23.5
35	17.8	18.5	19.0	19.6	20.5	21.6	22.2	22.7	23.2	23.5	23.8	24.1	24.4
36	18.5	19.2	19.7	20.3	21.3	22.4	23.1	23.6	24.0	24.4	24.7	25.0	25.3
37	19.2	20.0	20.5	21.1	22.1	23.2	23.9	24.4	24.8	25.2	25.6	25.9	26.1
38	19.9	20.7	21.2	21.9	22.9	24.0	24.7	25.2	25.7	26.1	26.4	26.7	27.0
39	20.6	21.5	22.0	22.6	23.7	24.8	25.5	26.1	26.5	26.9	27.3	27.6	27.9
40	21.4	22.2	22.7	23.4	24.4	25.6	26.3	26.9	27.4	27.8	28.1	28.5	28.7
41	22.1	23.0	23.5	24.2	25.2	26.4	27.2	27.8	28.2	28.6	29.0	29.3	29.6
42	22.8	23.7	24.2	25.0	26.0	27.2	28.0	28.6	29.1	29.5	29.9	30.2	30.5
43	23.6	24.5	25.0	25.7	26.8	28.1	28.8	29.4	29.9	30.4	30.7	31.1	31.4
44	24.3	25.2	25.8	26.5	27.6	28.9	29.7	30.3	30.8	31.2	31.6	31.9	32.3
45	25.1	26.0	26.6	27.3	28.4	29.7	30.5	31.1	31.7	32.1	32.5	32.8	33.1
46	25.8	26.8	27.3	28.1	29.3	30.5	31.4	32.0	32.5	33.0	33.4	33.7	34.0
47	26.6	27.5	28.1	28.9	30.1	31.4	32.2	32.9	33.4	33.8	34.2	34.6	34.9
48	27.3	28.3	28.9	29.7	30.9	32.2	33.1	33.7	34.2	34.7	35.1	35.5	35.8
49	28.1	29.1	29.7	30.5	31.7	33.0	33.9	34.6	35.1	35.6	36.0	36.4	36.7
50	28.9	29.9	30.5	31.3	32.5	33.9	34.8	35.4	36.0	36.5	36.9	37.2	37.6
	0.01%	0.02%	0.03%	0.05%	0.1%	0.2%	0.3%	0.4%	0.5%	0.6%	0.7%	0.8%	0.9%
N						B						

TABLE 8.1 *(Continued)*

					A in Erl							
					B							
1.0%	*1.2%*	*1.5%*	*2%*	*3%*	*5%*	*7%*	*10%*	*15%*	*20%*	*30%*	*40%*	*50%*
0.0101	0.0121	0.0152	0.0204	0.0309	0.0526	0.753	0.111	0.176	0.250	0.429	0.667	1.00
0.153	0.168	0.190	0.223	0.282	0.381	0.470	0.595	0.796	1.00	1.45	2.00	2.73
0.455	0.489	0.535	0.602	0.715	0.899	1.06	1.27	1.60	1.93	2.63	3.48	4.59
0.869	0.922	0.992	1.09	1.26	1.52	1.75	2.05	2.50	2.95	3.89	5.02	6.50
1.36	1.43	1.52	1.66	1.88	2.22	2.50	2.88	3.45	4.01	5.19	6.60	8.44
1.91	2.00	2.11	2.28	2.54	2.96	3.30	3.76	4.44	5.11	6.51	8.19	10.4
2.50	2.60	2.74	2.94	3.25	3.74	4.14	4.67	5.46	6.23	7.86	9.80	12.4
3.13	3.25	3.40	3.63	3.99	4.54	5.00	5.60	6.50	7.37	9.21	11.4	14.3
3.78	3.92	4.09	4.34	4.75	5.37	5.88	6.55	7.55	8.52	10.6	13.0	16.3
4.46	4.61	4.81	5.08	5.53	6.22	6.78	7.51	8.62	9.68	12.0	14.7	18.3
5.16	5.32	5.54	5.84	6.33	7.08	7.69	8.49	9.69	10.9	13.3	16.3	20.3
5.88	6.05	6.29	6.61	7.14	7.95	8.61	9.47	10.8	12.0	14.7	18.0	22.2
6.61	6.80	7.05	7.40	7.97	8.83	9.54	10.5	11.9	13.2	16.1	19.6	24.2
7.35	7.56	7.82	8.20	8.80	9.73	10.5	11.5	13.0	14.4	17.5	21.2	26.2
8.11	8.33	8.61	9.01	9.65	10.6	11.4	12.5	14.1	15.6	18.9	22.9	28.2
8.88	9.11	9.41	9.83	10.5	11.5	12.4	13.5	15.2	16.8	20.3	24.5	30.2
9.65	9.89	10.2	10.7	11.4	12.5	13.4	14.5	16.3	18.0	21.7	26.2	32.2
10.4	10.7	11.0	11.5	12.2	13.4	14.3	15.5	17.4	19.2	23.1	27.8	34.2
11.2	11.5	11.8	12.3	13.1	14.3	15.3	16.6	18.5	20.4	24.5	29.5	36.2
12.0	12.3	12.7	13.2	14.0	15.2	16.3	17.6	19.6	21.6	25.9	31.2	38.2
12.8	13.1	13.5	14.0	14.9	16.2	17.3	18.7	20.8	22.8	27.3	32.8	40.2
13.7	14.0	14.3	14.9	15.8	17.1	18.2	19.7	21.9	24.1	28.7	34.5	42.1
14.5	14.8	15.2	15.8	16.7	18.1	19.2	20.7	23.0	25.3	30.1	36.1	44.1
15.3	15.6	16.0	16.6	17.6	19.0	20.2	21.8	24.2	26.5	31.6	37.8	46.1
16.1	16.5	16.9	17.5	18.5	20.0	21.2	22.8	25.3	27.7	33.0	39.4	48.1
17.0	17.3	17.8	18.4	19.4	20.9	22.2	23.9	26.4	28.9	34.4	41.1	50.1
17.8	18.2	18.6	19.3	20.3	21.9	23.2	24.9	27.6	30.2	35.8	42.8	52.1
18.6	19.0	19.5	20.2	21.2	22.9	24.2	26.0	28.7	31.4	37.2	44.4	54.1
19.5	19.9	20.4	21.0	22.1	23.8	25.2	27.1	29.9	32.6	38.6	46.1	56.1
20.3	20.7	21.2	21.9	23.1	24.8	26.2	28.1	31.0	33.8	40.0	47.7	58.1
21.2	21.6	22.1	22.8	24.0	25.8	27.2	29.2	32.1	35.1	41.5	49.4	60.1
22.0	22.5	23.0	23.7	24.9	26.7	28.2	30.2	33.3	36.3	42.9	51.1	62.1
22.9	23.3	23.9	24.6	25.8	27.7	29.3	31.3	34.4	37.5	44.3	52.7	64.1
23.8	24.2	24.8	25.5	26.8	28.7	30.3	32.4	35.6	38.8	45.7	54.4	66.1
24.6	25.1	25.6	26.4	27.7	29.7	31.3	33.4	36.7	40.0	47.1	56.0	68.1
25.5	26.0	26.5	27.3	28.6	30.7	32.3	34.5	37.9	41.2	48.6	57.7	70.1
26.4	26.8	27.4	28.3	29.6	31.6	33.3	35.6	39.0	42.4	50.0	59.4	72.1
27.3	27.7	28.3	29.2	30.5	32.6	34.4	36.6	40.2	43.7	51.4	61.0	74.1
28.1	28.6	29.2	30.1	31.5	33.6	35.4	37.7	41.3	44.9	52.8	62.7	76.1
29.0	29.5	30.1	31.0	32.4	34.6	36.4	38.8	42.5	46.1	54.2	64.4	78.1
29.9	30.4	31.0	31.9	33.4	35.6	37.4	39.9	43.6	47.4	55.7	66.0	80.1
30.8	31.3	31.9	32.8	34.3	36.6	38.4	40.9	44.8	48.6	57.1	67.7	82.1
31.7	32.2	32.8	33.8	35.3	37.6	39.5	42.0	45.9	49.9	58.5	69.3	84.1
32.5	33.1	33.7	34.7	36.2	38.6	40.5	43.1	47.1	51.1	59.9	71.0	86.1
33.4	34.0	34.6	35.6	37.2	39.6	41.5	44.2	48.2	52.3	61.3	72.7	88.1
34.3	34.9	35.6	36.5	38.1	40.5	42.6	45.2	49.4	53.6	62.8	74.3	90.1
35.2	35.8	36.5	37.5	39.1	41.5	43.6	46.3	50.6	54.8	64.2	76.0	92.1
36.1	36.7	37.4	38.4	40.0	42.5	44.6	47.4	51.7	56.0	65.6	77.7	94.1
37.0	37.6	38.3	39.3	41.0	43.5	45.7	48.5	52.9	57.3	67.0	79.3	96.1
37.9	38.5	39.2	40.3	41.9	44.5	46.7	49.6	54.0	58.5	68.5	81.0	98.1
1.0%	*1.2%*	*1.5%*	*2%*	*3%*	*5%*	*7%*	*10%*	*15%*	*20%*	*30%*	*40%*	*50%*

B

TABLE 8.1 *(Continued)*

N	0.01%	0.02%	0.03%	0.05%	0.1%	0.2%	0.3%	0.4%	0.5%	0.6%	0.7%	0.8%	0.9%
						A in Erl *B*							
50	28.9	29.9	30.5	31.3	32.5	33.9	34.8	35.4	36.0	36.5	36.9	37.2	37.6
51	29.6	30.6	31.3	32.1	33.3	34.7	35.6	36.3	36.9	37.3	37.8	38.1	38.5
52	30.4	31.4	32.0	32.9	34.2	35.6	36.5	37.2	37.7	38.2	38.6	29.0	39.4
53	31.2	32.2	32.8	33.7	35.0	36.4	37.3	38.0	38.6	39.1	39.5	39.9	40.3
54	31.9	33.0	33.6	34.5	35.8	37.2	38.2	38.9	39.5	40.0	40.4	40.8	41.2
55	32.7	33.8	34.4	35.3	36.6	38.1	39.0	39.8	40.4	40.9	41.3	41.7	42.1
56	33.5	34.6	35.2	36.1	37.5	38.9	39.9	40.6	41.2	41.7	42.2	42.6	43.0
57	34.3	35.4	36.0	36.9	38.3	39.8	40.8	41.5	42.1	42.6	43.1	43.5	43.9
58	35.1	36.2	36.8	37.8	39.1	40.6	41.6	42.4	43.0	43.5	44.0	44.4	44.8
59	35.8	37.0	37.6	38.6	40.0	41.5	42.5	43.3	43.9	44.4	44.9	45.3	45.7
60	36.6	37.8	38.5	39.4	40.8	42.4	43.4	44.1	44.8	45.3	45.8	46.2	46.6
61	37.4	38.6	39.3	40.2	41.6	43.2	44.2	45.0	45.6	46.2	46.7	47.1	47.5
62	38.2	39.4	40.1	41.0	42.5	44.1	45.1	45.9	46.5	47.1	47.6	48.0	48.4
63	39.0	40.2	40.9	41.9	43.3	44.9	46.0	46.8	47.4	48.0	48.5	48.9	49.3
64	39.8	41.0	41.7	42.7	44.2	45.8	46.8	47.6	48.3	48.9	49.4	49.8	50.2
65	40.6	41.8	42.5	43.5	45.0	46.6	47.7	48.5	49.2	49.8	50.3	50.7	51.1
66	41.4	42.6	43.3	44.4	45.8	47.5	48.6	49.4	50.1	50.7	51.2	51.6	52.0
67	42.2	43.4	44.2	45.2	46.7	48.4	49.5	50.3	51.0	51.6	52.1	52.5	53.0
68	43.0	44.2	45.0	46.0	47.5	49.2	50.3	51.2	51.9	52.5	53.0	53.4	53.9
69	43.8	45.0	45.8	46.8	48.4	50.1	51.2	52.1	52.8	53.4	53.9	54.4	54.8
70	44.6	45.8	46.6	47.7	49.2	51.0	52.1	53.0	53.7	54.3	54.8	55.3	55.7
71	45.4	46.7	47.5	48.5	50.1	51.8	53.0	53.8	54.6	55.2	55.7	56.2	56.6
72	46.2	47.5	48.3	49.4	50.9	52.7	53.9	54.7	55.5	56.1	56.6	57.1	57.5
73	47.0	48.3	49.1	50.2	51.8	53.6	54.7	55.6	56.4	57.0	57.5	58.0	58.5
74	47.8	49.1	49.9	51.0	52.7	54.5	55.6	56.5	57.3	57.9	58.4	58.9	59.4
75	48.6	49.9	50.8	51.9	53.5	55.3	56.5	57.4	58.2	58.8	59.3	59.8	60.3
76	49.4	50.8	51.6	52.7	54.4	56.2	57.4	58.3	59.1	59.7	60.3	60.8	61.2
77	50.2	51.6	52.4	53.6	55.2	57.1	58.3	59.2	60.0	60.6	61.2	61.7	62.1
78	51.1	52.4	53.3	54.4	56.1	58.0	59.2	60.1	60.9	61.5	62.1	62.6	63.1
79	51.9	53.2	54.1	55.3	56.9	58.8	60.1	61.0	61.8	62.4	63.0	63.5	64.0
80	52.7	54.1	54.9	56.1	57.8	59.7	61.0	61.9	62.7	63.3	63.9	64.4	64.9
81	53.5	54.9	55.8	56.9	58.7	60.6	61.8	62.8	63.6	64.2	64.8	65.4	65.8
82	54.3	55.7	56.6	57.8	59.5	61.5	62.7	63.7	64.5	65.2	65.7	66.3	66.8
83	55.1	56.6	57.5	58.6	60.4	62.4	63.6	64.6	65.4	66.1	66.7	67.2	67.7
84	56.0	57.4	58.3	59.5	61.3	63.2	64.5	65.5	66.3	67.0	67.6	68.1	68.6
85	56.8	58.2	59.1	60.4	62.1	64.1	65.4	66.4	67.2	67.9	68.5	69.1	69.6
86	57.6	59.1	60.0	61.2	63.0	65.0	66.3	67.3	68.1	68.8	69.4	70.0	70.5
87	58.4	59.9	60.8	62.1	63.9	65.9	67.2	68.2	69.0	69.7	70.3	70.9	71.4
88	59.3	60.8	61.7	62.9	64.7	66.8	68.1	69.1	69.9	70.6	71.3	71.8	72.3
89	60.1	61.6	62.5	63.8	65.6	67.7	69.0	70.0	70.8	71.6	72.2	72.8	73.3
90	60.9	62.4	63.4	64.6	66.5	68.6	69.9	70.9	71.8	72.5	73.1	73.7	74.2
91	61.8	63.3	64.2	65.5	67.4	69.4	70.8	71.8	72.7	73.4	74.0	74.6	75.1
92	62.6	64.1	65.1	66.3	68.2	70.3	71.7	72.7	73.6	74.3	75.0	75.5	76.1
93	63.4	65.0	65.9	67.2	69.1	71.2	72.6	73.6	74.5	75.2	75.9	76.5	77.0
94	64.2	65.8	66.8	68.1	70.0	72.1	73.5	74.5	75.4	76.2	76.8	77.4	77.9
95	65.1	66.6	67.6	68.9	70.9	73.0	74.4	75.5	76.3	77.1	77.7	78.3	78.9
96	65.9	67.5	68.5	69.8	71.7	73.9	75.3	76.4	77.2	78.0	78.7	79.3	79.8
97	66.8	68.3	69.3	70.7	72.6	74.8	76.2	77.3	78.2	78.9	79.6	80.2	80.7
98	67.6	69.2	70.2	71.5	73.5	75.7	77.1	78.2	79.1	79.8	80.5	81.1	81.7
99	68.4	70.0	71.0	72.4	74.4	76.6	78.0	79.1	80.0	80.8	81.4	82.0	82.6
100	69.3	70.9	71.9	73.2	75.2	77.5	78.9	80.0	80.9	81.7	82.4	83.0	83.5
N	0.01%	0.02%	0.03%	0.05%	0.1%	0.2%	0.3%	0.4%	0.5%	0.6%	0.7%	0.8%	0.9%
							B						

TABLE 8.1 *(Continued)*

					A in Erl							
					B							
1.0%	1.2%	1.5%	2%	3%	5%	7%	10%	15%	20%	30%	40%	50%
37.9	38.5	39.2	40.3	41.9	44.5	46.7	49.6	54.0	58.5	68.5	81.0	98.1
38.8	39.4	40.1	41.2	42.9	45.5	47.7	50.6	55.2	59.7	69.9	82.7	100.1
39.7	40.3	41.0	42.1	43.9	46.5	48.8	51.7	56.3	61.0	71.3	84.3	102.1
40.6	41.2	42.0	43.1	44.8	47.5	49.8	52.8	57.5	62.2	72.7	86.0	104.1
41.5	42.1	42.9	44.0	45.8	48.5	50.8	53.9	58.7	63.5	74.2	87.6	106.1
42.4	43.0	43.8	44.9	46.7	49.5	51.9	55.0	59.8	64.7	75.6	89.3	108.1
43.3	43.9	44.7	45.9	47.7	50.5	52.9	56.1	61.0	65.9	77.0	91.0	110.1
44.2	44.8	45.7	46.8	48.7	51.5	53.9	57.1	62.1	67.2	78.4	92.6	112.1
45.1	45.8	46.6	47.8	49.6	52.6	55.0	58.2	63.3	68.4	79.8	94.3	114.1
46.0	46.7	47.5	48.7	50.6	53.6	56.0	59.3	64.5	69.7	81.3	96.0	116.1
46.9	47.6	48.4	49.6	51.6	54.6	57.1	60.4	65.6	70.9	82.7	97.6	118.1
47.9	48.5	49.4	50.6	52.5	55.6	58.1	61.5	66.8	72.1	84.1	99.3	120.1
48.8	49.4	50.3	51.5	53.5	56.6	59.1	62.6	68.0	73.4	85.5	101.0	122.1
49.7	50.4	51.2	52.5	54.5	57.6	60.2	63.7	69.1	74.6	87.0	102.6	124.1
50.6	51.3	52.2	53.4	55.4	58.6	61.2	64.8	70.3	75.9	88.4	104.3	126.1
51.5	52.2	53.1	54.4	56.4	59.6	62.3	65.8	71.4	77.1	89.8	106.0	128.1
52.4	53.1	54.0	55.3	57.4	60.6	63.3	66.9	72.6	78.3	91.2	107.6	130.1
53.4	54.1	55.0	56.3	58.4	61.6	64.4	68.0	73.8	79.6	92.7	109.3	132.1
54.3	55.0	55.9	57.2	59.3	62.6	65.4	69.1	74.9	80.8	94.1	111.0	134.1
55.2	55.9	56.9	58.2	60.3	63.7	66.4	70.2	76.1	82.1	95.5	112.6	136.1
56.1	56.8	57.8	59.1	61.3	64.7	67.5	71.3	77.3	83.3	96.9	114.3	138.1
57.0	57.8	58.7	60.1	62.3	65.7	68.5	72.4	78.4	84.6	98.4	115.9	140.1
58.0	58.7	59.7	61.0	63.2	66.7	69.6	73.5	79.6	85.8	99.8	117.6	142.1
58.9	59.6	60.6	62.0	64.2	67.7	70.6	74.6	80.8	87.0	101.2	119.3	144.1
59.8	60.6	61.6	62.9	65.2	68.7	71.7	75.6	81.9	88.3	102.7	120.9	146.1
60.7	61.5	62.5	63.9	66.2	69.7	72.7	76.7	83.1	89.5	104.1	122.6	148.0
61.7	62.4	63.4	64.9	67.2	70.8	73.8	77.8	84.2	90.8	105.5	124.3	150.0
62.6	63.4	64.4	65.8	68.1	71.8	74.8	78.9	85.4	92.0	106.9	125.9	152.0
63.5	64.3	65.3	66.8	69.1	72.8	75.9	80.0	86.6	93.3	108.4	127.6	154.0
64.4	65.2	66.3	67.7	70.1	73.8	76.9	81.1	87.7	94.5	109.8	129.3	156.0
65.4	66.2	67.2	68.7	71.1	74.8	78.0	82.2	88.9	95.7	111.2	130.9	158.0
66.3	67.1	68.2	69.6	72.1	75.8	79.0	83.3	90.1	97.0	112.6	132.6	160.0
67.2	68.0	69.1	70.6	73.0	76.9	80.1	84.4	91.2	98.2	114.1	134.3	162.0
68.2	69.0	70.1	71.6	74.0	77.9	81.1	85.5	92.4	99.5	115.5	135.9	164.0
69.1	69.9	71.0	72.5	75.0	78.9	82.2	86.6	93.6	100.7	116.9	137.6	166.0
70.0	70.9	71.9	73.5	76.0	79.9	83.2	87.7	94.7	102.0	118.3	139.3	168.0
70.9	71.8	72.9	74.5	77.0	80.9	84.3	88.8	95.9	103.2	119.8	140.9	170.0
71.9	72.7	73.8	75.4	78.0	82.0	85.3	89.9	97.1	104.5	121.2	142.6	172.0
72.8	73.7	74.8	76.4	78.9	83.0	86.4	91.0	98.2	105.7	122.6	144.3	174.0
73.7	74.6	75.7	77.3	79.9	84.0	87.4	92.1	99.4	106.9	124.0	145.9	176.0
74.7	75.6	76.7	78.3	80.9	85.0	88.5	93.1	100.6	108.2	125.5	147.6	178.0
75.6	76.5	77.6	79.3	81.9	86.0	89.5	94.2	101.7	109.4	126.9	149.3	180.0
76.6	77.4	78.6	80.2	82.9	87.1	90.6	95.3	102.9	110.7	128.3	150.9	182.0
77.5	78.4	79.6	81.2	83.9	88.1	91.6	96.4	104.1	111.9	129.7	152.6	184.0
78.4	79.3	80.5	82.2	84.9	89.1	92.7	97.5	105.3	113.2	131.2	154.3	186.0
79.4	80.3	81.5	83.1	85.8	90.1	93.7	98.6	106.4	114.4	132.6	155.9	188.0
80.3	81.2	82.4	84.1	86.8	91.1	94.8	99.7	107.6	115.7	134.0	157.6	190.0
81.2	82.2	83.4	85.1	87.8	92.2	95.8	100.8	108.8	116.9	135.5	159.3	192.0
82.2	83.1	84.3	86.0	88.8	93.2	96.9	101.9	109.9	118.2	136.9	160.9	194.0
83.1	84.1	85.3	87.0	89.8	94.2	97.9	103.0	111.1	119.4	138.3	162.6	196.0
84.1	85.0	86.2	88.0	90.8	95.2	99.0	104.1	112.3	120.6	139.7	164.3	198.0
1.0%	1.2%	1.5%	2%	3%	5%	7%	10%	15%	20%	30%	40%	50%

B

TABLE 8.1 *(Continued)*

N	A in Erl												
	B												
	0.01%	0.02%	0.03%	0.05%	0.1%	0.2%	0.3%	0.4%	0.5%	0.6%	0.7%	0.8%	0.9%
100	69.3	70.9	71.9	73.2	75.2	77.5	78.9	80.0	80.9	81.7	82.4	83.0	83.5
102	70.9	72.6	73.6	75.0	77.0	79.3	80.7	81.8	82.7	83.5	84.2	84.8	85.4
104	72.6	74.3	75.3	76.7	78.8	81.1	82.5	83.7	84.6	85.4	86.1	86.7	87.3
106	74.3	76.0	77.1	78.5	80.5	82.8	84.3	85.5	86.4	87.2	87.9	88.6	89.2
108	76.0	77.7	78.8	80.2	82.3	84.6	86.2	87.3	88.3	89.1	89.8	90.5	91.1
110	77.7	79.4	80.5	81.9	84.1	86.4	88.0	89.2	90.1	90.9	91.7	92.3	92.9
112	79.4	81.1	82.2	83.7	85.8	88.3	89.8	91.0	92.0	92.8	93.5	94.2	94.8
114	81.1	82.9	84.0	85.4	87.6	90.1	91.6	92.8	93.8	94.7	95.4	96.1	96.7
116	82.8	84.6	85.7	87.2	89.4	91.9	93.5	94.7	95.7	96.5	97.3	98.0	98.6
118	84.5	86.3	87.4	89.0	91.2	93.7	95.3	96.5	97.5	98.4	99.2	99.9	100.5
120	86.2	88.0	89.2	90.7	93.0	95.5	97.1	98.4	99.4	100.3	101.0	101.7	102.4
122	87.9	89.8	90.9	92.5	94.7	97.3	98.9	100.2	101.2	102.1	102.9	103.6	104.3
124	89.6	91.5	92.7	94.2	96.5	99.1	100.8	102.1	103.1	104.0	104.8	105.5	106.2
126	91.3	93.2	94.4	96.0	98.3	100.9	102.6	103.9	105.0	105.9	106.7	107.4	108.1
128	93.1	95.0	96.2	97.8	100.1	102.7	104.5	105.8	106.8	107.7	108.5	109.3	109.9
130	94.8	96.7	97.9	99.5	101.9	104.6	106.3	107.6	108.7	109.6	110.4	111.2	111.8
132	96.5	98.5	99.7	101.3	103.7	106.4	108.1	109.5	110.5	111.5	112.3	113.1	113.7
134	98.2	100.2	101.4	103.1	105.5	108.2	110.0	111.3	112.4	113.4	114.2	115.0	115.6
136	100.0	101.9	103.2	104.9	107.3	110.0	111.8	113.2	114.3	115.2	116.1	116.8	117.5
138	101.7	103.7	105.0	106.6	109.1	111.9	113.7	115.0	116.2	117.1	118.0	118.7	119.4
140	103.4	105.4	106.7	108.4	110.9	113.7	115.5	116.9	118.0	119.0	119.9	120.6	121.4
142	105.1	107.2	108.5	110.2	112.7	115.5	117.4	118.7	119.9	120.9	121.8	122.5	123.3
144	106.9	109.0	110.2	112.0	114.5	117.4	119.2	120.6	121.8	122.8	123.6	124.4	125.2
146	108.6	110.7	112.0	113.8	116.3	119.2	121.1	122.5	123.6	124.6	125.5	126.3	127.1
148	110.4	112.5	113.8	115.5	118.1	121.0	122.9	124.3	125.5	126.5	127.4	128.2	129.0
150	112.1	114.2	115.6	117.3	119.9	122.9	124.8	126.2	127.4	128.4	129.3	130.1	130.9
152	113.8	116.0	117.3	119.1	121.8	124.7	126.6	128.1	129.3	130.3	131.2	132.0	132.8
154	115.6	117.8	119.1	120.9	123.6	126.5	128.5	129.9	131.2	132.2	133.1	133.9	134.7
156	117.3	119.5	120.9	122.7	125.4	128.4	130.3	131.8	133.0	134.1	135.0	135.9	136.6
158	119.1	121.3	122.7	124.5	127.2	130.2	132.2	133.7	134.9	136.0	136.9	137.8	138.5
160	120.8	123.1	124.4	126.3	129.0	132.1	134.0	135.6	136.8	137.9	138.8	139.7	140.4
162	122.6	124.8	126.2	128.1	130.8	133.9	135.9	137.4	138.7	139.8	140.7	141.6	142.4
164	124.3	126.6	128.0	129.9	132.7	135.8	137.8	139.3	140.6	141.7	142.6	143.5	144.3
166	126.1	128.4	129.8	131.7	134.5	137.6	139.6	141.2	142.5	143.5	144.5	145.4	146.2
168	127.9	130.2	131.6	133.5	136.3	139.4	141.5	143.1	144.3	145.4	146.4	147.3	148.1
170	129.6	131.9	133.4	135.3	138.1	141.3	143.4	144.9	146.2	147.3	148.3	149.2	150.0
172	131.4	133.7	135.2	137.1	139.9	143.1	145.2	146.8	148.1	149.2	150.2	151.1	151.9
174	133.1	135.5	136.9	138.9	141.8	145.0	147.1	148.7	150.0	151.1	152.1	153.0	153.9
176	134.9	137.3	138.7	140.7	143.6	146.9	149.0	150.6	151.9	153.0	154.0	155.0	155.8
178	136.7	139.0	140.5	142.5	145.4	148.7	150.8	152.4	153.8	154.9	156.0	156.9	157.7
180	138.4	140.8	142.3	144.3	147.3	150.6	152.7	154.3	155.7	156.8	157.9	158.8	159.6
182	140.2	142.6	144.1	146.1	149.1	152.4	154.6	156.2	157.6	158.7	159.8	160.7	161.6
184	142.0	144.4	145.9	147.9	150.9	154.3	156.4	158.1	159.5	160.6	161.7	162.6	163.5
186	143.7	146.2	147.7	149.8	152.8	156.1	158.3	160.0	161.4	162.5	163.6	164.5	165.4
188	145.5	148.0	149.5	151.6	154.6	158.0	160.2	161.9	163.3	164.4	165.5	166.5	167.3
190	147.3	149.8	151.3	153.4	156.4	159.8	162.1	163.8	165.2	166.4	167.4	168.4	169.3
192	149.1	151.6	153.1	155.2	158.3	161.7	163.9	165.6	167.0	168.3	169.3	170.3	171.2
194	150.8	153.4	154.9	157.0	160.1	163.6	165.8	167.5	168.9	170.2	171.2	172.2	173.1
196	152.6	155.2	156.7	158.8	161.9	165.4	167.7	169.4	170.8	172.1	173.2	174.1	175.0
198	154.4	156.9	158.5	160.7	163.8	167.3	169.6	171.3	172.7	174.0	175.1	176.1	177.0
200	156.2	158.7	160.3	162.5	165.6	169.2	171.4	173.2	174.6	175.9	177.0	178.0	178.9
N	0.01%	0.02%	0.03%	0.05%	0.1%	0.2%	0.3%	0.4%	0.5%	0.6%	0.7%	0.8%	0.9%
							B						

TABLE 8.1 *(Continued)*

| | | | | | A in Erl | | | | | | | |
| | | | | | B | | | | | | | |
1.0%	1.2%	1.5%	2%	3%	5%	7%	10%	15%	20%	30%	40%	50%
84.1	85.0	86.2	88.0	90.8	95.2	99.0	104.1	112.3	120.6	139.7	164.3	198.0
85.9	86.9	88.1	89.9	92.8	97.3	101.1	106.3	114.6	123.1	142.6	167.6	202.0
87.8	88.8	90.1	91.9	94.8	99.3	103.2	108.5	116.9	125.6	145.4	170.9	206.0
89.7	90.7	92.0	93.8	96.7	101.4	105.3	110.7	119.3	128.1	148.3	174.2	210.0
91.6	92.6	93.9	95.7	98.7	103.4	107.4	112.9	121.6	130.6	151.1	177.6	214.0
93.5	94.5	95.8	97.7	100.7	105.5	109.5	115.1	124.0	133.1	154.0	180.9	218.0
95.4	96.4	97.7	99.6	102.7	107.5	111.7	117.3	126.3	135.6	156.9	184.2	222.0
97.3	98.3	99.7	101.6	104.7	109.6	113.8	119.5	128.6	138.1	159.7	187.6	226.0
99.2	100.2	101.6	103.5	106.7	111.7	115.9	121.7	131.0	140.6	162.6	190.9	230.0
101.1	102.1	103.5	105.5	108.7	113.7	118.0	123.9	133.3	143.1	165.4	194.2	234.0
103.0	104.0	105.4	107.4	110.7	115.8	120.1	126.1	135.7	145.6	168.3	197.6	238.0
104.9	105.9	107.4	109.4	112.6	117.8	122.2	128.3	138.0	148.1	171.1	200.9	242.0
106.8	107.9	109.3	111.3	114.6	119.9	124.4	130.5	140.3	150.6	174.0	204.2	246.0
108.7	109.8	111.2	113.3	116.6	121.9	126.5	132.7	142.7	153.0	176.8	207.6	250.0
110.6	111.7	113.2	115.2	118.6	124.0	128.6	134.9	145.0	155.5	179.7	210.9	254.0
112.5	113.6	115.1	117.2	120.6	126.1	130.7	137.1	147.4	158.0	182.5	214.2	258.0
114.4	115.5	117.0	119.1	122.6	128.1	132.8	139.3	149.7	160.5	185.4	217.6	262.0
116.3	117.4	119.0	121.1	124.6	130.2	134.9	141.5	152.0	163.0	188.3	220.9	266.0
118.2	119.4	120.9	123.1	126.6	132.3	137.1	143.7	154.4	165.5	191.1	224.2	270.0
120.1	121.3	122.8	125.0	128.6	134.3	139.2	145.9	156.7	168.0	194.0	227.6	274.0
122.0	123.2	124.8	127.0	130.6	136.4	141.3	148.1	159.1	170.5	196.8	230.9	278.0
123.9	125.1	126.7	128.9	132.6	138.4	143.4	150.3	161.4	173.0	199.7	234.2	282.0
125.8	127.0	128.6	130.9	134.6	140.5	145.6	152.5	163.8	175.5	202.5	237.6	286.0
127.7	129.0	130.6	132.9	136.6	142.6	147.7	154.7	166.1	178.0	205.4	240.9	290.0
129.7	130.9	132.5	134.8	138.6	144.6	149.8	156.9	168.5	180.5	208.2	244.2	294.0
131.6	132.8	134.5	136.8	140.6	146.7	151.9	159.1	170.8	183.0	211.1	247.6	298.0
133.5	134.8	136.4	138.8	142.6	148.8	154.0	161.3	173.1	185.5	214.0	250.9	302.0
135.4	136.7	138.4	140.7	144.6	150.8	156.2	163.5	175.5	188.0	216.8	254.2	306.0
137.3	138.6	140.3	142.7	146.6	152.9	158.3	165.7	177.8	190.5	219.7	257.6	310.0
139.2	140.5	142.3	144.7	148.6	155.0	160.4	167.9	180.2	193.0	222.5	260.9	314.0
141.2	142.5	144.2	146.6	150.6	157.0	162.5	170.2	182.5	195.5	225.4	264.2	318.0
143.1	144.4	146.1	148.6	152.7	159.1	164.7	172.4	184.9	198.0	228.2	267.6	322.0
145.0	146.3	148.1	150.6	154.7	161.2	166.8	174.6	187.2	200.4	231.1	270.9	326.0
146.9	148.3	150.0	152.6	156.7	163.3	168.9	176.8	189.6	202.9	233.9	274.2	330.0
148.9	150.2	152.0	154.5	158.7	165.3	171.0	179.0	191.9	205.4	236.8	277.6	334.0
150.8	152.1	153.9	156.5	160.7	167.4	173.2	181.2	194.2	207.9	239.7	280.9	338.0
152.7	154.1	155.9	158.5	162.7	169.5	175.3	183.4	196.6	210.4	242.5	284.2	342.0
154.6	156.0	157.8	160.4	164.7	171.5	177.4	185.6	198.9	212.9	245.4	287.6	346.0
156.6	158.0	159.8	162.4	166.7	173.6	179.6	187.8	201.3	215.4	248.2	290.9	350.0
158.5	159.9	161.8	164.4	168.7	175.7	181.7	190.0	203.6	217.9	251.1	294.2	354.0
160.4	161.8	163.7	166.4	170.7	177.8	183.8	192.2	206.0	220.4	253.9	297.5	358.0
162.3	163.8	165.7	168.3	172.8	179.8	185.9	194.4	208.3	222.9	256.8	300.9	362.0
164.3	165.7	167.6	170.3	174.8	181.9	188.1	196.6	210.7	225.4	259.6	304.2	366.0
166.2	167.7	169.6	172.3	176.8	184.0	190.2	198.9	213.0	227.9	262.5	307.5	370.0
168.1	169.6	171.5	174.3	178.8	186.1	192.3	201.1	215.4	230.4	265.4	310.9	374.0
170.1	171.5	173.5	176.5	180.8	188.1	194.5	203.3	217.7	232.9	268.2	314.2	378.0
172.0	173.5	175.4	178.2	182.8	190.2	196.6	205.5	220.1	235.4	271.1	317.5	382.0
173.9	175.4	177.4	180.2	184.8	192.3	198.7	207.7	222.4	237.9	273.9	320.9	386.0
175.9	177.4	179.4	182.2	186.9	194.4	200.8	209.9	224.8	240.4	276.8	324.2	390.0
177.8	179.3	181.3	184.2	188.9	196.4	203.0	212.1	227.1	242.9	279.6	327.5	394.0
179.7	181.3	183.3	186.2	190.9	198.5	205.1	214.3	229.4	245.4	282.5	330.9	398.0
1.0%	1.2%	1.5%	2%	3%	5%	7%	10%	15%	20%	30%	40%	50%
					B							

TABLE 8.1 *(Continued)*

N	A in Erl B												
	0.01%	0.02%	0.03%	0.05%	0.1%	0.2%	0.3%	0.4%	0.5%	0.6%	0.7%	0.8%	0.9%
200	156.2	158.7	160.3	162.5	165.6	169.2	171.4	173.2	174.6	175.9	177.0	178.0	178.9
202	158.0	160.5	162.1	164.3	167.5	171.0	173.3	175.1	176.5	177.8	178.9	179.9	180.8
204	159.7	162.3	164.0	166.1	169.3	172.9	175.2	177.0	178.4	179.7	180.8	181.8	182.8
206	161.5	164.1	165.8	167.9	171.2	174.8	177.1	178.9	180.4	181.6	182.7	183.8	184.7
208	163.3	165.9	167.6	169.8	173.0	176.6	179.0	180.8	182.3	183.5	184.7	185.7	186.6
210	165.1	167.7	169.4	171.6	174.8	178.5	180.9	182.7	184.2	185.4	186.6	187.6	188.6
212	166.9	169.5	171.2	173.4	176.7	180.4	182.7	184.6	186.1	187.4	188.5	189.5	190.5
214	168.7	171.3	173.0	175.2	178.5	182.2	184.6	186.5	188.0	189.3	190.4	191.5	192.4
216	170.5	173.2	174.8	177.1	180.4	184.1	186.5	188.4	189.9	191.2	192.3	193.4	194.4
218	172.3	175.0	176.6	178.9	182.2	186.0	188.4	190.2	191.8	193.1	194.3	195.3	196.3
220	174.0	176.8	178.5	180.7	184.1	187.8	190.3	192.1	193.7	195.0	196.2	197.2	198.2
222	175.8	178.6	180.3	182.6	185.9	189.7	192.2	194.0	195.6	196.9	198.1	199.2	200.2
224	177.6	180.4	182.1	184.4	187.8	191.6	194.1	195.9	197.5	198.8	200.0	201.1	202.1
226	179.4	182.2	183.9	186.2	189.6	193.5	195.9	197.8	199.4	200.8	202.0	203.0	204.0
228	181.2	184.0	185.7	188.1	191.5	195.3	197.8	199.7	201.3	202.7	203.9	205.0	206.0
230	183.0	185.8	187.6	189.9	193.3	197.2	199.7	201.6	203.2	204.6	205.8	206.9	207.9
232	184.8	187.6	189.4	191.7	195.2	199.1	201.6	203.5	205.1	206.5	207.7	208.8	209.8
234	186.6	189.4	191.2	193.6	197.1	201.0	203.5	205.4	207.1	208.4	209.7	210.8	211.8
236	188.4	191.3	193.0	195.4	198.9	202.8	205.4	207.4	209.0	210.4	211.6	212.7	213.7
238	190.2	193.1	194.9	197.2	200.8	204.7	207.3	209.3	210.9	212.3	213.5	214.6	215.7
240	192.0	194.9	196.7	199.1	202.6	206.6	209.2	211.2	212.8	214.2	215.4	216.6	217.6
242	193.8	196.7	198.5	200.9	204.5	208.5	211.1	213.1	214.7	216.1	217.4	218.5	219.5
244	195.6	198.5	200.3	202.8	206.3	210.4	213.0	215.0	216.6	218.0	219.3	220.4	221.5
246	197.4	200.3	202.2	204.6	208.2	212.2	214.9	216.9	218.5	220.0	221.2	222.4	223.4
248	199.2	202.2	204.0	206.4	210.1	214.1	216.8	218.8	220.4	221.9	223.2	224.3	225.4
250	201.0	204.0	205.8	208.3	211.9	216.0	218.7	220.7	222.4	223.8	225.1	226.2	227.3
	0.908	*0.914*	*0.920*	*0.926*	*0.934*	*0.944*	*0.950*	*0.956*	*0.960*	*0.964*	*0.968*	*0.972*	*0.974*
300	246.4	249.7	251.8	254.6	258.6	263.2	266.2	268.5	270.4	272.0	273.5	274.8	276.0
	0.918	*0.924*	*0.928*	*0.932*	*0.942*	*0.952*	*0.958*	*0.962*	*0.966*	*0.970*	*0.972*	*0.976*	*0.978*
350	292.3	295.9	298.2	301.2	305.7	310.8	314.1	316.6	318.7	320.5	322.1	323.6	324.9
	0.922	*0.928*	*0.932*	*0.938*	*0.946*	*0.954*	*0.960*	*0.966*	*0.970*	*0.972*	*0.976*	*0.978*	*0.982*
400	338.4	342.3	344.8	348.1	353.0	358.5	362.1	364.9	367.2	369.1	370.9	372.5	374.0
	0.928	*0.934*	*0.938*	*0.942*	*0.950*	*0.958*	*0.964*	*0.968*	*0.972*	*0.976*	*0.978*	*0.982*	*0.984*
450	384.8	389.0	391.7	395.2	400.5	406.4	410.3	413.3	415.8	417.9	419.8	421.6	423.2
	0.932	*0.938*	*0.942*	*0.946*	*0.954*	*0.962*	*0.968*	*0.972*	*0.974*	*0.978*	*0.982*	*0.982*	*0.984*
500	431.4	435.9	438.8	442.5	448.2	454.5	458.7	461.9	464.5	466.8	468.9	470.7	472.4
	0.938	*0.943*	*0.946*	*0.951*	*0.957*	*0.965*	*0.970*	*0.974*	*0.978*	*0.981*	*0.983*	*0.986*	*0.989*
600	525.2	530.2	533.4	537.6	543.9	551.0	555.7	559.3	562.3	564.9	567.2	569.3	571.3
	0.943	*0.948*	*0.951*	*0.956*	*0.962*	*0.969*	*0.974*	*0.978*	*0.981*	*0.984*	*0.986*	*0.989*	*0.990*
700	619.5	625.0	628.5	633.2	640.1	647.9	653.1	657.1	660.4	663.3	665.8	668.2	670.3
	0.948	*0.953*	*0.955*	*0.959*	*0.965*	*0.972*	*0.976*	*0.980*	*0.983*	*0.985*	*0.989*	*0.990*	*0.993*
800	714.3	720.3	724.0	729.1	736.6	745.1	750.7	755.1	758.7	761.8	764.7	767.2	769.6
	0.951	*0.955*	*0.959*	*0.962*	*0.967*	*0.974*	*0.979*	*0.982*	*0.985*	*0.988*	*0.990*	*0.993*	*0.994*
900	809.4	815.8	819.9	825.3	833.3	842.5	848.6	853.3	857.2	860.6	863.7	866.5	869.0
	0.954	*0.959*	*0.961*	*0.964*	*0.970*	*0.976*	*0.980*	*0.984*	*0.987*	*0.989*	*0.991*	*0.993*	*0.996*
1000	904.8	911.7	916.0	921.7	930.3	940.1	946.6	951.7	955.9	959.5	962.8	965.8	968.6
	0.962	*0.963*	*0.960*	*0.963*	*0.977*	*0.979*	*0.984*	*0.983*	*0.991*	*0.995*	*0.992*	*0.992*	*0.994*
1100	1001.	1008.	1012.	1018.	1028.	1038.	1045.	1050.	1055.	1059.	1062.	1065.	1068.
N	*0.01%*	*0.02%*	*0.03%*	*0.05%*	*0.1%*	*0.2%*	*0.3%*	*0.4%*	*0.5%*	*0.6%*	*0.7%*	*0.8%*	*0.9%* B

TABLE 8.1 *(Continued)*

					A in Erl							
					B							
1.0%	*1.2%*	*1.5%*	*2%*	*3%*	*5%*	*7%*	*10%*	*15%*	*20%*	*30%*	*40%*	*50%*
179.7	181.3	183.3	186.2	190.9	198.5	205.1	214.3	229.4	245.4	282.5	330.9	398.0
181.7	183.2	185.2	188.1	192.9	200.6	207.2	216.5	231.8	247.9	285.4	334.2	402.0
183.6	185.2	187.2	190.1	194.9	202.7	209.4	218.7	234.1	250.4	288.2	337.5	406.0
185.5	187.1	189.2	192.1	196.9	204.7	211.5	221.0	236.5	252.9	291.1	340.9	410.0
187.5	189.1	191.1	194.1	199.0	206.8	213.6	223.2	238.8	255.4	293.9	344.2	414.0
189.4	191.0	193.1	196.1	201.0	208.9	215.8	225.4	241.2	257.9	296.8	347.5	418.0
191.4	193.0	195.1	198.1	203.0	211.0	217.9	227.6	243.5	260.4	299.6	350.9	422.0
193.3	194.9	197.0	200.0	205.0	213.0	220.0	229.8	245.9	262.9	302.5	354.2	426.0
195.2	196.9	199.0	202.0	207.0	215.1	222.2	232.0	248.2	265.4	305.3	357.5	430.0
197.2	198.8	201.0	204.0	209.1	217.2	224.3	234.2	250.6	267.9	308.2	360.9	434.0
199.1	200.8	202.9	206.0	211.1	219.3	226.4	236.4	252.9	270.4	311.1	364.2	438.0
201.1	202.7	204.9	208.0	213.1	221.4	228.6	238.6	255.3	272.9	313.9	367.5	442.0
203.0	204.7	206.8	210.0	215.1	223.4	230.7	240.9	257.6	275.4	316.8	370.9	446.0
204.9	206.6	208.8	212.0	217.1	225.5	232.8	243.1	260.0	277.8	319.6	374.2	450.0
206.9	208.6	210.8	213.9	219.2	227.6	235.0	245.3	262.3	280.3	322.5	377.5	454.0
208.8	210.5	212.8	215.9	221.2	229.7	237.1	247.5	264.7	282.8	325.3	380.9	458.0
210.8	212.5	214.7	217.9	223.2	231.8	239.2	249.7	267.0	285.3	328.2	384.2	462.0
212.7	214.4	216.7	219.9	225.2	233.8	241.4	251.9	269.4	287.8	331.1	387.5	466.0
214.7	216.4	218.7	221.9	227.2	235.9	243.5	254.1	271.7	290.3	333.9	390.9	470.0
216.6	218.3	220.6	223.9	229.3	238.0	245.6	256.3	274.1	292.8	336.8	394.2	474.0
218.6	220.3	222.6	225.9	231.3	240.1	247.8	258.6	276.4	295.3	339.6	397.5	478.0
220.5	222.3	224.6	227.9	233.3	242.2	249.9	260.8	278.8	297.8	342.5	400.9	482.0
222.5	224.2	226.5	229.9	235.3	244.3	252.0	263.0	281.1	300.3	345.3	404.2	486.0
224.4	226.2	228.5	231.8	237.4	246.3	254.2	265.2	283.4	302.8	348.2	407.5	490.0
226.3	228.1	230.5	233.8	239.4	248.4	256.3	267.4	285.8	305.3	351.0	410.9	494.0
228.3	230.1	232.5	235.8	241.4	250.5	258.4	269.6	288.1	307.8	353.9	414.2	498.0
0.976	*0.982*	*0.988*	*0.998*	*1.014*	*1.042*	*1.070*	*1.108*	*1.176*	*1.250*	*1.428*	*1.666*	*2.000*
277.1	279.2	281.9	285.7	292.1	302.6	311.9	325.0	346.9	370.3	425.3	497.5	598.0
0.982	*0.984*	*0.990*	*1.000*	*1.016*	*1.044*	*1.070*	*1.108*	*1.174*	*1.248*	*1.428*	*1.668*	*2.000*
326.2	328.4	331.4	335.7	342.9	354.8	365.4	380.4	405.6	432.7	496.7	580.9	698.0
0.982	*0.988*	*0.994*	*1.004*	*1.020*	*1.046*	*1.070*	*1.108*	*1.176*	*1.250*	*1.430*	*1.666*	*2.000*
375.3	377.8	381.1	385.9	393.9	407.1	418.9	435.8	464.4	495.2	568.2	664.2	798.0
0.986	*0.990*	*0.996*	*1.004*	*1.018*	*1.046*	*1.072*	*1.110*	*1.176*	*1.250*	*1.428*	*1.666*	*2.000*
424.6	427.3	430.9	436.1	444.8	459.4	472.5	491.3	523.2	557.7	639.6	747.5	898.0
0.988	*0.994*	*0.998*	*1.006*	*1.022*	*1.048*	*1.070*	*1.108*	*1.176*	*1.250*	*1.428*	*1.668*	*2.000*
474.0	477.0	480.8	486.4	495.9	511.8	526.0	546.7	582.0	620.2	711.0	830.9	998.0
0.991	*0.994*	*1.000*	*1.008*	*1.022*	*1.047*	*1.073*	*1.110*	*1.176*	*1.249*	*1.429*	*1.666*	*2.000*
573.1	576.4	580.8	587.2	598.1	616.5	633.3	657.7	699.6	745.1	853.9	997.5	1198.
0.993	*0.997*	*1.002*	*1.010*	*1.024*	*1.049*	*1.073*	*1.110*	*1.176*	*1.250*	*1.428*	*1.665*	*2.00*
672.4	676.1	681.0	688.2	700.5	721.4	740.6	768.7	817.2	870.1	996.7	1164.	1398.
0.994	*0.998*	*1.004*	*1.011*	*1.025*	*1.050*	*1.073*	*1.110*	*1.176*	*1.250*	*1.433*	*1.67*	*2.00*
771.8	775.9	781.4	789.3	803.0	826.4	847.9	879.7	934.8	995.1	1140.	1331.	1598.
0.997	*1.000*	*1.004*	*1.013*	*1.025*	*1.050*	*1.074*	*1.111*	*1.172*	*1.249*	*1.42*	*1.67*	*2.00*
871.5	875.9	881.8	890.6	905.5	931.4	955.3	990.8	1052.	1120.	1282.	1498.	1798.
0.997	*1.001*	*1.006*	*1.013*	*1.025*	*1.046*	*1.077*	*1.112*	*1.18*	*1.25*	*1.43*	*1.66*	*2.00*
971.2	976.0	982.4	991.9	1008.	1036.	1063.	1102.	1170.	1245.	1425.	1664.	1998.
0.998	*1.000*	*1.006*	*1.011*	*1.03*	*1.05*	*1.07*	*1.11*	*1.18*	*1.25*	*1.43*	*1.67*	*2.00*
1071.	1076.	1083.	1093.	1111.	1141.	1170.	1213.	1288.	1370.	1568.	1831.	2198.
1.0%	*1.2%*	*1.5%*	*2%*	*3%*	*5%*	*7%*	*10%*	*15%*	*20%*	*30%*	*40%*	*50%*

B

TABLE 8.2 Erlang C Model—Blocked Calls Held

P(B) = 0.010					
servers (N)	erlangs (A)	servers (N)	erlangs (A)	servers (N)	erlangs (A)
10	4.08	410	363.51	810	744.01
20	10.97	420	372.93	820	753.59
30	18.59	430	382.35	830	763.18
40	26.58	440	391.78	840	772.77
50	34.80	450	401.22	850	782.36
60	43.20	460	410.66	860	791.95
70	51.73	470	420.11	870	801.55
80	60.36	480	429.57	880	811.15
90	69.07	490	439.03	890	820.75
100	77.85	500	448.49	900	830.35
110	86.69	510	457.96	910	839.96
120	95.58	520	467.44	920	849.56
130	104.52	530	476.92	930	859.17
140	113.50	540	486.41	940	868.78
150	122.51	550	495.90	950	878.40
160	131.56	560	505.40	960	888.01
170	140.63	570	514.90	970	897.63
180	149.74	580	524.40	980	907.25
190	158.86	590	533.91	990	916.87
200	168.01	600	543.42	1000	926.50
210	177.18	610	552.94	1010	936.12
220	186.37	620	562.46	1020	945.75
230	195.58	630	571.99	1030	955.38
240	204.81	640	581.52	1040	965.01
250	214.05	650	591.05	1050	974.64
260	223.30	660	600.58	1060	984.28
270	232.57	670	610.12	1070	993.91
280	241.86	680	619.67	1080	1003.55
290	251.15	690	629.21	1090	1013.19
300	260.46	700	638.76	1100	1022.83
310	269.78	710	648.32	1110	1032.47
320	279.11	720	657.87	1120	1042.12
330	288.46	730	667.43	1130	1051.76
340	297.81	740	676.99	1140	1061.41
350	307.17	750	686.56	1150	1071.06
360	316.54	760	696.13	1160	1080.71
370	325.92	770	705.70	1170	1090.36
380	335.30	780	715.27	1180	1100.02
390	344.70	790	724.85	1190	1109.67
400	354.10	800	734.42	1200	1119.33

TABLE 8.2   (*Continued*)

servers (N)	erlangs (A)	servers (N)	erlangs (A)	servers (N)	erlangs (A)
	P(B) = 0.020				
10	4.54	410	368.36	810	750.98
20	11.77	420	377.84	820	760.61
30	19.64	430	387.33	830	770.24
40	27.84	440	396.82	840	779.87
50	36.26	450	406.32	850	789.51
60	44.83	460	415.82	860	799.15
70	53.52	470	425.33	870	808.79
80	62.30	480	434.85	880	818.43
90	71.15	490	444.37	890	828.07
100	80.06	500	453.89	900	837.72
110	89.03	510	463.42	910	847.37
120	98.04	520	472.95	920	857.02
130	107.09	530	482.49	930	866.67
140	116.18	540	492.03	940	876.32
150	125.30	550	501.58	950	885.98
160	134.45	560	511.13	960	895.64
170	143.63	570	520.69	970	905.30
180	152.83	580	530.24	980	914.96
190	162.05	590	539.81	990	924.62
200	171.29	600	549.37	1000	934.28
210	180.55	610	558.94	1010	943.95
220	189.83	620	568.52	1020	953.62
230	199.12	630	578.09	1030	963.29
240	208.43	640	587.67	1040	972.96
250	217.76	650	597.26	1050	982.63
260	227.09	660	606.84	1060	992.31
270	236.44	670	616.43	1070	1001.98
280	245.80	680	626.02	1080	1011.66
290	255.18	690	635.62	1090	1021.34
300	264.56	700	645.22	1100	1031.02
310	273.96	710	654.82	1110	1040.70
320	283.36	720	664.42	1120	1050.38
330	292.77	730	674.03	1130	1060.07
340	302.19	740	683.64	1140	1069.75
350	311.62	750	693.25	1150	1079.44
360	321.06	760	702.87	1160	1089.13
370	330.51	770	712.48	1170	1098.82
380	339.96	780	722.10	1180	1108.51
390	349.42	790	731.73	1190	1118.20
400	358.89	800	741.35	1200	1127.90

TABLE 8.2   (Continued)

\multicolumn{6}{c}{P(B) = 0.050}					
servers (N)	erlangs (A)	servers (N)	erlangs (A)	servers (N)	erlangs (A)
10	5.29	410	375.53	810	761.24
20	13.00	420	385.10	820	770.94
30	21.25	430	394.68	830	780.63
40	29.77	440	404.27	840	790.33
50	38.47	450	413.85	850	800.03
60	47.29	460	423.45	860	809.74
70	56.21	470	433.04	870	819.44
80	65.21	480	442.64	880	829.15
90	74.26	490	452.25	890	838.85
100	83.37	500	461.86	900	848.56
110	92.52	510	471.47	910	858.27
120	101.71	520	481.09	920	867.99
130	110.93	530	490.71	930	877.70
140	120.18	540	500.33	940	887.42
150	129.46	550	509.96	950	897.13
160	138.76	560	519.59	960	906.85
170	148.08	570	529.22	970	916.57
180	157.42	580	538.86	980	926.29
190	166.78	590	548.50	990	936.02
200	176.16	600	558.14	1000	945.74
210	185.55	610	567.79	1010	955.47
220	194.96	620	577.44	1020	965.19
230	204.38	630	587.09	1030	974.92
240	213.81	640	596.74	1040	984.65
250	223.25	650	606.40	1050	994.38
260	232.71	660	616.06	1060	1004.11
270	242.17	670	625.72	1070	1013.85
280	251.65	680	635.39	1080	1023.58
290	261.13	690	645.06	1090	1033.32
300	270.63	700	654.73	1100	1043.06
310	280.13	710	664.40	1110	1052.79
320	289.64	720	674.08	1120	1062.53
330	299.16	730	683.75	1130	1072.27
340	308.68	740	693.43	1140	1082.02
350	318.21	750	703.11	1150	1091.76
360	327.75	760	712.79	1160	1101.50
370	337.29	770	722.48	1170	1111.25
380	346.85	780	732.17	1180	1120.99
390	356.40	790	741.86	1190	1130.74
400	365.96	800	751.55	1200	1140.49

TABLE 8.2 (*Continued*)

P(B) = 0.100					
servers (N)	erlangs (A)	servers (N)	erlangs (A)	servers (N)	erlangs (A)
10	5.99	410	381.69	810	770.03
20	14.12	420	391.34	820	779.78
30	22.68	430	401.00	830	789.53
40	31.48	440	410.66	840	799.28
50	40.42	450	420.32	850	809.04
60	49.46	460	429.99	860	818.80
70	58.57	470	439.66	870	828.56
80	67.75	480	449.33	880	838.32
90	76.98	490	459.01	890	848.08
100	86.25	500	468.69	900	857.84
110	95.56	510	478.37	910	867.60
120	104.90	520	488.06	920	877.37
130	114.26	530	497.75	930	887.13
140	123.65	540	507.44	940	896.90
150	133.06	550	517.14	950	906.67
160	142.49	560	526.84	960	916.44
170	151.93	570	536.54	970	926.21
180	161.40	580	546.24	980	935.99
190	170.87	590	555.95	990	945.76
200	180.37	600	565.66	1000	955.53
210	189.87	610	575.37	1010	965.31
220	199.38	620	585.08	1020	975.09
230	208.91	630	594.80	1030	984.87
240	218.45	640	604.52	1040	994.65
250	227.99	650	614.24	1050	1004.43
260	237.55	660	623.96	1060	1014.21
270	247.11	670	633.68	1070	1023.99
280	256.68	680	643.41	1080	1033.77
290	266.26	690	653.14	1090	1043.56
300	275.85	700	662.87	1100	1053.34
310	285.44	710	672.60	1110	1063.13
320	295.04	720	682.34	1120	1072.92
330	304.65	730	692.07	1130	1082.70
340	314.26	740	701.81	1140	1092.49
350	323.88	750	711.55	1150	1102.28
360	333.50	760	721.29	1160	1112.08
370	343.13	770	731.04	1170	1121.87
380	352.76	780	740.78	1180	1131.66
390	362.40	790	750.53	1190	1141.45
400	372.04	800	760.28	1200	1151.25

TABLE 8.2 *(Continued)*

P(B) = 0.200					
servers (N)	erlangs (A)	servers (N)	erlangs (A)	servers (N)	erlangs (A)
10	6.85	410	388.69	810	779.97
20	15.45	420	398.43	820	789.79
30	24.38	430	408.17	830	799.60
40	33.48	440	417.92	840	809.42
50	42.69	450	427.67	850	819.23
60	51.97	460	437.42	860	829.05
70	61.31	470	447.17	870	838.87
80	70.70	480	456.93	880	848.69
90	80.12	490	466.69	890	858.52
100	89.57	500	476.45	900	868.34
110	99.06	510	486.21	910	878.16
120	108.56	520	495.98	920	887.99
130	118.09	530	505.75	930	897.81
140	127.63	540	515.52	940	907.64
150	137.19	550	525.29	950	917.46
160	146.76	560	535.06	960	927.29
170	156.35	570	544.84	970	937.12
180	165.95	580	554.62	980	946.95
190	175.56	590	564.40	990	956.78
200	185.18	600	574.18	1000	966.61
210	194.80	610	583.97	1010	976.45
220	204.44	620	593.75	1020	986.28
230	214.09	630	603.54	1030	996.11
240	223.74	640	613.33	1040	1005.95
250	233.40	650	623.12	1050	1015.79
260	243.07	660	632.91	1060	1025.62
270	252.74	670	642.71	1070	1035.46
280	262.42	680	652.50	1080	1045.30
290	272.11	690	662.30	1090	1055.14
300	281.80	700	672.10	1100	1064.98
310	291.50	710	681.90	1110	1074.82
320	301.20	720	691.70	1120	1084.66
330	310.90	730	701.50	1130	1094.50
340	320.61	740	711.31	1140	1104.34
350	330.33	750	721.11	1150	1114.19
360	340.04	760	730.92	1160	1124.03
370	349.77	770	740.73	1170	1133.87
380	359.49	780	750.54	1180	1143.72
390	369.22	790	760.35	1190	1153.56
400	378.95	800	770.16	1200	1163.41

CASE 2. Channel-Sharing Case

N_1 is the number of nominal channels assigned in a cell.
N_2 is number of channels sharing with other cell.
$\Delta N = N_2 - N_1$.
A is the resultant offered load.

$$A \simeq \frac{1}{2}[A(N_1, B) + A(N_2, B) - A(\Delta N, B)] \qquad (8.4.7)$$

Substituting $N_1 = 45$, $\Delta N = 15$, and $N_2 = 60$ into Eq. (8.4.7), and obtaining each value of $A(N, B)$ from Table 8.1, the resultant offered load A becomes

$$A \simeq \frac{A(45, 0.02) + A(60, 0.02) - A(15, 0.02)}{2}$$

$$= \frac{35.6 + 49.6 - 9.01}{2} = 38.10 \qquad (8.4.8)$$

Then the number of mobile users that can be served is

$$M = \frac{38.1 \times 60}{1.76} = 1299 \text{ users} \qquad (8.4.9)$$

Comparing Eq. (8.4.6) with Eq. (8.4.9), we can see that the channel-sharing scheme always serves more users than the no-channel-sharing scheme. However, the drawback to the channel-sharing scheme is that the hardware of the additional 15 channels has to be provided at each cell site. Also system control is more complicated for the channel-sharing scheme.

Example 8.1. If all of the 45 nominal channels of a cell are shared with those of an adjacent cell, then how many users can be served in a cell?

SOLUTION. Assume that the blocking probability $B = 0.02$ and that the average calling time $t = 1.75$ min. Then applying Eq. (8.4.7) with $N_1 = 45$, $N_2 = 90$, and $N = 45$, we obtain

$$A = \frac{A(45, 0.02) + A(90, 0.02) - A(45, 0.02)}{2}$$

$$= \frac{78.3}{2} = 39.15$$

$$M = \frac{39.15 \times 60}{1.76} = 1334.66 \text{ users} \qquad (8.4.10)$$

This result indicates that sharing a total of 90 channels between two cells will always provide service to the largest number of users.

Channel Sharing in Directional Antenna Cells

Assume that three directional antennas are used in three sectors of each cell site. Then a total of 45 channels assigned at each site becomes 15 channels for each sector. The strategy of sharing channels between sectors at a cell is always recommended. However, in the system there may not be a great deal of freedom in sharing channels. Suppose that channels may always be shared counterclockwise to avoid adjacent-channel interference, as shown In Fig. 8.9B. Assume also that the number of users that can be served is indicated in two cases, no-channel-sharing case and channel-sharing case. The conditions given here are the same as mentioned earlier.

CASE 1. No-Channel-Sharing Case

$N = 15$ channels/sector.
$A (15, 0.02) = 9.01$.
$M = (9.01 \times 60)/1.76 = 307.16$ users.

CASE 2. Channel-Sharing Case
Substituting N_1 and 15, $N_2 = 30$ and $\Delta N = 15$ into Eq. (8.4.7) yields

$$A = \frac{A(15, 0.02) + A(30, 0.02) - A(15, 0.02)}{2}$$

$$= \frac{21.9}{2} = 10.95$$

$$M = \frac{10.95 \times 60}{1.76} = 373.30 \text{ users}$$

In comparing the number of the mobile users in these two cases, it is apparent that the channel-sharing scheme always provides more users than the no-channel-sharing scheme. Equation (8.4.11) sums up the results:

$$A(N, B) > 2 \times A\left(\frac{N}{2}, B\right) \tag{8.4.11}$$

8.4.4   Channel Borrowing

Channel borrowing is usually performed from a permanent base. Since traffic density is not uniformly distributed over an entire coverage area, some areas need more channels to provide necessary service. Clearly, borrowing is a long-term commitment; it makes no difference if it is in an omnidirectional antenna cell or a directional antenna cell. Channel borrowing will be illustrated in an omnidirectional antenna cell. Assume that 45 channels are normally assigned to each cell site. However, to meet special conditions, 15 channels are needed from an adjacent cell, as shown in Fig. 8.9C. The performance should be based on the total number of users in these two cells.

First, we define the notations as

$$N_1 = \text{the nominal channels}$$

$$\Delta N = \text{borrowing channels}$$

Then, we consider the resultant offered load in the two cells:

$$A' = A(N_1 + \Delta N, B) + A(N_1 - \Delta N, B) \text{ (in two cells)} \qquad (8.4.12)$$

Suppose that $N_1 = 45$, $\Delta N = 15$, $B = 0.02$. We substitute these values into Eq. (8.4.12). The values of $A(N, B)$ can be found from Table 8.1. Then

$$A' = A(60, 0.02) + A(30, 0.02)$$
$$= 49.6 + 21.9 - 71.5 \qquad (8.4.13)$$

which is the same as the no-borrowing case whereby $A(60, 0.02) = A(45, 0.125)$ and $A(30, 0.02) = A(45, <0.001)$ found from Table 8.1:

$$A = A(45, 0.125) + A(45, <0.001) = 71.5 \qquad (8.4.14)$$

The number of users provided in two cells is

$$M = \frac{71.50 \times 60}{1.76} = 2437.50/\text{two cells}$$

Equation (8.4.14) shows that the channel borrowing scheme lowers the blocking probability in one cell and increases that in the other cell. If the expected number of users in two cells is always different, then the channel borrowing scheme becomes effective.

From the previous analysis, the conclusion is that the number of users provided by the channel-borrowing scheme is the same as that provided by the no-borrowing scheme, but it is less than that provided by the channel-sharing scheme. The following equation is always valid:

$$A(2N, B) \geq A(N + \Delta N, B) + A(N - \Delta N, B) \qquad (8.4.15)$$

8.5 SWITCHING CAPACITY CONSIDERATION

In calculating switching capacity for handling the traffic of L cell sites, the analysis given in Section 8.4 may be used as well. Since a switching system has to handle all the traffic load dynamically, the system itself builds in the load-sharing feature. The following notations are defined:

N is the number of nominal channels/cell.
L is the number of cell sites/system.
B is the blocking probability.
$\bar{t}$ is the average calling time.

Then the offered load of the switching system is

$$A_s = A(NL, B) \tag{8.5.1}$$

For $N = 45$ channels/cells and $L = 20$ cells, Eq. (8.5.1) becomes

$$A_s = a(900, 0.02) = 890.6 \text{ Erlangs} \quad \text{(from Table 8.1)}$$

For an average calling time of $\bar{t} = 1.76$ min, the switching system will handle

$$M_s = \frac{890.6 \times 60}{1.76} = 30361 \text{ users/20 cells} \tag{8.5.2}$$

which is higher than either 20 times 1215 users/cell for a no-channel-sharing scheme [Eq. (8.4.6)] or 20 times 1299 users/cell for a partial-channel-sharing scheme [Eq. (8.4.9)], or 20 times 1335 users/cell for a full-channel-sharing scheme [Eq. (8.4.10)] in an omnidirectional antenna system. From this observation it can be concluded that system traffic congestion is more likely the fault of certain cell sites than the switching system. In planning a mobile telecommunication system, it is very important that a proper frequency assignment plan is determined for each cell site.[5]

REFERENCES

1. Langseth, R. E., and Y. S. Yeh, "Some Results on Digital Signaling over the Mobile Radio Channel" (Microwave Mobile Symposium, Boulder, CO, 1973).
2. Lee, W. C. Y., *Mobile Communications Engineers* (McGraw-Hill, 1982): 394.
3. Lee, W. C. Y., "Elements of Mobile Cellular System Design," *IEEE Trans. Veh. Tech.* (May 1986).
4. Siemens Corp., "Telephone Traffic Theory Tables and Charts" (Part I., by Siemens Telephone and Switching Division, Munich, May 1970).
5. Lee, W. C. Y., *Mobile Cellular Telecommunications Systems* (McGraw Hill, 1989), p. 256.

PROBLEMS

8.1 If the length of a word is 20 bits, the bit error rate P_e is 10^{-2}, and the Hamming distance d is 4, then what is the false-alarm rate P_f? What would be the difference if we just use

$$P_f \approx P_e^d$$

8.2 Explain why the transmission performance without the repetition scheme is worse in the fast-fading case than in the slow-fading case. After using the repetition scheme, the transmission performance is reversed.

8.3 In a seven-cell frequency-reuse cellular system, there are 395 voice channels to be evenly assigned to each cell. If the blocking probability is 2%, what is the offered load in Erlang? If the average talking time (holding time) is 100 s, how many calls can be served in a busy hour?

8.4 In a three-sector cell the voice frequency channels assigned to the cell are 57 channels. One-third of the channels will be assigned to each sector. What is the offered load in Erlang if the probability is 2%.

8.5 In a 5-km cell we would like to handle 2000 calls per hour. Assume that the blocking probability is 1% and that the average holding time is 1.7 min, how many radio channels are needed?

8.6 Suppose that there are 60 channels assigned to a cell. If the cell radius is 4 km, the blocking probability 2%, the average holding time 100 s, how many customers can be served per 100 ft²?

8.7 Consider the system described in Fig. 8.8. The inner cell uses 15 kHz channels, and the outer cell uses 30 kHz channels. What would be the capacity increase as compared with the cells using all of the 30 kHz channels?

8.8 In a channel-sharing scheme the number of nominal channels assigned in a cell is 40. The number of channels sharing with other cells is 20; the blocking probability is 2%. Find the resultant offered load and compare the offered loads between the sharing scheme and the no-sharing scheme.

8.9 Derive Eq. (8.4.7). If $A = 40$ Erlangs and $B = 0.02$, what are the estimated values of N_1 and N_2?

8.10 Derive Eq. (8.4.12), and prove that Eq. (8.4.15) is always true.

9

CELLULAR CDMA

9.1 WHY CDMA?

The code division multiple access (CDMA) scheme was developed mainly to increase capacity. The development of digital cellular systems for increasing capacity came just as the analog cellular system faced a capacity limitation in 1987. In digital systems there are three basic multiple-access schemes: frequency division multiple access (FDMA), time division multiple access (TDMA), and code division multiple access (CDMA). In theory, it does not matter whether the spectrum is divided into frequencies, time slots, or codes; the capacity provided from these three multiple-access schemes is the same. However, in the cellular system we might find that one is better suited in certain communication media than another. In the North American cellular system, in particular, no additional spectrum is allocated for digital cellular, so the analog and digital systems coexist in the spectrum. The problem of transition from analog to digital is another consideration. Although the CDMA has been used in satellite communications, the same system cannot be directly applied to the mobile cellular system. To design a cellular CDMA system, we first need to understand the mobile radio environment; then we can study whether the characteristics of CDMA are suitable for the mobile radio environment.

In a CDMA system the propagation of a wideband carrier signal is used, whereas the propagation of a narrowband carrier signal is a conventional means of communication. Therefore we begin our discussion of wave propagation with the more conventional narrowband wave; then we discuss the wideband wave.

9.2 NARROWBAND (NB) WAVE PROPAGATION

A signal is transmitted from a cell site, received by either a mobile unit or a portable unit, and then propagated over a terrain configuration between two

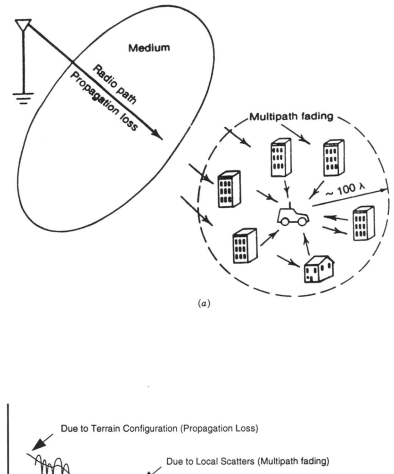

(a)

(b)

Figure 9.1. (a) A mobile radio environment—propagation loss and multipath fading, (b) mobile radio environment, (c) a typical fading received from moving mobile unit.

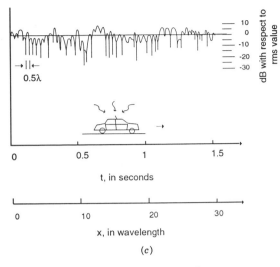

(c)

Figure 9.1. *(Continued)*

ends. The effect of the terrain configuration generates a different long-term fading characteristic that follows a log-normal variation appearing on the envelope of the received signal as shown in Fig. 9.1. Since the antenna height of a mobile or portable unit is close to the ground, three effects are observed:[1] (1) excessive path loss, (2) multipath fading, and (3) the time delay spread phenomenon.

9.2.1 Excessive Path Loss of a CW (Narrowband) Propagation in a Mobile Radio Environment

Suppose that the transmitting power is P_t; then the poynting vector (or the transmitted power density) U_t is

$$U_t = \frac{P_t}{4\pi r^2} \tag{9.2.1}$$

At the receiving end, the signal arrived after passing through the mobile radio environment. The received power can be expressed as

$$P_r = U_t \cdot C(d, f)A_e(f) \tag{9.2.2}$$

where $C(d, f)$ is the medium characteristic, $A_e(f)$ is the effective aperture of the receiving antenna at the receiving end, and it can be expressed as

$$A_e(f) = \frac{c^2 G}{4\pi f^2} \tag{9.2.3}$$

where c is the speed of light and G is the gain of the receiving antenna. Then substituting Eqs. (9.2.1) and (9.2.3) into Eq. (9.2.2), we have

$$P_r = \frac{P_t}{4\pi r^2} C(d, f) \frac{c^2 G}{4\pi f^2} \qquad (9.2.4)$$

Since the antenna height of mobile unit is close to the ground, the signal received is not only from the direct path but also from the strong reflected path. These two paths create an excessive path loss which is 40 dB/dec (fourth power law applied), doubling the path loss in decibels of the free-space path loss [see Eq. (2.3.19)].

In Eq. (9.2.4) P_r is the received power that has been found from the experimental data shown in Section 2.3 as

$$P_r \alpha \frac{1}{r^4} \cdot \frac{1}{f^3} \qquad (9.2.5)$$

Comparing Eq. (9.2.5) with Eq. (9.2.4), we see that the medium characteristic $C(d, f)$ is

$$C(d, f) \alpha \frac{1}{r^2 f}$$

or

$$C(d, f) = \frac{k}{r^2 f} \qquad (9.2.6)$$

where k is a constant. Equation 9.2.4 can be rewritten as

$$P_r = \frac{kc^2 GP_t}{(4\pi r^2)^2} \cdot \frac{1}{f^3} \qquad (9.2.7)$$

Once the medium characteristic $C(d, f)$ is found as shown in Eq. (9.2.6), the wideband propagation path loss can be derived.

9.2.2 Multipath Fading Characteristic

Because of low antenna height on mobile units, the human-made structures surrounding antennas cause multipath fading in the received signals. This is called *Rayleigh fading,* as shown in Fig. 9.1 (see also Section 1.3). The multipath fading creates the burst error in digital transmission. The average

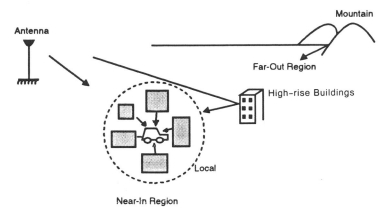

Figure 9.2. Distribution of scatters in mobile radio environment.

duration of fades t as well as the level crossing rates n at 10 dB below the average power of a signal is a function of vehicle speed V and wavelength λ.

$$\bar{t} = 0.132\left(\frac{\lambda}{V}\right) \qquad \text{seconds} \qquad (9.2.8)$$

$$\bar{n} = 0.75\left(\frac{V}{\lambda}\right) \qquad \text{crossings per second} \qquad (9.2.9)$$

For a frequency of 850 MHz and a speed of 15 mph then $\bar{t} = 6$ ms and $\bar{n} = 16$ crossings per second. Equations (9.2.8) and (9.2.9) can be derived from Section 3.1.

9.2.3 Time Delay Spread

The time delay spread phenomenon occurs if a time dispersive medium is present. In a mobile radio environment a single symbol transmitted from one end and received at the other end receives not only its own symbol but also many echos of its symbol. The time delay spread intervals, which are measured from the first symbol to the last detectable echo, are different in different built environments. The average time delay spread due to the local scatterers in suburban areas is 0.5 μs and in urban areas is 3 μs, as described in Section 1.5.6. These local scatterers are in the near-end region illustrated in Fig. 9.2, and the time delay spread corresponding to this region is illustrated in Fig. 9.3. As Fig. 9.2 shows, there are other types of time delay spreads. One kind of delayed wave is caused by the reflection of high-rise buildings (far-out region), and another kind by the reflection from mountains. Their corresponding time delays are given in Fig. 9.3. In certain mountain areas the time delay spread can be up to 100 μs. These time delay spreads would cause

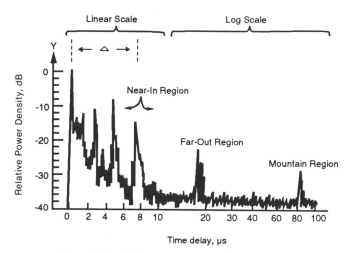

Figure 9.3. An illustration on time-delay spread.

intersymbol interference (ISI) for data transmission.[2] To avoid ISI, the trans-mission rate R_b should not exceed the inverse value of the delay spread Δ if the mobile unit is at a standstill (nonfading case),

$$R_b < \frac{1}{\Delta} \qquad (9.2.10)$$

or R_b should not exceed the inverse value of $2\pi\Delta$ if the mobile unit is in motion (fading case)

$$R_b < \frac{1}{(2\pi\Delta)} \qquad (9.2.11)$$

If the transmission rate R_b is higher than given in Eq. (9.2.10) or Eq. (9.2.11), both FDMA and TDMA need equalizers capable of reducing the ISI to some degree depending on the complication of the time delay spread length and the wave arrival distribution.[3,4,5] An FDMA system always requires a slower transmission rate than a TDMA system when both systems offer the same radio capacity. Usually an FDMA system can get away with not using an equalizer as long as its transmission rate does not rise too high above 10 kilo-samples per second. The CDMA system does not need an equalizer, but a simpler device called a correlator can be used. It will be described later.

9.3   WIDEBAND (WB) SIGNAL PROPAGATION

Wideband (WB) signal transmission is often used in a mobile radio environ-ment. The advantages of using wideband transmission are (1) to reduce fade (i.e., by applying to the frequency diversity concept), (2) to avoid jamming (i.e., by spreading transmitted energy over a wideband so that jamming is ineffective), and (3) to increase cellular system capacity.

What is the wideband signal propagation characteristics? This section will try to answer this question after a simple analysis of wideband propagation path loss and signal fading. To do this, the characteristics of wideband must first be derived from a CW or narrowband signal.

9.3.1 Wideband Signal Path Loss in a Mobile Radio Environment

Suppose that a transmitted power P_t in watts is used to send a wideband (WB) signal with its bandwidth B in Hz along a mobile radio path. The power spectrum of each frequency component of the WB signal over the bandwidth B is $S_t(f)$. Then the relation between $S_t(f)$ and P_t is

$$P_t = \int_{f_0 - B/2}^{f_0 + B/2} S_t(f) \, df \tag{9.3.1}$$

and the poynting vector at the transmitting end is

$$U_t = \frac{P_t}{4\pi r^2} = \frac{\int_{f_0 - B/2}^{f_0 + B/2} S_t(f) \, df}{4\pi r^2} \tag{9.3.2}$$

At the receiving end, the received power of a wideband signal after passing through a mobile radio environment can be found as in Eq. (9.2.2):

$$P_r = \frac{1}{4\pi r^2} \int_{f - B/2}^{f + B/2} S_t(f) \cdot C(d, f) \cdot A_e(f) \, df \tag{9.3.3}$$

Substituting Eqs. (9.2.3) and (9.2.6) into Eq. (9.3.3) yields

$$
\begin{aligned}
P_r &= \frac{1}{4\pi r^2} \int_{f_0 - B/2}^{f_0 + B/2} S_t(f) \frac{k}{r^2 f} \cdot \frac{c^2 G}{4\pi f^2} \, df \\
&= \frac{kc^2 G}{(4\pi r^2)^2} \int_{f_0 - B/2}^{f_0 + B/2} S_t(f) \frac{1}{f^3} \, df
\end{aligned}
\tag{9.3.4}
$$

Equation (9.3.4) can be solved if $S_t(f)$ is known. For simplicity, let

$$S_t(f) = \text{constant} \qquad f_0 - \frac{B}{2} \le f \le f_0 + \frac{B}{2} \tag{9.3.5}$$

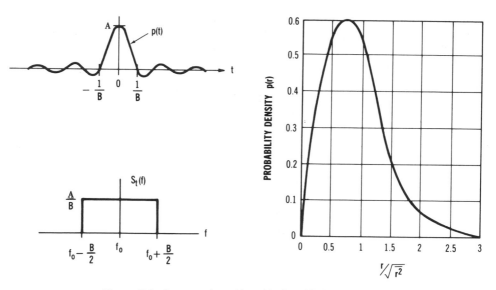

Figure 9.4. A sync pulse $p(t)$ and its band limited spectrum.

This condition can be realized by designing the waveforms of the sending pulses as sinc pulses

$$p(t) = A \operatorname{sinc}(t \cdot B) \tag{9.3.6}$$

then

$$S_t(f) = \frac{A}{B} \qquad f_0 - \frac{B}{2} \le f \le f_0 + \frac{B}{2} \tag{9.3.7}$$

where A is the amplitude of the pulse and B is the total bandwidth. Equations (9.3.6) and (9.3.7) are shown in Fig. 9.4. Equation (9.3.7) can be substituted into Eq. (9.3.4) to yield the received power of a wideband signal:

$$
\begin{aligned}
P_r &= \frac{kc^2GA}{(4\pi r^2)^2 B} \int_{f_0 - B/2}^{f_0 + B/2} \frac{1}{f^3} \, df \\[2mm]
&= \frac{kc^2GA}{(4\pi r^2)^2 B} \cdot \frac{1}{2} \left\{ \frac{1}{(f_0 - B/2)^2} - \frac{1}{(f_0 + B/2)^2} \right\} \\[2mm]
&= K \frac{f_0}{[f_0^2 - (B/2)^2]^2} \\[2mm]
&= K \frac{1}{f_0^3[1 - (B/2f_0)^2]^2} \tag{9.3.8}
\end{aligned}
$$

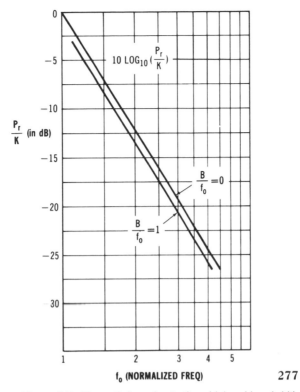

Figure 9.5. The path loss due to the wideband bandwidth.

where

$$K = \frac{kc^2 GA}{(4\pi r^2)^2} \tag{9.3.9}$$

When the bandwidth B approaches zero, Eq. (9.3.8) becomes

$$P_r = K \frac{1}{f_0^3} \quad \text{(narrow-band)} \tag{9.3.10}$$

that agrees with Eq. (9.2.7). Equation (9.3.8) is plotted in Fig. 9.5. The received power is a function of both the carrier frequency and the bandwidth. In principle, the larger the bandwidth, the greater is the received power. If the bandwidth B is half of the carrier frequency, then substituting $B = f_0/2$ into Eq. (9.3.8) yields

$$P_r = K \frac{f_0}{[f_0^2 - (f_0/4)^2]^2} = \frac{K}{(15/16)f_0^3} \tag{9.3.11}$$

Equation (9.3.11) shows that the received power of a wideband signal is $B = f_0/2$, which is merely 0.28 dB higher than the received power of a CW wave. If the bandwidth is extremely wide, $B = f_0$, Eq. (9.3.8) gives

$$P_r = \frac{K}{(9/16)f_0^3} \qquad (9.3.12)$$

that is only 2.5 dB higher than the received power of a CW wave. Therefore the propagation path loss of a wideband signal can be estimated by the path-loss rule of a CW wave in the mobile radio environment.

9.3.2 Wideband Signal Fading

The wideband signal propagation can be described in terms of path loss and the signal fading. The path losses of both the wideband signal and the narrowband signal can be observed from the measured data and proved by theory (as shown in Section 9.3.1). However, the characteristics of wideband signal fading are different from narrowband signal fading characteristics (which are described in Chapter 1). The wideband signal fading is not as severe as the narrowband signal fading. The wideband signal has less fading because its reception takes advantage of the natural frequency diversity over the wideband signal.

Multipath Fading Characteristic on Wideband
The wideband pulse signaling $S_0(t)$ can be expressed as[7]

$$S_0(t) = A \frac{\sin(\pi B t)}{\pi t} \qquad (9.3.13)$$

where A is the pulse amplitude shown in Fig. 9.4. The received signal can be represented as

$$S(t) = \left(\frac{A}{B}\right) \sum_{m=-\infty}^{\infty} b_m(t) \frac{\sin \pi B(t - m/B)}{\pi(t - m/B)} \qquad (9.3.14)$$

The pulse width of $1/B$ is the time interval of the pulse occupied. We count all b_m that are not vanishing over a range of a finite number of m corresponding to a time delay spread Δ. Then the effective number of diversity branches M can be approximated by

$$M = \frac{\Delta + 1/B}{1/B} = B \cdot \Delta + 1. \qquad (9.3.15)$$

The effective number of diversity varies for built structures. The M is larger in the urban area than in the suburban area. If $\Delta = 0.5$ μs for suburban and $\Delta = 3$ μs for urban, and $B = 30$ kHz for narrowband and 1.25 MHz for wideband, we find the effective number of diversity M as follows:

M DIVERSITY BRANCHES

Human-Made environment	$B = 30$ kHz	$B = 1.25$ MHz
$\Delta = 0.5$ μs Suburban	1.015	1.625
$\Delta = 3$ μs Urban	1.09	4.75

The wider the bandwidth, the less fading there is. For $B = 1.25$ MHz, the fading of its received signal is reduced as if a diversity-branch receiver $M = 1.625$ (between a single branch and two branches) were applied in suburban areas and $M = 4.75$ (between four branches and five branches) in urban areas. The wideband signal would provide more diversity gain in urban areas than in suburban areas. For $B = 30$ kHz, no effective diversity gain is noticeable on its narrowband received signal.

9.4   KEY ELEMENTS IN DESIGNING CELLULAR

As described in Section 5.5, the frequency reuse concept guides the cellular system design for obtaining high system capacity.

1. Co-channel interference reduction factor (CIRF). The minimum separation between two co-channel cells D_s is based on a co-channel interference reduction factor (CIRF) q that is expressed as

$$q = \frac{D_s}{R} \qquad (9.4.1)$$

   where R is the cell radius. The value of q is different for each system. For analog cellular systems, $q = 4.6$ is based on the channel bandwidth $B_c = 30$ kHz and the carrier-to-interference ratio C/I equal 18 dB.

2. Hand-offs. The hand-off is a unique feature in cellular. It switches the call to a new frequency channel in a new cell site without interrupting the call or alerting the user. Reducing unnecessary hand-offs and making the necessary hand-off successfully are very important tasks for cellular system operators in analog systems or in future FDMA or TDMA digital systems.

3. Frequency management and frequency assignment. Based on the minimum distance D_s, the number of cells K in a cell reuse pattern can be obtained,

$$K = \frac{(D_s/R)^2}{3} = \frac{q^2}{3} \qquad (9.4.2)$$

*The notation of CIRF is changed from a to q in this chapter to gain clarity since the letter a is a commonly used letter.

The total number of allocated channels is divided by K. There are K sets of frequencies; each cell operates its own set of frequencies managed by the system operator. This is the frequency management task. During a call process different frequencies are assigned to different calls. This is the frequency assignment task. Both tasks are critically impacted to interference and capacity.

4. Reverse-link power control. The reverse-link power control is for reducing near-end to far-end interference. The interference occurs when a mobile unit close to the cell site masks the received signal at the cell site so that the signal from a far-end mobile unit cannot be received by the cell site. This is a unique type of interference for the mobile radio environment.

5. Forward-link power control. The forward-link power control is used to reduce the necessary interference outside its own cell boundary.

6. Capacity enhancement. The capacity of cellular systems can be increased by handling q in two ways:

   a. Within standard cellular equipment the value of q shown in Eq. (9.4.1) remains a constant. As the cell radius R is reduced, D_s reduces. As D_s gets smaller, the same frequency can be repeated over and over in the same geographical area. That is why small cells (sometimes called *microcells* or *pico cells*) are used to increase capacity.

   b. Among different cellular systems many different types of radio equipment can be chosen. The idea is to search for those cellular systems that provide ever smaller values of q. When q in Eq. (9.4.1) is small, D_s will be small, even if the cell radius remains unchanged. The q in properly designed digital cellular systems is smaller than the q in analog systems. By choosing a small q of a new system based on the q of an old system, we can increase the capacity of the system without reducing the size of the cell. That is why the new digital system is preferred to the old analog system.

Reducing the size of cells in a system requires using more cells. This is always costly. Therefore it is important that digital cellular systems are used properly to achieve an optimum value of q.

9.5   SPREAD TECHNIQUES IN MODULATION

Spreading techniques in modulation are generally used in antijamming military operations. There are two spreading techniques: spectrum spreading (spread spectrum), and time spreading (time hopping).

9.5.1   Spread Spectrum Techniques

Spectrum spreading can be accomplished in either of two techniques: by direct sequencing or by frequency hopping.

1. Direct sequence (DS) method. Each information bit is symbolized by a large number of coded bits called *chips*. For example, if an information bit rate R = 10 kbps is used and it needs an information bandwidth B = 10 kHz, and if each bit of 10 kbps is coded by 100 chips, then the chip rate is 1 Mbps, which needs a DS bandwidth B_{ss} = 1 MHz. The bandwidth is thus spreading from 10 kHz to 1 MHz. The spectrum spreading in DS is measured by the processing gain (PG) in dB

$$PG = 10 \log \frac{B_{ss}}{B} \quad \text{(in dB)} \quad (9.5.1)$$

Then the PG of this example is 20 dB. Or we say that this SS system has 20 dB processing gain. The first DS experiment was carried out in 1949 by L. A. DeRosa and M. Rogoff who established a link between New Jersey and California.

2. Frequency hopping (FH) method. A frequency hopping receiver would equip N frequency channels for an active call in order to hop over those N frequencies in some determined hopping pattern. If the information channel width is 10 kHz and there are 100 channels to hop, N = 100, the FH bandwidth B_{ss} = 1 MHz. The spectrum is spreading from 10 kHz (no hopping) to 1 MHz (frequency hopping). The spectrum spreading in FH is measured by PG as

$$PG = 10 \log N \quad \text{(in dB)} \quad (9.5.2)$$

Then the PG of this example is 20 dB. The total hopping frequency channels in FH are also called *chips*. There can be either fast hopping, which makes two or more hops for each symbol, or slow hopping, which makes two or more symbols for each hop. In general, the transmission data rate is the symbol rate. The symbol rate is equal to the bit rate at a binary transmission. Due to the limitation of today's technology, the FH is using a slow hopping pattern.

9.5.2 Time Hopping—Spread Time Technique

A message transmitted with a data rate R requiring a transmit time interval T is now allocated at a longer transmission time interval T_s the data are sent in bursts dictated by a hopping pattern. The time interval between bursts t_n also can be varied. The time spreading data rate R_s is always less than the information bit rate R. Assume that N bursts occurred in time T, then

$$R_s = \left(\frac{T_s}{T}\right)R = \left(1 - \frac{\sum_1^N t_n}{T}\right)R \quad (9.5.3)$$

9.6 DESCRIPTION OF DS MODULATION

The spread spectrum (DS and FH) was used for reducing intentional interference (enemy jamming). We are interested in using it for increasing capacity
instead of reducing the intentional interference. Immediately we recognize
that slow frequency hopping does not increase capacity. Slow hopping lets
good channels downgrade and bad channels upgrade. For a system to be
designed for capacity, all the channels have to be deployed only marginally
well. If bad channels occur in a high-capacity SS system, the system does not
provide normal channels with excessive signal levels that can average with
the poor signal levels from those bad channels to within an acceptable quality
level. It just pulls down all the channels to an unacceptable level. The way
to improve this situation is either to drop the back channels or to correct the
bad channels by other means. Fast hopping would help increase the capacity
because it would provide diversity but the technology for fast hopping at 800
MHz is not available.

9.6.1 Basic DS Technique

The basic DS technique is illustrated in Fig. 9.6. The data $x(t)$ transmitted at
a rate R are modulated first by a carrier f_0 and then by a spreading code $G(t)$
to form a DS signal $S_t(t)$ with a chip rate R_p which takes a DS bandwidth B_{ss}.
The DS signal $S_t(t - T)$ after a propagation time delay T is received and
goes through a correlator using the same spreading code $G(t)$ pre-stored in
it to despread the DS signal. Then the despread signal $S(t - T)$ is obtained.
After demodulating it by f_0, $x(t)$ is recovered. Take a constant envelop signal
modulated on a carrier f_0 at transmitting end shown in Fig. 9.6. Let $x(t)$ be
a data stream modulated by a binary phase shift keying (BPSK) so that

$$x(t) = \pm 1 \qquad (9.6.1)$$

modulated by a binary shift keying

$$S(t) = x(t)\cos(2\pi f_0 t). \qquad (9.6.2)$$

At the transmitting end the spreading sequence $G(t)$ modulation also uses
BPSK

$$G(t) = \pm 1 \qquad (9.6.3)$$

then

$$S_t(t) = x(t)G(t)\cos(2\pi f_0 t). \qquad (9.6.4)$$

At the receiving end, the $S_t(t - T)$ is received after T seconds of propa-

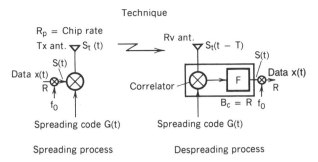

Figure 9.6. Basic spread—spectrum technique. Tx Ant.—Transmitting antenna; Rv Ant.—Receiving antenna.

gation delay. The despreading processing then takes place. The signal $S(t - T)$ coming out from the correlator is

$$S(t - T) = x(t - T) \cdot G(t - T)G(t - \hat{T})\cos[2\pi f_0(t - T)] \qquad (9.6.5)$$

where $\hat{T}$ is the estimated propagation delay generated in the receiver. Since $G(t) = \pm 1$,

$$G(t - T)G(t - \hat{T}) = 1 \qquad (9.6.6)$$

For a good correlator $T = \hat{T}$. Then

$$S(t - T) = x(t - T)\cos[2\pi f_0(t - t)] \qquad (9.6.7)$$

The data $x(t - T)$ are recovered, after being modulated by the carrier frequency f_0, as shown in Fig. 9.6.

9.6.2 Pseudonoise (PN) Code Generator

Pseudonoise (PN) code coming from a PN sequence is a deterministic signal. For example, the sequence 000100110101111 is a pseudonoise sequence. It contains three properties:

1. Balance property—7 zeros and 8 ones. The numbers of zeros and ones of a PN code are different only by one.
2. Run property—4 "zero" runs (or "one" runs):
    Runs = 4
    1/2 of runs (i.e., 2) of length 1 [i.e., two single "zeros (or ones)"]
    1/4 of runs (i.e., 1) of length 2 [i.e., one "2 consecutive zeros (or ones)"]
    1/8 of runs (i.e., 0.5) of length 3 [i.e., one "3 consecutive zeros (or

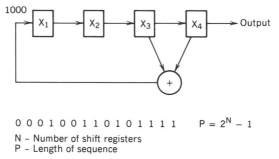

0 0 0 1 0 0 1 1 0 1 0 1 1 1 1   P = 2^N - 1

N - Number of shift registers
P - Length of sequence

Figure 9.7. PN code (linear maximal length sequence) generator.

ones)"]. (In the preceding example, 1/8 of the runs cannot be counted because the code is too short.)

3. Correlation property. Let D denote the "difference," S denote the "same." The two PN codes are compared as follows:

0	0	0	1	0	0	1	1	0	1	0	1	1	1	1
1	0	0	0	1	0	0	1	1	0	1	1	1	1	1
D	*S*	*S*	*D*	*D*	*S*	*D*	*S*	*D*	*D*	*D*	*D*	*S*	*S*	*S*

The value of the correlation of two N-bit sequences can be obtained by counting the number N_d of D's and the number N_s or S's and inserting them into the following equation:

$$P = \frac{1}{N} (N_s - N_d) = \frac{1}{15} (7 - 8) = -\frac{1}{15} \qquad (9.6.8)$$

Then the correlation of a 15-bit PN code is $-1/15$. The PN code generator of a four-shift register is shown in Fig. 9.7. The modulo 2 adder is summing the shift register X_3 and the shift register X_4. The summing signal then feeds back to the shift register X_1. Suppose that a 4-bit sequence 1000 is fed into the shift register X_1. The output PN sequence from this PN code generator is 00010011010111. The code length L of any PN code generator is dependent upon the number of shift registers N.

$$L = 2^N - 1 \qquad (9.6.9)$$

The PN sequence generated in Fig. 9.7 is also called the *linear maximal length sequence*. For $N = 4$, L is 15.

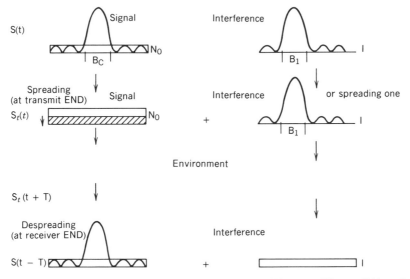

Figure 9.8. Spread spectrum. The interference source could have a different $G_i(t)$ to do the spreading and end the same result.

9.6.3 Reduction of Interference by a DS Signal

Before the spreading process the signal $S(t)$ of Fig. 9.6 is illustrated in both the frequency and time domains, as shown in Fig. 9.8. After spreading $(S)t$ with a given $G(t)$, the output $S_t(t - T)$ is transmitted out while the interference in the air is a narrowband signal or a DS signal with a different $G_I(t)$. When $S_t(t - T)$ is received after a propagation delay T, it is despreaded by the same $G(t)$, resulting in $S(t - T)$. The interference signal can spread to an SS signal by the $G(t)$ if it is a narrowband signal, or it can remain an SS signal because $G(t)$ and $G_I(t)$ do not agree with each other. Consequently a low level of interference within the desired signal bandwidth B_c can be achieved.

9.7 CAPACITIES OF MULTIPLE-ACCESS SCHEMES

Multiple-access schemes are used to provide resources for establishing calls. There are five multiple-access schemes: FDMA serves the calls with different frequency channels. TDMA serves the calls with different time slots. CDMA serves the calls with different code sequences. PDMA (polarization division multiple access) serves the calls with different polarization; PDMA is not applicable to mobile radio communications.[6] SDMA (space division multiple access) serves the calls by spot beam antennas. The calls in different areas covered by the spot beams can be served by the same frequency—the frequency reuse concept. The first three multiple-access schemes can be applied in the cellular systems. The illustration of the differences among three mul-

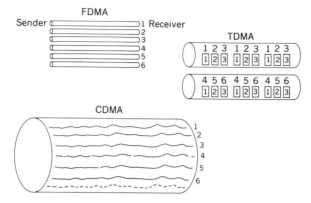

Figure 9.9. Illustration of different multiple-access systems.

tiple-access schemes are shown in Fig. 9.9. Assume that a set of six channels is assigned to a cell. In FDMA six frequency channels serve six calls. In TDMA the channel bandwidth is three times wider than that of FDMA channel bandwidth. Thus two TDMA channel bandwidths equal six FDMA channel bandwidths, with each TDMA channel providing three time slots. The total of six time slots serves six calls. In CDMA one big channel has a bandwidth equal to six FDMA channels. The CDMA radio channel can provide six code sequences and serve six calls. Also CDMA can squeeze additional code sequences into the same radio channel, but the other two multiple access schemes cannot. Adding more code sequences of course degrades the voice quality.

In analog systems, only FDMA can be applied. The C/I received at the RF is closely related to the S/N at the baseband which is related to the voice quality. In digital systems, all three, FDMA, TDMA, and CDMA can be applied. The C/I received at the RF is closely related to the E_b/I_0 at the baseband.

$$\frac{C}{I} = \left(\frac{E_b}{I_0}\right)\left(\frac{R_b}{B_c}\right)$$

$$= \left(\frac{E_b}{I_0}\right) \bigg/ \left(\frac{B_c}{R_b}\right) \tag{9.7.1}$$

where E_b is the energy per bit; I_0 is the interference power per hertz, R_b is the bit per second, and B_c is the radio channel bandwidth in hertz. In digital FDMA or TDMA there are designated channels or time slots for calls. Thus if $R_b \geq B_c$, and E_b/I_0 at the baseband is always greater than one; then C/I is also greater than one (i.e., has a positive value in decibels). In CDMA all the coded sequences, say, N, share one radio channel; thus B_c is much greater than R_b. The notation B_c is often replaced by B_{ss} which is the spread spectrum channel. Within the radio channel any one code sequence is interfered with

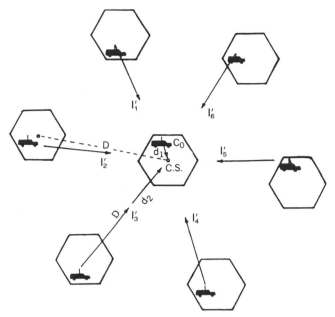

Figure 9.10. Co-channel interference.

$N - 1$ of other code sequences. As a result the interference level is always higher than the signal level, and C/I is less than one (i.e., has a negative value in decibels).

9.7.1 Capacity of Cellular FDMA and TDMA[9]

In FDMA or TDMA each frequency channel or each time slot is assigned to one call. During the call period, no other calls can share the same channel or slot. The co-channel interference comes from a distance $D_s = qR$. Assume that the worst case of having six cochannel interferers (see Fig. 9.10) and the fourth power law path loss are applied. The capacity of the cellular FDMA and cellular TDMA can be found by the radio capacity m, expressed as

$$m = \frac{B_t/B_c}{K} = \frac{M}{\sqrt{2/3(C/I)_s}} \qquad \text{number of channels per cell} \qquad (9.7.2)$$

where

$$
\begin{aligned}
B_t &= \text{total bandwidth (transmitted or received)} \\
B_c &= \text{channel bandwidth (transmitted or received) or equivalent} \\
&\quad \text{channel bandwidth} \\
M = B_t/B_c &= \text{total number of channels or equivalent channels} \\
(C/I)_s &= \text{minimum required carrier-to-interference ratio per channel} \\
&\quad \text{or time slot}
\end{aligned}
$$

Equation (9.7.2) can be directly applied to both analog FDMA and digital FDMA systems. In TDMA systems B_c is an equivalent channel bandwidth. For example, a TDMA radio channel bandwidth of 30 kHz with three time slots can have an equivalent channel bandwidth of 10 kHz (B_c = 10 kHz). The minimum required $(C/I)_s$ of each time slot turns out to be the same as $(C/I)_s$ of the TDMA equivalent channel. The radio capacity is based on two parameters, B_c and $(C/I)_s$ as indicated in Eq. (9.7.2). It has the same two parameters that appear in Shannon's channel capacity formula. The difference between Eq. (9.7.2) and Shannon's is that in the former the two parameters are related and in the latter they are independent. The $(C/I)_s$ of radio capacity can be found using a standard voice quality as soon as the channel bandwidth B_c is known.

9.7.2 Radio Capacity of Cellular CDMA

Cellular CDMA is uniquely designed to work in cellular systems. The primary purpose of using this CDMA is for high capacity. In cellular CDMA there are two CIRF values. One CIRF is called *adjacent CIRF*, $q_a = D_s/R = 2$. It means that the same radio channel can be reused in all neighboring cells. The other CIRF is called *self-CIRF*, $q_s = 1$. It means that different code sequences use the same radio channel to carry different traffic channels. The two CIRF's are shown in Fig. 9.11. With the smallest value of CIRF, the CDMA system is the most efficient frequency reuse system we can find.

1. Required $(C/I)_s$ in cellular CDMA. $(C/I)_s$ can be found from Eq. (9.7.1) depending on the value of E_b/I_0, which is measured at the baseband and determined by the voice quality. For example, if the vocoder rate is R_b = 8 kb/s and the total wideband channel bandwidth B_t = 1.25 MHz, then E_b/I_0 is determined as follows:

$$\frac{E_b}{I_0} = 7 \text{ dB}$$

then

$$(C/I)_s = 0.032 \ (=) -15 \text{ dB}$$

$$\frac{E_b}{I_0} = 4.5 \text{ dB}$$

then

$$(C/I)_s = 0.01792 \ (=) -17.5 \text{ dB}$$

Next the radio capacity of this system can be derived, with calculations

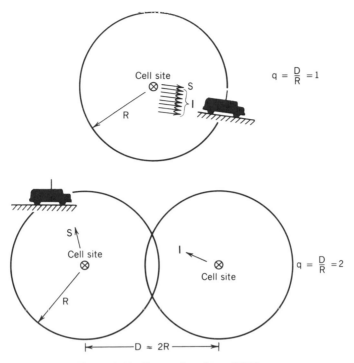

Figure 9.11. Expression of two CIRF's.

based on the forward link, and it can be improved by the power control schemes.

2. Without power control scheme. The radio capacity is calculated from the forward-link C/I ratio. The $(C/I)_s$ received by a mobile unit at the boundary of a CDMA cell shown in Fig. 9.12 can be obtained from the nine interfering cells as follows:

$$\left(\frac{C}{I}\right)_s = \frac{\alpha \cdot R^{-4}}{\underbrace{\alpha(M-1) \cdot R^{-4}}_{\text{Within the cell}} + \underbrace{\alpha \cdot 2M \cdot R^{-4}}_{\substack{\text{Two closest} \\ \text{adjacent channels}}}}$$

$$+ \underbrace{\alpha \cdot 3M \cdot (2R)^{-4}}_{\substack{\text{Three intermediate-} \\ \text{range channels}}} + \underbrace{\alpha \cdot 6M(2.633)^{-4}}_{\text{Six distant cells}}$$

$$= \frac{1}{3.3123M - 1} \tag{9.7.3}$$

where α is a constant factor, M is the number of traffic channels. $(C/I)_s$

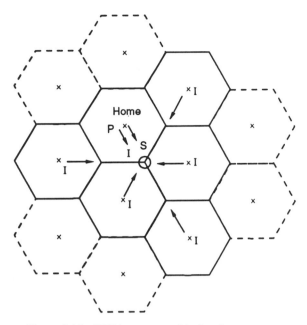

Figure 9.12. CDMA system and its interference.

can be determined from E_b/I_0 and R_b/B_s in Eq. (9.7.1). M is found from Eq. (9.7.3):

$$\left(\frac{C}{I}\right)_s = 0.032 \qquad M = 9.736$$

$$\left(\frac{C}{I}\right)_s = 0.01792 \quad M = 17.15$$

The radio capacity of CDMA defined in Eq. (9.7.2) is

$$m = \frac{M}{K} \qquad \text{number of traffic channels per cell} \qquad (9.7.4)$$

In this case $K = q_a^2/3 = 4/3 = 1.33$.

The difference in radio capacity between CDMA and FDMA or TDMA is shown in Eq. (9.7.4). That is, K is fixed but M is not for CDMA; however M is fixed but K is not for FDMA or TDMA. Therefore

$$m = \frac{M}{1.33} = 7.32 \qquad \text{traffic channels per cell for } E_b/I_0 = 7 \text{ dB}$$

$$= 12.9 \qquad\qquad \text{traffic channels/cell for } E_b/I_0 = 4.5 \text{ dB}$$

9.7.3 Power Control Scheme in CDMA

We can increase the radio capacity by using a suitable power control scheme. The power control scheme used at the forward link of each cell can reduce the interference to the other adjacent cells. The less interference there is generated in a cell, the more the value of M increases. Equation (9.7.3) indicates that if we can neglect all the interference, then (see Fig. 9.12)

$$\left(\frac{C}{I}\right)_s = \frac{R^{-4}}{(M-1)R^{-4}} = \frac{1}{M-1} \tag{9.7.5}$$

for

$$\left(\frac{C}{I}\right)_s = 0.032 \qquad M = 30.25$$

$$\left(\frac{C}{I}\right)_s = 0.01792 \quad M = 54.8$$

By comparing Eq. (9.7.3) with Eq. (9.7.5), we see that the total number of traffic channels M is drastically reduced due to the existence of interference. However, since interference is always existing in the adjacent cells, we can only reduce it by using a power control scheme. A power control scheme for the total power should be considered after combining all traffic channels and determining as follows the necessary power delivery to the close-in mobile unit and the total power reduced at the boundary:

CASE 1: FINDING THE NECESSARY POWER DELIVERY TO A CLOSE-IN MOBILE UNIT
Suppose that the transmitted power at the cell site for the jth mobile unit is P_j which is proportional to r_j^n:

$$P_j \propto r_j^n \tag{9.7.6}$$

where r_j is the distance between the cell site and the jth mobile unit and n is a number. In examining the number n we find that the power control scheme of using $n = 2$ in Eq. (9.7.6) can provide the optimum capacity and also meet the requirement that the forward-link signal can still reach the near-end mobile unit at distance r_j from the cell site with a reduced power

$$P_j = P_R \left(\frac{r_j}{R}\right)^2 \tag{9.7.7}$$

where P_R is the power required to reach those mobile units at the cell boundary

R. The M mobile units served by M traffic channels are assumed uniformly distributed in a cell. Then

$$p(M_l) = kr_l \qquad 0 \le r_l \le R \tag{9.7.8}$$

where $M = \sum_{l=1}^{L} M_l$. There are L groups of mobile units; each L is equally circled around the cell site. M_l is the number of mobile units in the lth group depending on its location. k is a constant. Equation (9.7.8) indicates that if fewer mobile units are closely circling around the cell site, more mobile units are at the outside ring of the cell site. Assume that the distance r_0 is from the cell site to a desired mobile unit; also assume that r_0 is a near-in distance between the close-by mobile unit and the cell site. With the help of Eqs. (9.7.6) and (9.7.7) the power transmitted from the cell site P_t is equal to

$$
\begin{aligned}
P_t &= \sum^{M_1} P_1 + \sum^{M_2} P_2 + \sum^{M_3} P_3 + \ldots + \sum^{M_L} P_L \\
&= P_r \left[\sum^{kr_1} \left(\frac{r_1}{R}\right)^2 + \sum^{kr_2} \left(\frac{r_2}{R}\right)^2 + \ldots + \sum^{kr_L} \left(\frac{r_L}{R}\right)^2 \right] \\
&= P_r \left[kr_1 \left(\frac{r_1}{R}\right)^2 + kr_2 \left(\frac{r_2}{R}\right)^2 + \ldots + kr_L \left(\frac{r_L}{R}\right)^2 \right]
\end{aligned}
\tag{9.7.9}
$$

Since r_L is the distance from the cell site to the cell boundary, $r_L = R$, then Eq. (9.7.9) becomes

$$P_t = P_R k \int_0^R \frac{r^3}{R^2} \, dr = P_R k \frac{R^2}{4} \tag{9.7.10}$$

The total number of mobile units M can be obtained as

$$
\begin{aligned}
M &= \sum_{l=1}^{L} M_l = k(r_1 + r_2 + \ldots + R) \\
&= k \int r \, dr = k \frac{R^2}{2}
\end{aligned}
\tag{9.7.11}
$$

Substituting Eq. (9.7.10) into Eq. (9.7.11) yields

$$p_t = P_R k \left[\frac{M}{2k}\right] = P_R \frac{M}{2} \tag{9.7.12}$$

If the full power P_R is applied to every channel, then

$$P_t = M P_R \tag{9.7.13}$$

Comparing Eqs. (9.7.12) and (9.7.13), we see that the total transmitted power reduces to one-half for the power control scheme of Eq (9.7.7). The $(C/I)_s$ of a mobile unit at a distance r_0 that is close to the cell site is

$$\left(\frac{C}{I}\right)_{s1} = \frac{P_R(r_0/R)^2 \cdot r_0^{-4}}{P_R(M/2) \cdot r_0^{-4}} = \frac{(r_0/R)^2}{(M/2)} \tag{9.7.14}$$

The interference from the adjacent cells can be neglected in Eq. (9.7.14) in this case.

CASE 2: REDUCING THE TOTAL POWER AT THE CELL BOUNDARY
The $(C/I)_s$ of a mobile unit at a distance R that is at the cell boundary can be obtained similarly to Eq. (9.7.3):

$$\left(\frac{C}{I}\right)_{s2} = \frac{P_R}{P_R[(M - 1)/2 + 2(M/2) + 3(M/2) \cdot (2)^{-4} + 6(M/2)(2.633^{-4})]}$$

$$= \frac{1}{1.656M} \tag{9.7.15}$$

The values of M and m can be found from Eq. (9.7.15) in applying the power control scheme:

$$\begin{array}{llll}
M = 18.87 & m = 14.19 & \text{for } C/I = 0.032 \ (-15 \text{ dB}) \\
M = 23.7 & m = 28.33 & \text{for } C/I = 0.01792 \ (-17 \text{ dB})
\end{array} \tag{9.7.16}$$

Now the value of $(C/I)_s$ which is received by the mobile unit at the distance r_0 computed in Eq. (9.7.14), should be checked in terms of Eq. (9.7.15) to see whether it is valid:

$$\left(\frac{C}{I}\right)_{s1} = \frac{(r_0/R)^2}{M/2} = \frac{3.3(r_0/R)^2}{3.3(M/2)} \geq \frac{1}{1.656M} \tag{9.7.17}$$

The power reduction ratio $(r_0/R)^2$ in Eq. (9.7.17) cannot be less than 0.302 for mobile units located less than the distance $r_0 = 0.55R$. If we set the lowest power at $0.302P_R$, then the total power has to be changed:

$$P_t = P_R k \left[\frac{r_0^2}{R^2} r_1 + \frac{r_2^3}{R^2} + \frac{r_3^3}{R^2} + \cdots\right]$$

$$= P_R k \left[\left(\frac{r_0}{R}\right)^2 \int_0^{r_0} r \, dr + \int_{r_0}^R \frac{r^3}{R^2} dr\right]$$

$$= P_R k \frac{R^2}{4}\left[1 + \left(\frac{r_0}{R}\right)^4\right] \tag{9.7.18}$$

For $r_0/R = 0.55$, then $(r_0/R)^4 = 0.0913$.

The transmitted power P_t in Eq. (9.7.18) has to be adjusted as

$$P_t = P_R k\left(\frac{R^2}{4}\right) \cdot 1.0913 = P_R(M/2) \cdot 1.0913 \qquad (9.7.19)$$

Equation (9.7.19) indicates that when the condition of the lowest power per traffic channel is $0.302P_R$ at the cell site that serves mobile units within and equal to the distance r_0, $r_0 = 0.55R$, the total power at the cell site is slightly increased by 1.0913 times compared with Eq. (9.7.10). Under the adjusted transmitted power P_t in Eq. (9.7.10), the actual values of M and m are reduced:

$$M = \frac{18.87}{1.0913} = 17.3 \quad m = 13 \qquad \text{for } (C/I)_s = 0.032$$

$$M = \frac{33.7}{1.0913} = 30.0 \quad m = 25.96 \qquad \text{for } (C/I)_s = 0.1792 \quad (9.7.20)$$

Comparing Eq. (9.7.20) with Eq. (9.7.16), we find no significant change of M and m when the adjusted transmitted power is applied.

9.7.4 Comparison of Different CDMA Cases

Table 9.1 lists the performance of five different cases:

Case 1: No adjacent cell interference is considered (this is not a real case)
Case 2: No power control, adjacent cell interference is considered
Case 3: Power control with $n = 1$, adjacent cell interference is considered
Case 4: Power control with $n = 2$, adjacent cell interference is considered
Case 5: Power control with $n = 3$, adjacent cell interference is considered

Case 1 in the table is not a real case. In case 2, without power control the performance is poor. Power control schemes are used in cases 3 to 5. In these cases, to provide the minimum transmitted power at the cell site for serving those mobile units within or equal to the distance r_0, the total transmitted power at the cell site is increased (as indicated under the healing "after adjusting the transmitted power"). In comparing the number of channels per cell m among cases 3 to 5, we found that case 4 has 2 channels more than case 3 but one channel less than case 5. However, case 5 is harder to implement than case 4. One channel gain in case 5 over case 4 can be washed out in the practical situation. When we used power control schemes of $n > 3$, we found no further improvement in radio capacity. We can conclude that $n = 2$ in case 4 is the best choice.

9.8 REDUCTION OF NEAR–FAR RATIO INTERFERENCE IN CDMA

In CDMA all traffic channels share one radio channel. Therefore a strong signal received from a near-in mobile unit will mask the weak signal from a far-end mobile unit at the cell site. A power control scheme applied on the reverse link would reduce this near-far ratio interference. Then the signals received at the cell site from all the mobile units within a cell would remain at the same level. The scheme is described as follows: The power transmitted from each mobile unit has to be adjusted based on its distance from the cell site as

$$P_j = P_R \left(\frac{r_j}{R}\right)^4 \tag{9.8.1}$$

where P_R, r, and R are as given previously, and a fourth power rule is applied in Eq. (9.8.1). Neglecting the interfering signals from adjacent cell, the C/I received from a mobile unit J at the cell site can be obtained as

$$\frac{C}{I} = \frac{P_R(r_j/R)^4(r_j)^{-4}}{\sum\limits_{1}^{M-1} P_R(r_j/R)^4(r_j)^{-4}} = \frac{1}{M-1} \tag{9.8.2}$$

The C/I of Eq. (9.8.2) has to be greater than or equal to the required $(C/I)_s$:

$$\frac{C}{I} \geq \left(\frac{C}{I}\right)_s \tag{9.8.3}$$

Applying $(C/I)_s$ in Eq. (9.8.2), we obtain

$$M = 30.25 \quad m = 22.74 \quad \text{for } (C/I)_s = 0.032 \ (-15 \text{ dB})$$

$$M = 54.5 \quad m = 4.2 \quad \text{for } (C/I)_s = 0.01792 \ (-17 \text{ dB})$$

The number of channels M obtained from the reverse link is much higher than that from the forward channel, as shown in Table 9.1. This indicates that for greater radio capacity the number of channels on the forward link has to be increased. The capacity of CDMA is determined from the forward link not from the reverse link.

9.9 NATURAL ATTRIBUTES OF CDMA

There are many advantages to using CDMA in the cellular system.

TABLE 9.1 Performance of Different Power Control Schemes

Performance in Different Cases	Adjacent Cell Interfering					No Adjacent Cell Interference Considered
	No Power Control	Power Control Schemes				
	Case 2 $N = 0$	Case 3 $N = 1$	Case 4 $N = 2$	Case 5 $N = 3$		Case 1
Power control due to the distance from the cell site	P_R	$P_R(r_j/R)$	$P_R(r_j/R)^2$	$P_R(r_j/R)^3$		P_R
R_o	N/A	0.303R	0.55R	0.7R		N/A
Before adjusting the Tx power Total transmitted power at the cell site	MP_R	$P_R(2\,M/3)$	$P_R(M/2)$	$P_R(2\,M/5)$		MP_R
The $(C/I)_s$ Received at R_0	$\dfrac{1}{M-1}$	$(r_o/R)/(2\,M/3)$	$(r_o/R)^2/(M/2)$	$(r_o/R)^3/(2\,M/5)$		$\dfrac{1}{M-1}$
At R (cell boundary)	$\dfrac{1}{3.3123M-1}$	$\dfrac{1}{2.2M}$	$\dfrac{1}{1.656M}$	$\dfrac{1}{1.32M}$		$\dfrac{1}{M-1}$

M at $(C/I)_s = 0.032$	9.736	14.2	18.87	23.67	30.25
$(C/I)_s = 0.0179$	17.15	25.36	33.7	42.27	54.8
M at $(C/I)_s = 0.032$	7.32	10.67	14.19	17.8	22.74
$= 0.0179$	12.9	19	28.33	31.78	41.2

After adjusting the Tx power
Total transmitted power at the cell site

The $(C/I)_s$ received

	$P_R(2\,M/3) \times 1.0139$	$P_R(M/2) \times 1.09$	$P_R(2\,M/5) \times 1.25$
At $R \leq R_o$	$(r_o/R)^2/[(2\,M/3) \times 1.0139]$	$(r_o/R)^2/[(M/2) \times 1.09]$	$(r_o/R)^3/[(2\,M/5) \times 1.25]$
At $R > R_o$	$(r/R)^2/[(2\,M/3) \times 1.0139]$	$(r/R)^2/[(M/2) \times 1.09]$	$(r/R)^3/[(2\,M/5) \times 1.25]$
At R	$\dfrac{1}{2.23M}$	$\dfrac{1}{1.8M}$	$\dfrac{1}{1.65M}$
M at $(C/I)_s = 0.032$ $(-15\ \text{dB})$	14.2	17.3	19
at $(C/I)_s = 0.010792$ $(-17.4\ \text{dB})$	25.36	31	33.8
m at $(C/I)_s = -15\ \text{dB}$	10.67	13	14
at $(C/I)_s = -17.4\ \text{dB}$	19	23.3	25.4

1. Voice activity cycles. CDMA can take the natural shape of human conversation. The human voice activity cycle is 35%. The rest of the time we are either listening or pausing. In CDMA all the users are sharing one radio channel. When users assigned to the channel are not talking, all others on the channel benefit with less interference in a single CDMA radio channel. Thus, the voice activity cycle reduces mutual interference by 65% tripling the true channel capacity. CDMA is the only technology that takes advantage of this phenomenon. Therefore due to the voice activity cycle the radio capacity shown in Eq. (9.7.20) can be three times higher. This means that the radio capacity is about 40 channels per cell for $C/I = -15$ dB or $E_b/I_0 = 7$ dB.

2. No equalizer needed. When the transmission rate is much higher than 10 kbps in both FDMA and TDMA, an equalizer is needed to reduce the intersymbol interference caused by time delay spread. However, in CDMA only a correlator is needed at the receiver end to recover the desired signal from despreading the SS signal. The correlator is simpler to install than the equalizer.

3. One radio per site. Only one radio is needed at each site or at each sector. It saves equipment space and is easy to install.

4. No hard hand-off. Since every cell uses the same CDMA radio, the only difference is in the code sequences. There is no hand-off from one frequency to another frequency while it moves from cell to cell. However a code sequence will be changed from one cell to another cell. This is called a *soft hand-off*.

5. No guard time in CDMA. Guard time is required in TDMA between time slots. Guard time does occupy the time periods of certain bits. The waste bits could be used to improve quality performance in TDMA. In CDMA guard time does not exist.

6. Sectorization for Capacity. In FDMA and TDMA each cell is divided into sectors in order to reduce interference. As a result the trunking efficiency of dividing channels in each sector decreases. In CDMA sectorization is used to increase capacity in a way that introduces three radios in three sectors and therefore obtains three times the capacity as compared with the theoretical one radio in a cell.

7. Less fading. Less fading is observed in the wideband signal while it propagates in a mobile radio environment. It is more advantageous to use a wideband signal in urban areas than in suburban areas for reasons noted in Section 9.3.2.

8. Easy transition. In a situation where two systems, analog and CDMA, have to share the same allocated spectrum, 10% of the bandwidth (1.25 MHz) will increase twice as much ($=0.1 \times 20$) as the full bandwidth of FM radio capacity (discussed below). Since only 5% (heavy users) of the total users take more than 30% of the total traffic, the

system providers can let the heavy users exchange their analog units for the dual mode (analog/CDMA) units and convert 30% of capacity to CDMA on the first day of CDMA operations.

9. Capacity advantage. Given that a 10% of 12.5 MHz spectrum, i.e. 1.25 MHz is used to compare the capacities of three multiple access schemes; FDMA, TDMA and CDMA.

B_t = 1.25 MHz, the total bandwidth

B_{ss} = 1.25 MHz the CDMA radio channel

B_c = 30 kHz for FM (i.e. FDMA)

B_c = 30 kHz and three time slots for TDMA

the capacity of FM is,

$$\text{Total number of channels} = \frac{1.25 \times 10^6}{30 \times 10^3} = 41.67$$

Cell reuse pattern $K = 7$

$$\text{Radio capacity } m = \frac{41.67}{7} = 6 \quad \text{channels per cell}$$

and the capacity of TDMA is,

$$\text{Total number of channels} = \frac{1.25 \times 10^2}{10 \times 10^3} = 125$$

Cell reuse pattern $K = 4$ (assumed)

$$\text{Radio capacity } m_{\text{TDMA}} = \frac{125}{4} = 31.25 \quad \text{channels per cell}$$

whereas the capacity of CDMA is,

Total of channels per cell $m = 13$

Cell reuse pattern $K = 1.33$

Radio capacity from Eq. (9.7.20) for $E_b/I = 7$ dB

with voice activity cycle and sectorization,

$$m_{\text{CDMA}} = 13 \times 3 \times 3 = 120 \quad \text{channels per cell}$$

Therefore

$$m_{\text{CDMA}} = 20 \times m_{\text{FM}}$$

$$= 4 \times m_{\text{TDMA}}$$

10. No frequency management or assignment needed. In FDMA and TDMA the frequency management is always a critical task. Since there is only one common radio channel in CDMA, no frequency management is needed. Also, to reduce real time interference, a dynamic frequency assignment has to be implemented in TDMA and FDMA. This calls for a linear broadband power amplifier which is hard to develop. CDMA does not need a dynamic frequency assignment.

11. Soft capacity. In CDMA all the traffic channels share one CDMA radio channel. We can add an additional user so that the voice quality is just slightly degraded as compared to that of the normal 40-channel cell. The difference in dB is only 10 log 41/40 which is 0.24 dB down in C/I ratio.

12. Coexistance. Both analog and CDMA systems can operate in two different spectra. CDMA only needs 10% of bandwidth to general 200% of capacity, so there is no interference between the two systems.

13. For microcell and in-building systems. CDMA is a natural waveform suitable for microcell and in-building because it is suceptable to noise and interference.

This summary of CDMA highlights the potential of increasing capacity in future cellular communications. Two papers[11,12] analyzed CDMA in depth; other interesting literature on CDMA can be found in references 13–21. The material of this chapter is mainly from References 20 and 21.

REFERENCES

1. Lee, W. C. Y., *Mobile Cellular Telecommunications Systems* (McGraw-Hill, 1989): ch. 4.

2. Lee, W. C. Y., *Mobile Communications Engineering* (McGraw-Hill, 1982): 340–399.

3. Proakis, John G., "Adaptive Equalization for a TDMA Digital Mobile Radio," *IEEE Trans. Veh. Tech.*: 333–341.

4. Crozier, S. N., D. D. Falconer, and S. Mahmond, "Short-Block Equalization Techniques Employing Channel Estimation for Fading Time-Dispersive Channels," *Proc. IEEE Vehicular Technology Conference* (San Francisco, CA, 1989): 142–146.

5. Monsen, P., "Theoretical and Measured Performance of a DEF Modem on a Fading Multipath Channel," *IEEE Trans. Commun.* COM-25, (Oct. 1977): 1144–1153.

6. Lee, W. C. Y., *Mobile Communications Engineering* (McGraw-Hill, 1982): 163.

7. Schwartz, M., W. R. Bennett, and S. Stein, *Communications Systems and Techniques* (McGraw-Hill, 1966): 561.

8. Sklar, B., *Digital Communications, Fundamentals and Applications* (Prentice Hall, 1988): 546.

9. Lee, W. C. Y., "Spectrum Efficiency in Cellular," *IEEE Trans. on Veh. Tech.* 38 (May 1989): 69–75.

10. PacTel Cellular & Qualcomm, *CDMA Cellular—The Next Generation* (Pamphlet distributed at CDMA demonstration, Qualcomm, San Diego, CA, Oct. 20–Nov. 7, 1989).

11. Gilhousen, K. S., I. M. Jacobs, R. Padovani, A. J. Viterbi, L. A. Weaver, and C. E. Wheatley, "On the Capacity of a Cellular CDMA System," *IEEE Trans. Veh. Tech.*: Vol. 40 May, 1991: 303–312.

12. Pickholtz, R. L., L. B. Milstein, and D. L. Schilling, "Spread Spectrum for Mobile Communications," *IEEE Trans. Veh. Tech.* Vol. 40, May 1991: 313–322.

13. Viterbi, A. J., "When Not to Spread Spectrum—A Sequel," *IEEE Commun.* 23 (Apr. 1985): 12–17.

14. Milstein, L. B., R. L. Pickholtz, and D. L. Schilling, "Optimization of the Processing Gain of an FSK-FH System," *IEEE Trans. Commun.* COM-28 (July 1980): 1062–1079.

15. Huth, G. K., "Optimization of Coded Spread Spectrum System Performance," *IEEE Trans. Commun.* COM-25 (Aug. 1977): 763–770.

16. Simon, M. K., J. K. Omira, R. A. Scholtz, and B. K. Levin, *Spread Spectrum Communications,* vol. 2 (Rockville, MD: Computer Science Press, 1985).

17. Pickholtz, R. L., D. L. Schilling, and L. B. Milstein, "Theory of Spread-Spectrum Communications—A Tutorial," *IEEE Trans. Commun.* COM-30 (May 1982): 855–884.

18. Scholtz, R. A., "The origins of Spread Spectrum Communications," *IEEE Trans. Commun.* COM-30 (May 1982): 882–854.

19. Viterbi, A. J., "Spread Spectrum Communications—Myths and Realities," *IEEE Commun.* (May 1979): 11–18.

20. Lee, W. C. Y., "Radio Access Technology—CDMA/Spread Spectrum," seminar notes used for the one-day seminar of IEEE San Francisco Section/Pacific Bell at San Ramon, Ca. on Jan 23, 1990, and the seminar of IEEE New Jersey Section/ Rutgers Univ. WINLAB at Piscataway, N.J. on April 25, 1990.

21. Lee, W. C. Y., "Overview of Cellular CDMA" IEEE Trans. on Vehicular Technology, Vol. 40, May 1991, pp 291–302.

PROBLEMS

9.1 If the carrier frequency is doubled, how much additional propagation loss would result from the increase of frequency?

9.2 If the bandwidth of a broadband signal is one-third of the carrier frequency, what could be the additional propagation loss?

9.3 Given a bandwidth of 1.23 MHz and a time delay spread in a suburban area of 0.5 μs, what is the effective number M of diversity branches affected by the received signal compared with a 30-kHz bandwidth signal?

9.4   A direct-sequence signal bandwidth spreads from 10 kHz to 10 MHz. What is the processing gain? Is the processing gain a real gain, or can it only be realized when the interference appears?

9.5   In a newly designed TDMA system based on the accepted voice quality the required C/I is 14 dB, and the channel bandwidth is 30 kHz with two time slots. What is the radio capacity of this system?

9.6   In a CDMA system the required $E_b/I_0 = 7$ dB. The processing gain is 22 dB. If there is no forward power control, how many traffic channels are available?

9.7   Compare the radio capacities of two systems, both occupying the same allocated band. One system has a required $C/I \geq 22$ dB and a channel bandwidth of 16 kHz, and the other system has a required $C/I \geq 15$ dB and a channel bandwidth of 35 kHz.

9.8   A band of 1.23 MHz is used for a 30-kHz channel analog system with $(C/I)_s \geq 18$ dB. The same band is used for a CDMA system with $E_b/I_0 \geq 8$ dB and a processing gain of 22 dB. The forward power control is not applied in the CDMA system. If both systems use the three-sector cells, what is the ratio of two radio capacities between these two systems?

9.9   If the time delay in an environment is very large, say, $\Delta = 30$ μs, and the bandwidth of a narrowband signal is $B = 30$ kHz, are the wideband effective diversity branches approximated in Eq. (9.3.15) still applicable? Under what condition is Eq. (9.3.15) valid?

9.10   Will the effective diversity branches in a CDMA system increase more in urban areas than in suburban areas?

10

MICROCELL SYSTEMS

10.1 DESIGN OF A CONVENTIONAL CELLULAR SYSTEM

The advanced mobile phone service (AMPS) cellular system at 850 MHz, as used in North America, is a high capacity system. Its spectrum utilization is based on the frequency reuse concept, whereby a frequency can be reused repeatedly in different geographical locations. Various locations using the same frequencies are called co-channel cells described in Section 5.5.

The minimal separation (D_s) required between two nearby co-channel cells is based on specifying a tolerable co-channel interference, which is measured by a required carrier-to-interference ratio $(C/I)_s$. The $(C/I)_s$ ratio also is a function of the minimum acceptable voice quality of the system. In an AMPS, $(C/I)_s$ is equal to about 18 dB (the level at which 75% of the users grade the system "good" or "excellent"), and the minimal required separation D_s based on $(C/I)_s = 18$ dB is about $4.6R$, where R is the radius of the cell. In a cellular system the number of cells K in a cell reuse pattern is a function of the co-channel separation D_s. For $D_s = 4.6R$, we have $K = 7$. This means that a cluster of seven cells can share the entire allocated spectrum. In each of the two bands allocated for cellular, there are 395 voice channels; each cell can have 57 channels on average. The detail is described in Section 5.6.

By 1991 the conventional cellular systems in use since 1984 began to reach their capacity in the larger markets. To increase system capacity, we may take one of three approaches based on the co-channel interference reduction factor (CIRF), q_s, which is defined as[1]

$$q_s = \frac{D_s}{R} = \sqrt{3K} \qquad (10.1.1)$$

where D_s is the minimum required distance between any two co-channel cells in a cellular system (see Fig. 10.1A) corresponding to the required carrier-to-interference ratio (C/I) received at both the cell site and the mobile unit in a cell, R is the cell radius, K is the number of cells in a cell reuse pattern.

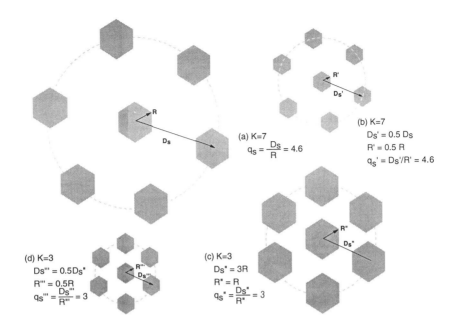

Figure 10.1. Four CIRF configurations.

K is also called the *cell reuse factor*. The first two approaches for increasing capacity are conventional approaches; the third one is the new approach. Equation (10.1.1) is derived from an idealized hexagonal cell layout and is commonly used. The three approaches for increasing capacity can be stated as follows:

1. Split the cells. Capacity can be increased by reducing R but keeping q_s unchanged as in Eq. (10.1.1) (see Fig. 10.1(b)), that is, by rescaling the system. When R is smaller than one mile or one kilometer, the cell is commonly called a *microcell*. In a first-order approximation every time R is reduced by one-half, capacity increases by four. The measurement of capacity in this approach is the number of channels per square kilometer. The cell-splitting approach leads to an increase in radio capacity.[2] Splitting cells is independent of the system's scale, that is, the value of q_s remains unchanged. This approach can be used in any analog or digital system. A conventional microcell takes this approach.

2. Reduce the cell reuse factor (also called *a reduction of the required D/R approach*). We can increase capacity by determining methods by which D can be reduced, that is, by forming a new configuration but keeping R unchanged in Eq. (10.1.1) (see Fig. 10.1(c)). As a result q_s is reduced, as is the cell reuse factor K, as is clear from Eq. (10.1.1). The value of D_s, however, is a function of the required $(C/I)_s$. For example, if a new cellular system can achieve a frequency reuse factor

of $K = 3$, then the capacity of the new system can be obtained by comparing it with the AMPS system of $K = 7$. Since K is reduced from $K = 7$ to $K = 3$, the capacity is increased by $7/3 = 2.33$ times. The measurement of radio capacity in this approach is the number of channels per cell:

$$m = \frac{\text{total voice channels}}{K} \qquad (10.1.2)$$

The reduction of the cell reuse factor approach would increase radio capacity m, as is clear from Eq. (10.1.2)

In the past, sectorization was used to reduce the value of K in an analog system. When the co-channel interference in a cell increases, either a three-sector or six-sector cell configuration should be used in order not to expand the required co-channel cell separation D_s. In other words, with a given interference, the sectorization seems to be able to reduce the value of D_s. However, when we take a good look at sectorization, the method of assigning a set of frequency channels in each sector is the same as if assigned in a cell. The hand offs occur as the vehicle passes among the sectors, the same as among cells. Therefore, if in a $K = 7$ system of three-sector cells, the number of channels per sector (assuming a total of 395 channels) is

$$\frac{395}{7 \times 3} = 19 \qquad \text{channels per sectors}$$

in a $K = 4$ system of six-sector cells the number of channels per sector is

$$\frac{395}{4 \times 6} = 16 \qquad \text{channels per sectors}$$

We see from these two values that there is not much difference in radio capacity between the two cellular configurations. Further capacity can be obtained by using the cell-splitting cells approach so that the size of each sector is reduced. The problem for the sectorization is that the trunking efficiency of the utilized channels decreases. The cell usage with the same number of channels in an omni cell is much higher than that of a sector cell. Therefore sectorization is not an effective method of reducing q_s.

3. Reduce the required D/R by a new microcell approach. The schematic of the new microcell system is shown in Fig. 10.1D. In this case not only the cell radius reduces, but also the CIRF reduces. Furthermore there is no degradation in trunking efficiency; it is a true $K = 3$ system. The advantages of this system include both reduction of co-channel

interference and confining the co-channel interference relative to the signal to a small area. It will be discussed in detail in the next section.

10.2 DESCRIPTION OF NEW MICROCELL SYSTEM DESIGN

Generally the new microcell consists of three zones, as shown in Figs. 10.2 and 10.3. (More than three zones can be incorporated if needed.) Each zone has a zone site, and one of the three zone sites usually is colocated with a base site. All radio transmitters and the receivers that serve a microcell are installed at the base site. Every zone site physically shares the same radio equipment installed at the base. To serve a vehicle from a zone site, an 800-MHz cellular signal can be converted up to a microwave or optical signal at the base and then converted down back to the 800-MHz signal at the zone site to serve the vehicle at that zone as if the vehicle were located at the base.

Conversely, after boasting with a low-noise amplifier at a zone site, the received cellular signal can be up-converted to either a microwave or optical signal and then down-converted at 800 MHz at the base. In this case the zone site only requires an up/down converter, power amplifier and a low-noise broadband pre-amplifier which is easy to install because of the small size and the light weight of the zone-site apparatus.

10.2.1 Signal Coming from Mobile Unit

A mobile unit driving in a microcell sends a signal. Each zone site receives the signal and passes it through its up/down converter, up-converting the signal and sending it through the microwave or optical signal medium and then down-converting the signal at the base site. Thus the mobile signals received from all zones are sent back to the base site. A zone selector located at the base site is used to select a suitable zone to serve the mobile unit, choosing the zone with the strongest signal strength. Then the base site delivers the cellular signal to the zone site through its up-converting processing.

10.2.2 Signal Coming from Base Site

The suitable zone site receives the cellular signal from the base site through a down-converting process and transmits to the mobile unit after amplification. Therefore, although the receivers at three zones are all active, only the transmitter of one zone is active in that particular frequency to serve that particular mobile unit. When the mobile unit is moved from zone to zone, the assigned channel frequency remains unchanged. The zone selector at the base site simple shifts the transmitting signal (base-to-mobile) from one zone to another zone according to the mobile unit's location. Only one active base-transmitting zone at one time serves a vehicle (one assigned frequency) in a cell. As a result, no hand off action is needed when the mobile unit is entering

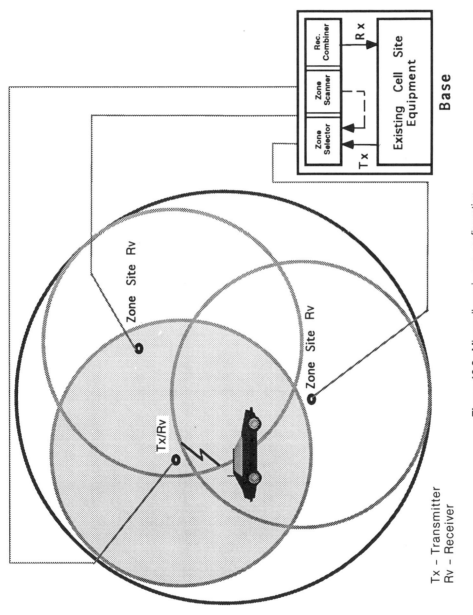

Base

Tx – Transmitter
Rv – Receiver

Figure 10.2. Microcell omni-zone configuration.

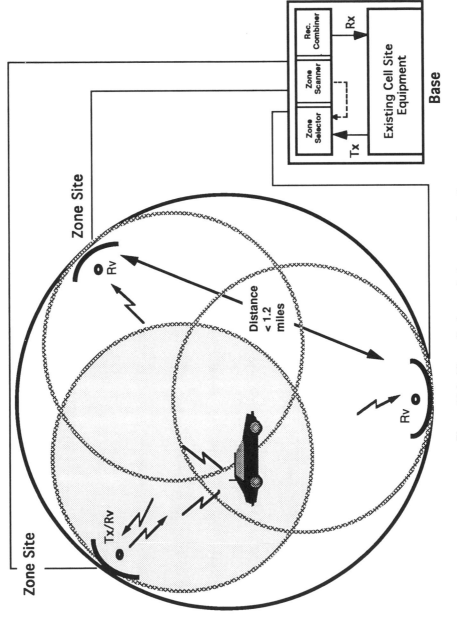

Figure 10.3. Microcell edge-excited zone configuration.

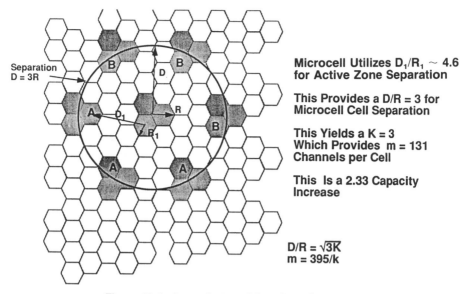

Microcell Utilizes $D_1/R_1 \sim 4.6$ for Active Zone Separation

This Provides a $D/R = 3$ for Microcell Cell Separation

This Yields a $K = 3$ Which Provides $m = 131$ Channels per Cell

This Is a 2.33 Capacity Increase

$D/R = \sqrt{3K}$

$m = 395/k$

Figure 10.4. An application of the microcell system.

a new active zone. A microcell can handle 60 frequencies assigned to 60 vehicles. On average, each zone handles 20 mobile calls associated with 20 frequencies at a time.

10.3   ANALYSIS OF CAPACITY AND VOICE QUALITY

The new microcell system can be implemented in three ways: by a selective omni-zone approach, by a selective edge-excited approach, or by a nonselective edge-excited approach or by a nonselective edge-excited approach.

10.3.1   Selective Omni-Zone Approach

We can place a zone site at the center of each zone as shown in Fig. 10.2. The transmit power of each zone site would be center excited. We would calculate the C/I ratio from the new system shown in Fig. 10.4.

To prove that capacity has increased and that voice quality in this new microcell system has improved, as shown in Fig. 10.1(d), we calculate the co-channel interference reduction factor (CIRF) q_s, which is a key element in the design of a cellular system. In the conventional microcell system, q_s is used for measuring both the voice quality and the capacity because they are related. In this microcell system, on the other hand, there are two CIRFs to be considered because voice quality and the capacity are measured differently. One CIRF q_{s1} is used to measure the voice quality, and another CIRF q_{s2}

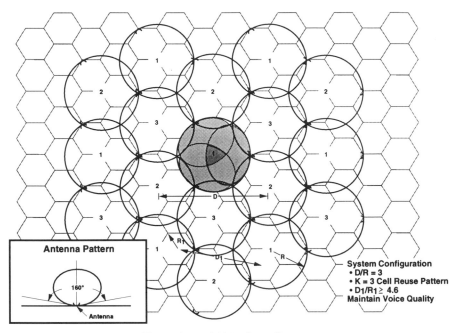

Figure 10.5. A K-3 microcell system.

is used to measure the radio capacity. The microcell system is shown in Fig. 10.5.

CIRF between Two Active Base Transmitting Co-Channel Zones

This CIRF has a new q_{s1} defined as $q_{s1} = D_1/R_1$, where D_1 is the distance between one active zone in one microcell and the corresponding active zone in the other microcell, as shown in Fig. 10.4. R_1 is the radius of each zone. The real coverage area of each zone is used for estimating interference. Therefore the radius R_1 of a real coverage area is used to confine the zone area.

There are many values of q_{s1}, depending on which two active co-channel zones are considered. Among them, the two closest co-channel active zones are the worst to be used for measuring CIRF q_{s1}. As we know, in an AMPS system, C/I has to be 18 dB, implying that q_s must be 4.6 in order to maintain an acceptable voice quality when using 30 kHz analog FM radios. In the AMPS system the earlier simulation shown that $q_{s1} = 4.6$ was adequate for omni-directional cells.[5]

When the cell site antenna height is normally 100 to 150 ft high, and the ground is not flat, the co-channel interference received on the reverse link (mobile-to-base) is larger than expected. Therefore a sectorization architecture was introduced for the macrocells. In a microcell system the microcell antenna height is always lower than 100 ft, normally 40 to 50 ft, and generally the ground in a small area around the antenna is flat. Under this condition the co-channel interference on the reverse link is reduced, and the sectorization arrangement becomes unnecessary for $K = 7$ system configuration. This is supported by measured data. Since the same radios are used in the microcell system, q_{s1} has to be at least the same as 4.6 in order to be back on the $K = 7$ configuration.

By construction it is shown that the q_{s1} of the two closest co-channel active zones in their corresponding microcells is 4.6, as shown in Fig. 10.4. In this microcell normally the q_{s1} between any two active zones in two corresponding co-channel microcells is always equal or greater than 4.6, as shown in Fig. 10.4. The voice quality in this microcell system based on $q_{s1} \geq 4.6$ has been proved to be equal or better than the voice quality of AMPS system. This q_{s1} is used to measure the voice quality only in the new microcell system.

CIRF between Two Co-Channel Microcells

Radio capacity depends on the separation of two neighboring co-channel cells. In the microcell system the CIRF q_{s2} is defined as $q_{s2} = D/R$, where D is the distance between two co-channel microcells and R is the microcell radius (see Fig. 10.4). In this case $q_{s2} = D/R = 3$ is equivalent to $K = 3$, shown in Eq. (10.1.1). The three zones per microcell and the $K = 3$ system is illustrated in Fig. 10.5. Thus the entire cell is covered. Since K is reduced from $K = 7$ of the AMPS system to $K = 3$, the microcell system has increased $7/3 = 2.33$ times, as is shown by Eq. (10.1.2). Therefore q_{s2} is used to measure the capacity.

The frequency assignment in a $K = 3$ system is illustrated in Fig. 10.5. The total allocated 395 channels can be divided into three groups. The first group consists of channels 1, 4, 7, 10, etc. The second group consists of channels 2, 5, 8, 11, etc. The channels of the third group are 3, 6, 9, 12, etc. As the figure shows, each group will be assigned to each cell according to the cell number.

Improvement of Carrier-to-Interference Ratio

The $q_s = D_s/R = 4.6$ in an AMPS system has to meet two requirements: all co-channel cells must be a distance of $4.6R$ away from the serving cell, and the value of $q_s = 4.6$ has to be based on $C/I = 18$ dB, where the interference is received from six co-channels cells at the first tier (see Fig. 10.1(a)).[3] The C/I of 18 dB at RF averaged over Rayleigh fading provides a good, and even excellent, signal to the user in an analog system.

In a microcell system the separation D_1 of any two nearest co-channel zones between two active zones in two corresponding microcells is $4.6R_1$, as shown in Fig. 10.4. All other co-channel zones in their microcells are separated further than $4.6R_1$. In the worst case scenario we can choose between an active zone in the center cell and individual co-channel zones in six corre-

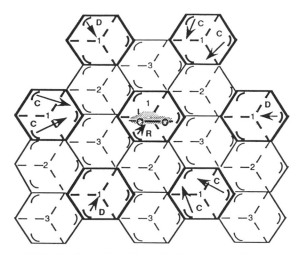

Figure 10.6. Configuration of the selective edge-excited zone cells.

sponding microcells, and calculate the C/I received at that zone in the center cell of the microcell system, as shown in the following (see Fig. 10.4):

$$\frac{C}{I} = \frac{R_1^{-4}}{\sum_{i=1}^{6} D_1^{-4}} = \frac{R_1^{-4}}{3(4.6R_1)^{-4} \cdot 3(5.75R_1)^{-4}}$$

$$= 105 \quad (= 20 \text{ dB}) \tag{10.3.1}$$

Equation (10.3.1) indicates that according to the worst case the C/I in a microcell system is 2 dB better than that of an AMPS system. In the worst case of Eq. (10.3.1) where all the co-channel zones are in either A or B zones, the C/I can be even better if all the active co-channel zones are other than A or B zones. The C/I in the normal case then is always greater than 20 dB. The $C/I \geq 20$ dB is at least 2 dB better than the required $(C/I)_s$ of the AMPS system. This proves that the voice quality of the microcell system is always better than that of the AMPS system. One remark is that this calculation is based on the signal coverage in each zone regardless of the type of antenna. This is because the shape of the coverage takes care of the antenna pattern.

10.3.2 Selective Edge-Excited Zone Approach

In the edge-excited zone approach all the zone sites are moved from the center to the edges of the zones on the perimeter of the cell boundary (see Fig. 10.3). The calculation of C/I in this edge-excited zone approach is based on the $K = 3$ configuration shown in Fig. 10.6. The center cell is the serving cell.

One selected zone is serving the mobile call. The center of the cell is the weak spot for receiving the signal from the zone site. There are six interfering cells around the serving cell. Among the six interfering cells, three cells may have two zones sites for interfering with the mobile call in the center cell. The other three cells have only one zone site in each cell to interfere with the mobile call. Since in a cell only one zone site is turned on at a time on any one frequency, the probability of interfering with the mobile call from each interfering zone site is one-third. The distance from each interfering zone site to the vehicle can be obtained from Fig. 10.6. Each of the three interfering cells has two C zones that can interfere with the mobile call. However, the probability of this occurring is only two-thirds. The probability that the remaining three interfering cells, each having one D zone that can interfere with the mobile call, is one-third. The C/I ratio is obtained at the vehicle from six co-channel cells (denoted "1") as shown in Fig. 10.6.

$$\frac{C}{I} = \frac{R^{-4}}{\underbrace{3[^2/_3(3.6R)^{-4}]}_{C \text{ zones}} + \underbrace{3[^1/_3(4R)^{-4}]}_{D \text{ zones}}} = 63 \quad (= 18 \text{ dB}) \qquad (10.3.3)$$

In the edge-excited zone approach the C/I can still be maintained at 18 dB, which is the level for acceptable voice quality. The $K = 3$ configuration shown in Fig. 10.6 confirms the increase of radio capacity. Still the omni-zone approach provides the best voice quality.

10.3.3   Nonselective Edge-Excited Zone Approach

Situations where all the zones have to be turned on call for a nonselective edge-excited zone configuration. In a nonselective edge-excited zone configuration, all the cells are treated as omni-cells because all zone sites are transmitting concurrently. In an analog system the regular center-excited omni-cells require the co-channel interference reduction factor q which is equivalent to $D/R = 4.6$, as mentioned previously.

In the edge-excited zone cells the D_1/R_1 has to be 4.6 in order to maintain the voice quality. Recall that D_1 is the co-channel zone separation and that R_1 is the distance from the zone transmitter to the zone boundary equal to the cell radius. The new q $(=D_1/R_1)$ is 3.6, as shown in Fig. 10.7. The frequency reuse factor K then becomes

$$K = \frac{(q)^2}{3} = \frac{(3.6)^2}{3} = 4.32 \approx 4 \qquad (10.3.4)$$

This result proves that the edge-excited approach can increase the radio capacity by $7/4 = 1.75$ times.

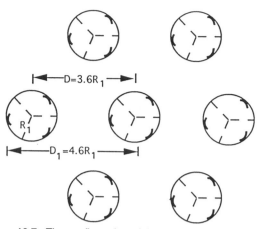

Figure 10.7. The configuration of the non-selective zone cells.

10.3.4 Summary

Radio capacity can be increased by 2.33 times if a selective zone approach is used either in an omni-zone configuration or in an edge-excited zone configuration. The radio capacity can be increased by 1.75 times if a nonselective zone approach is used. The efficiency of using this microcell configuration reaches a maximum because $K = 3$ is the smallest number in a frequency reuse system. For an analog system with a nonselective zone configuration, radio capacity can be increased by 1.75 times. For microcells to CDMA systems, the nonselective zone configuration can be used to further reduce the interference.

10.4 REDUCTION OF HAND-OFFS

Technically *hand-offs* means to hand off one frequency to another frequency while a vehicle enters a new cell or a new sector. Within each microcell no hand-offs from zone to zone are needed; zone-to-zone switchings are handled by a zone selector. The active zone follows the mobile unit as it moves from one zone to another. The channel frequency assigned to the mobile unit remains unchanged.

In this section we roughly estimate how many hand-offs can be eliminated relative to a microcell plan in which three zones are used. In a regular cell there are three sectors. The car can move in any one of three directions, as shown in Fig. 10.8. When a car moves through the other two sectors, it needs the hand-offs operation. As it enters and as it moves out of a cell, a hand-off occurs. In a microcell a car moving to either of the other two zones does not need a hand off. A hand-off only occurs when the car moves in or out of the cell. Since the shape of a zone is based on the hexagonal cell, the zone

❏ **Reducing up to One-half the Number of Hand-offs (HO)**

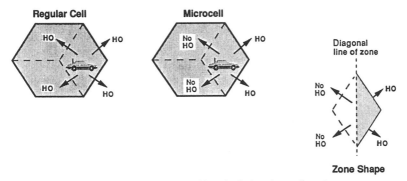

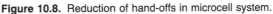

Figure 10.8. Reduction of hand-offs in microcell system.

❏ **Three Bottlenecks**

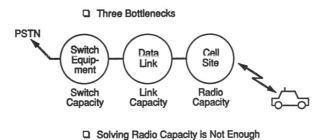

❏ **Solving Radio Capacity is Not Enough**

Figure 10.9. Cellular systems.

is diamond shaped. Symmetry to the diagonal line of travel can be observed in that at the left side of the diagonal line no hand-offs are required and on the right side hand-offs are needed. Consequently we estimate that only one-half of the hand-offs required in a regular cell configuration will occur in a microcell configuration. This reduction in number of hand-offs in the microcell system makes a great contribution to the capacity of the system.

10.5 SYSTEM CAPACITY

In any cellular design the overall capacity of each system can differ from system to system. System capacity may be capped by three limiting elements: radio capacity, control link capacity, and switch capacity (see Fig. 10.9).

Radio capacity is the element most often addressed in the literature.[3] Link capacity, or switch capacity are often overlooked in measurements of system capacity. Control link capacity measures the capacity of transmitting data over a linkage between the cell site and the switches. If the number of microwave linkages or T1 carrier lines are not sufficient, a bottleneck results.

Switch capacity measures the traffic capacity at the switching office. Again, if the switch is not big enough to handle the radio capacity, a bottleneck problem occurs. Among the three capacities, the weakest is used to gauge the system's capacity. Therefore improving the radio capacity in the system is not enough. Improving system capacity requires upgrading the lowest capacity of the three. With this in mind, every system operator should be aware that radio capacity is not the entire problem nor the entire solution.

Because in a microcell system design fewer intercell hand-offs are needed as compared to a regular system, both the switching load and the control link load are cut roughly in half, leaving two times the load to be handled by the present capacity. The easing of half the load means that twice the load can be added back onto the system. This roughly two times (2.33 times to be exact) radio capacity is exactly what the microcell system will offer without changing the present switching equipment.

10.6 ATTRIBUTES OF MICROCELL[7,8]

The new microcell design has many attractive features:

1. Increased system capacity. Based on the cell reuse pattern (reduced from $K = 7$ to $K = 3$), it offers 2.33 times the AMPS system capacity.
2. Voice quality improved. The voice quality of the microcell system always is better than the voice quality of AMPS.
3. Interference reduced. (a) In the selective omni-zone configuration, the power of three zone sites is reduced as compared to the power of the center site in a cell. In the selective edge-excited zone configuration all antennas in a cell face each other so that the interference signal has to cross the cell before interfering with a neighboring cell. In both configurations the coverage is only served in one active zone; therefore the interference is very weak compared with the interference from a transmitter from the center of a regular cell. (b) The three zone sites receiving the mobile signal simultaneously form a three-branch path diversity scheme which is suitable for low power portable units. The diversity scheme increases the probability of a signal being received at the base. (c) The microcell system offers the best arrangement for controlling interference. The active zone follows the vehicles, and cell coverage can be easily engineered by using three different transmit powers at three zone sites.
4. Adaptability. This microcell design can be added to any vendors' system without modifying the vendor's hardware or software.
5. Size of the zone apparatus. The zone up/down converters are small, so they can be mounted on the side of a building or on poles. The system therefore is a PCS (personal communications service) type that enables

tight control of interference. It is easy to remount from pole to pole when the signal coverage requirement is changed.

6. The microcell is an intelligent cell. The new microcell knows where to locate the mobile unit in a particular zone of the cell and deliver the power to that zone. Since the signal power is reduced, the two microcells can be closer, therefore increasing the capacity.

REFERENCES

1. Lee, W. C. Y., *Mobile Cellular Telecommunications Systems* (McGraw-Hill, 1989): 57.
2. MacDonald, V. H., "The Cellular Concept," *Bell Sys. Tech. J.* 58 (Jan. 1979): 15.
3. Lee, W. C. Y., "Spectrum Efficiency in Cellular," *IEEE Trans. Veh. Tech.* (May 1989): 69–75.
4. Lee, W. C. Y., "Cellular Telephone System" (U.S. Patent 4,932,049, June 5, 1990).
5. Lee, W. C. Y., "Microcell System for Cellular Telephone System" (U.S. Patent 5,067,147, Nov. 19, 1991).
6. Ott, Gary D., "Vehicle Location in Cellular a Mobile Radio Systems," *IEEE Trans. Veh. Tech.,* VT-26 (Feb. 1977): 43–46.
7. Lee, W. C. Y., "Microcell Architecture—Smaller Cells for Greater Performance" *IEEE Communications Magazine* 29 (Nov. 1991): 19–23.
8. Lee, W. C. Y., "An Innovative Microcell System," *Cellular Business* (Dec. 1991): 42–44.

PROBLEMS

10.1 What is the difference between the conventional microcell system and the new microcell system?

10.2 The new microcell system consists of three zones. Based on the selective omni-zone approach shown in Fig. 10.2, find the C/I in the system.

10.3 In the new microcell system why can both voice quality and the frequency-reuse factor be improved at the same time?

10.4 Why is the microcell suitable for portable units?

10.5 What is the microcell's signal delivery system?

10.6 In the selective edge-excited zone approach shown in Fig. 10.3, find the C/I in the system.

10.7 Will the radio capacity be constrained if the overall switching capacity is limited? Will the new microcell system be constrained?

10.8 The signal strength P_r is received from an edge-excited zone site P_{t1} at the boundary of the zone with the distance d. If the site is replaced by the omni-zone site, the zone radius $R = d/2$. The same signal strength P_r will be received at the boundary R. What is the power relationship between P_{t1} and P_{t2}. At point A (a distance $2d$ away from P_{t1} and $3R$ from P_{t2}), which site (P_{t1} or P_{t2}) will interfere more than the other. (See Fig. P10.1.)

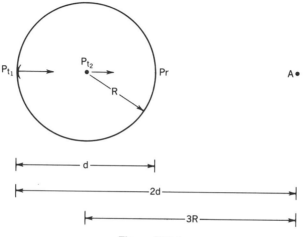

Figure P10.1

10.9 Why can the new microcell system be said to be transparent to any cell site equipment?

10.10 Given 60 channels allocated to either the microcell system or the macrocell system, calculate the radio capacities of both systems, including trunking efficiency.

11

MISCELLANEOUS RELATED SYSTEMS

11.1 PCS (PERSONAL COMMUNICATIONS SERVICE)

Over the last decade cellular systems providers have convinced the FCC that cellular systems promote high spectrum efficiencies. In addition the general public has become interested in cellular systems, and as a result there has been an unexpected increase in the rate of subscription even though marketing predictions at the time had taken a pessimistic view of cellular's expansion. Now the entire communications industry has begun to recognize that mobile cellular has a strong business potential, and everyone is eager to get established in the field. In mid-1991 the Federal Communications Commission released a Notice of Inquire (NOI) into how to develop and implement a new personal communications service (PCS). Since then the wireless communications community has begun to take even greater action.

11.1.1 Requirements of PCS

The requirements of PCS can be listed as follows:

PCS Subscriber Unit
The actual definition of a personal communications service is not clear. What is evident, however, is that in the 1990s, with lifestyle changes and people on the move, any subscriber unit for a PCS has to be carried easily, must be user-friendly, and should be operable anywhere in the world.

Mobility
Traditional marketing analysis based on market segments is no longer applicable for developing personal communications services. PCS is a dynamically mobile market: A subscriber can be in the office, at home, or on the street— anywhere, and at any time. The PCS subscriber unit carried by a user should work for this individual at all times. There can be no distinction between

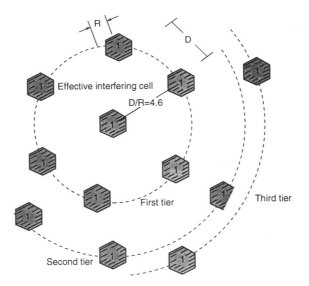

Figure 11.1. Six effective interfering cells of cell 1.

office use and home use. Therefore traditional market segments or demographic data would become irrelevant for studying people on the move.

Spectrum Efficiency

In choosing a PCS system, one should consider its spectrum efficiency. The spectrum is a precious and limited natural resource. A PCS system needs to serve an enormous number of customers in a limited, allocated spectrum. If we forecast handling 10,000 calls per busy hour per square kilometer (km²), then a PCS system needs roughly 350 radio channels in that 1 km². Can the industry design a system with that specification? In determining this, we must answer two questions: Do we have enough allocated spectrum? Will the system we choose be spectrum-efficient?

Radio channels cannot be deployed in the field in the same way as wireless channels. The co-channels and adjacent channels must be located far apart in order to reduce their mutual interference. To use the same channel again and again in order to increase channel capacity in cellular systems, the minimum required separation D is specified for two co-channel cells by how many cell radii R away, or by the D/R ratio. In today's analog systems (see Fig. 11.1), the co-channel separation D is 4.6 times R, or $D/R = 4.6$. The number is based on system performance for a desired voice quality. For achieving higher system capacity, a new system with a smaller required D/R should be chosen. From this required D/R, the number of channels per cell is obtained. Then the number of channels per square kilometer can be figured out. The larger the number of channels per square kilometer, the higher will be the

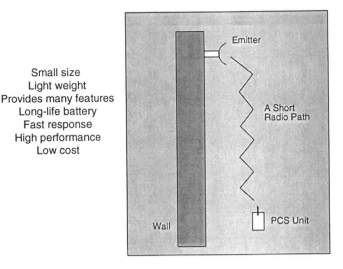

Small size
Light weight
Provides many features
Long-life battery
Fast response
High performance
Low cost

Emitter

A Short
Radio Path

PCS Unit

Wall

Figure 11.2. PCS subscriber unit.

system capacity. Increasing the system capacity means enhancing spectrum efficiency.

One-Unit Concept

Ideally the PCS subscriber appliance should be designed as either a one-for-all unit (i.e., it works in one system) or an all-in-one unit (i.e., it works in multiple systems). A one-system unit would be in the interest of end users; the one unit would contain low performance or high performance options, all capable of operating within the same system.

Why the one-system unit? Because we must place the interests of the end user above those of the manufacturing companies or systems providers. Ultimately, multiple systems—one for in-house use, one for vehicles, one for urban travel, etc.—would become burdensome to the end user.

Since the trend in communications technology is toward an increase in mobility, it is not practical for personal communications to maintain separate communications access lines for different systems. Any PCS subscriber unit should be small, lightweight, provide many features, and have a long-life battery. Today's technology can fulfill these requirements if the PCS infrastructure can take more of the burden and deliver the signal through a shorter radio path to the unit (Fig. 11.2). The cellular system is being designed for this purpose.

Intelligent Network

Managing a PCS system requires an intelligent network (IN) that has a centralized database and provides customer services through a service control point (SCP). The IN will only recognize the subscriber's identification num-

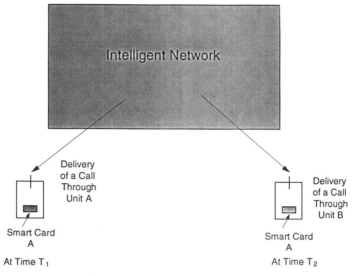

Figure 11.3. Intelligent network.

ber, not the subscriber's unit. For example, a call can be delivered to a subscriber anywhere and through any subscriber unit. The network can track a subscriber's locations and be updated by the subscriber's assigned personal ID numbers, regardless of the units they use. Sometimes the assigned personal number will be provided on a smart card. The IN will recognize the smart card, but not the unit (Fig. 11.3).

The smart card will be able to be used for any PCS subscriber unit. As soon as the smart card is plugged into a unit, that unit becomes that individual's subscriber unit. The network will update the information in the database and track on the new one. All the features and billings will be carried over to the new unit.

Multiple Access

To serve a call, the system provider will either assign a radio channel to a user (this is called *frequency division multiple access,* or FDMA), a time slot to a user (*time division multiple access,* or TDMA), or a code sequence to a user (*code division multiple access,* or CDMA). In satellite communications, all three multiple access methods can be used. But before choosing one of the three multiple access schemes for PCS, we have to understand the environment in which the PCSs will operate.

11.1.2 PCS Environment

To design a system for the PCS subscriber unit, system providers must understand the PCS environment. It is a human-made environment that includes cities, buildings, streets, highways, vehicles, and traffic lights. Because of these factors, propagation properties such as path loss, fading, and time delay spread of PCS waves are different from those in other media. Human-made noise heavily dominates the PCS environment. In a human-made environment, the excessive path loss, the sever signal fading, and the long time delay spread due to the multipath radio phenomena will make the PCS system a difficult one to build.

From recent studies, the hot issues concern wideband signal application, CDMA, and how to share the existing allocated spectrum. It seems that the wideband signal has some advantages for implementation in the PCS environment:

- Less signal fading is received, when compared to narrowband signals.
- There is less signal fading received in urban than in suburban regions.
- The wideband signal is suitable for CDMA.
- It is the output of spread spectrum modulation.

For spectrum efficiency, the study of sharing with existing allocated spectrum also can be a challenging task.

11.1.3 Some Concerns

Cost
The cost consideration for both future PCS subscriber units and the PCS infrastructure cannot override the quality of the personal communications service and the degree of risks involved. Furthermore the estimated cost of a newly developed system is based on many unrealized assumptions. Therefore, rather than relying on the estimated starting cost for evaluating the work of a newly designed system, we should make predictions based on PCS system unit cost as if the service already had been provided for three years. This analysis would consider radio and LSI (large scale integration) technology improvements, increasing the volume of units in use, and the ever-expanding expectations from the marketplace. Estimates on a fourth-year cost of a new unit is a fair indicator for evaluating a new PCS system.

Cellular PCS
Let's conclude by asking the following questions:

- Can a newly proposed PCS system be designed for the unique PCS environment as described?
- Can the entire industry plan PCS's future as a team?
- Can a new PCS system be successfully promoted after a sound verification on a prototype?

By answering these questions and examining the capability of cellular systems, we find that cellular can be a good candidate for the PCS system. It can reach everywhere, inside or outside any building.

The recent development of pocket-sized and six-ounce hand-held units makes cellular subscriber units highly attractive for future PCSs. To achieve low power consumption and increased talking time for a unit, cellular systems would require only the installation of cell site antennas or enhancers closer to the end users. The cellular system can serve the PCS one-for-all units. The idea of today's cellular system evolving into a PCS system is a viable idea whose time has come.

11.2   PORTABLE TELEPHONE SYSTEMS

A portable telephone system should provide service to all users with no restrictions regarding geographical areas, buildings, or moving conditions. A portable telephone system is available at the present with limit capacity, but it may take a decade before this system becomes a PCS system. The portable telephone system is a three-dimensional system and a dynamic system in which every portable telephone set can be moved around. In the future portable telephone units will be even used on aircraft. This system is a more complicated system than the land-mobile system (or the land-mobile system can be considered a subsystem of the portable telephone system). We could start from scratch to design a portable telephone system and take the land-mobile telephone to be a subsystem. Land-mobile systems were designed before portable systems because of their relative simplicity. Many of the design rules presented in previous chapters were developed just for land-mobile systems. Those rules cannot be used to design the complicated portable telephone system, or used to overlay the portable telephone system on a land-mobile system. Until enough data have been gathered and a mature land-mobile system has been established, it is not wise to set up two systems on the same frequency spectrum at the same time. Since cellular systems grow to become mature land-mobile systems, it will be a real challenge to solve the co-channel interference problem in its own system.

Of course, the portable telephone system can be allocated a frequency band different from the land-mobile frequency band, in which case, it can be started up without delay. Then the designers need only to avoid any possible interference with the existing systems in that same band. However, from motives of economy (of spectrum allocation and radio hardware), it seems natural to have two systems—portable and mobile—sharing the same cellular system if possible.

In 1992, cellular land-mobile telephone systems have been fully developed in all U.S. cities. Most systems are converted from a startup system which is a noise-limited system (Section 4.1) to a mature system which is an interference-limited system (Section 4.1). Users experience many co-channel, adjacent-channel, and near-end-to-far-end interference problems to be solved.

Although possible solutions for the mature system are added, the calling failure rate is still high, the hand-off performance is poor, and the voice qualities often become unacceptable. For achieving a successful mobile cellular system the portable telephone system cannot be added on at the same time the cellular system is being operable, or it would be too hard to rule out causes for the interference. A portable telephone system must wait until the interference problems in the land-mobile system are controlled. Then it can be added after implementing solutions to eliminate the interference problems inherent in two systems. In the following section several topics concerned with the portable telephone system, such as propagation path loss, the effect of the human body, and system control considerations are introduced.

11.2.1 Propagation Path Loss

The portable telephone system is a three-dimensional system in that individuals can use the portable phone not only at ground level but also inside buildings and on different floors. The signal strength of the portable unit increases with elevation and is severely attenuated by the metal structure of buildings. The losses of mean signal strength in decibels outside the building at the street level are measured for different floor heights, as shown in Fig. 11.4. (These data were collected in Japan.[2]) The floor-height gain follows a slope of 2.75 dB per floor from the first to the fifteenth floor on a linear scale. This means a 2.75-dB increase in strength per floor. After that the strength increases following a log scale based on a 6- to 7-dB/oct rule applying to floor heights. It is the same height-gain rule used in the land-mobile system discussed in Section 2.3.5. The level of reception inside the tenth floor is the same as that on the outside ground level. The received signal strength inside the first floor shows 26 dB more loss than that outside at ground level. Since building penetration depends on building structure, it can be said that building penetration is about 25 to 30 dB in general. Data collected in Chicago[3] indicate a slope of 2.67 dB per floor as shown in Fig. 11.5. The slopes from the two figures are very close. The loss measured at the first floor inside the building (Fig. 11.5) is 15 dB more than the reception level outside the building. The reception at the seventh floor is the same as that at street level outside the building.

The difference between the two building penetration losses in Figs. 11.4 and 11.5 can be attributed to different building construction in the two countries. U.S. buildings use heavy metal main frames and few supported metal secondary frames. There is less building penetration because there are fewer metal frames to block the incoming signal. Japanese buildings always have many supported metal frames to allow buildings to survive during earthquakes.

Shadow loss is defined as the loss due to a building standing in the radio path. It is the same as the difference between the total loss and the loss due to building penetration. The shadow loss is about 27 dB, as shown in

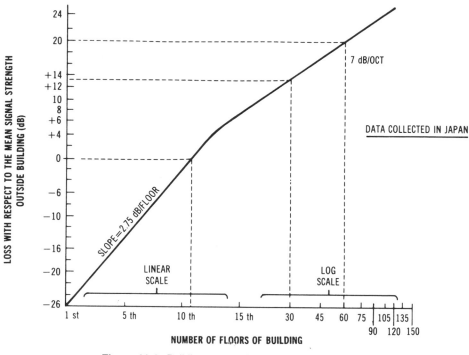

Figure 11.4. Building penetration loss in the Tokyo area.

Fig.11.5, regardless of the floor at which the measurement is taken. A summary is given in Table 11.1.

11.2.2 Body Effects

Since the portable unit is always carried by the user, the way the user positions the antenna affects the reception level of the unit. A study of the user's body's effects on the portable telephone unit is not restricted to the types of environment the user frequents. Therefore the body's effect on the portable unit at street level outside buildings will be described in this section.[4] If the user raises his or her portable unit overhead, the signal level received by the vertical antenna of the unit will be set at a reference level, a 0-dB level. When the user holds the antenna vertically at shoulder level, there is a 1- to 2-dB loss; at the waist level there is about a 5-dB loss. (These figures are obtained from Table 11.2.) If the antenna is placed horizontally above the head, it always receives a weaker signal than the same antenna placed vertically above the head. However, there is no difference in reception levels between two polarizations received at waist position.

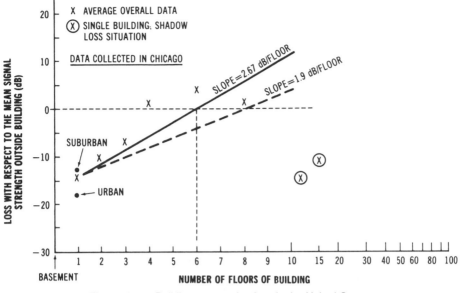

Figure 11.5. Building penetration loss in the United States

TABLE 11.1   **Building Penetration Loss**

Condition	Building Penetration Loss	Shadow Loss
Building penetration	+27 dB (Tokyo) +15 dB (Chicago)	27 dB (Chicago)
Window area	−6 dB	Regardless of floor height
1st−15th floors	2.75 dB/floor (Tokyo) 2.67 dB/floor (Chicago)	
15th−30th floors	7 dB/oct (Tokyo)	

11.2.3   Radio Phenomenon of Portable Units[5]

The portable communications unit can be developed as a future personal communications service (PCS). Engineers traditionally thought that a cellular portable unit, which is stationary while in use, could be treated as a fixed station and could be implemented more easily in the communications network than in a mobile unit. But cellular portable units are operating in mobile radio environments far more complicated than other communications environments. Their signal reception depends on not only natural terrain configurations but also human-made structures. Because human-made structures vary from one geographical area to another, a system that works well in one

TABLE 11.2   Portable Radio—Body Effect and Antenna Loss in Multipath Area

Antenna Position	Waist		Head		
Direction of Street	Vertical (dB)	Horizontal (dB)	Vertical (dB)	Horizontal (dB)	
In-line course in-wide street	5.5	3.3	2.0	5.8	— 0 dB
Crosswise course in wide street	5.2	3.9	1.0	5.7	— 1–2 dB (loss)
In-line course In narrow street	4.2	3.7	—	—	— 5 dB (loss)
Crosswise course in narrow street	3.5	6.0	—	—	
Average	4.9	4.2	1.5	5.8	

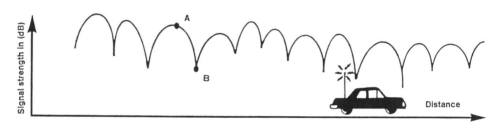

Figure 11.6. Fading signal from single frequency.

situation can totally fail in another. The radio characteristics of the environment must be considered when designing a portable-unit system.

Because of structures such as office buildings and houses, radio waves often do not follow a direct path from one point to another. What exists instead is a multipath radio phenomenon (Fig. 9.1). In this scenario the scattered radio waves are reflected and refracted from a single transmitted source through different paths before reaching the receiver unit, which can be placed either inside a building or on the street. The received signal strength at the unit is no longer a constant when the user carries the unit up and down inside a building or along the road.

When all the scattered waves are added in phase at particular spot, the resultant wave (signal strength) is strong (Fig. 11.6, spot A). When out of phase, the resultant wave becomes weak (spot B). This can be compared to a sea wave picture, even though a radio wave is invisible. At some point the user may be at the peak of the radio wave, while in other places the user may be in the valley of the radio wave. The distance between the peak and the valley is about one-half of the wavelength of the operating signal. At a frequency of 850 MHz the average distance is about 6 in. This drastic variation is called signal fading, which is difficult to handle in the mobile radio environment.

Each cellular operator uses 416 different channels. Each channel consists of two frequencies—forward-link frequency (base-to-mobile) and reverse-link frequency (mobile-to-base)—for a total of 832 different frequencies with 832 corresponding wavelengths. Imagine then that there are 832 different sea wave pictures. Since the wavelengths are different in each wave, one position can be one radio wave's peak but another radio wave's null. This phenomenon is called *selective fading*.

Selective fading does not affect the signal reception of a mobile unit because mobile units are always moving along a street. The signal strength of a frequency always varies up and down, creating a fading signal. In a fading signal only the average, or local-mean, received signal strength over a period of time counts as the signal arriving at the mobile unit. The local-mean signal strength of the 832 frequencies measured along a given street are apparently the same. The frequency chosen makes no difference in average signal strength.

Portable-unit signals do not have an average signal strength. Because the portable unit is normally stationary, the signal strength received depends on the location of the spot due to multipath fading. The signal is strong when the spot is out of the fade of the signal strength. The signal is weak when the spot is in the fade of the signal strength. If the portable telephone is located as at spot B in Fig. 11.6, no signal can be received. Therefore no average signal strength can be used to represent the portable-unit signal.

Selective fading also affects the signal reception of portable units. In a cellular system a set of 4 out of 832 channel frequencies is used to complete a call. The picture of these 832 different sea waves can be visualized for the signal reception of portable units. Since at one spot there is only one real-time value of signal strength of a channel received, either strong or weak, there is no chance to take advantage of the average value over time or locations. Therefore the real-time signal strengths are different from 832 different frequencies.

Fading cannot always be overcome by selecting a strong signal spot for portable units because the setup channels in cellular systems are used to set up calls. Each setup channel consists of a forward-link frequency and a reverse-link frequency to link back the information. A new voice channel is assigned as soon as a call is set up. It also consists of two frequencies so that both parties can talk on their own frequencies. This is called a *deplexing system*.

In a cellular portable unit, the unit indicator shows either "no service" or "in service" at a weak or strong spot. Although the forward-link setup channel can serve a strong signal at one spot, the same spot may not be served by a strong reverse-link setup channel or the forward and reverse links of the new assigned voice channel because of selective fading. A completed call setup cannot be guaranteed by pushing the send button while the in-service light is on, based on a strong setup channel at the portable unit. The same situation

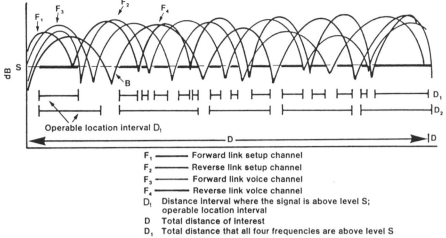

F₁ ─────── Forward link setup channel
F₂ ─────── Reverse link setup channel
F₃ ─────── Forward link voice channel
F₄ ─────── Reverse link voice channel
Dₜ   Distance interval where the signal is above level S;
     operable location interval
D   Total distance of interest
D₁   Total distance that all four frequencies are above level S
D₂   Total distance that only F₄ is above level S

Figure 11.7. Portable cellular—four-frequency selective fading system.

applies when a portable user, expecting an incoming call, chooses a spot where the signal strength of a forward setup channel appears strong. The user may not receive a call because the strong signal spot of the forward setup channel does not coincide with that of the reverse setup channel or the voice channel. This situation happens more often at the cell boundary where signal strength generally is weaker.

The four-frequency selective fading phenomenon is illustrated in Fig. 11.7. Assume first that if only one frequency F_1 is used for conducting a call, the total distance of interest is D. The distance intervals in which the signal is above level S are denoted as D_I. The percentage of the summed operable distance intervals D_I, as opposed to the total distance D in a single-frequency system, is large. But when four frequencies are used for completing a call as in today's analog cellular systems, the percentage of the summed operable distance intervals D_I versus D becomes much smaller (Fig. 11.7).

A probability calculation can be used to find the percentage of the signal above the level S in comparing a single-frequency system and a four-frequency system (Fig. 11.8). The x-axis is the signal level S, and the y-axis is the percentage of the signal exceeding the level S. In the figure, at a signal level of -5 dB (below average power), 72% of total location that a received signal is above that level is for a single-frequency system, but only 27% of total location that the received signal is above the level is for a four-frequency system. At a signal level of -10 dB, 91% of location is accepted for a single-frequency system, but only 69% of location is accepted for a four-frequency system. This phenomenon can be remedied by using something called a *diversity scheme*.

In a diversity scheme the portable becomes a two-antenna spaced diversity

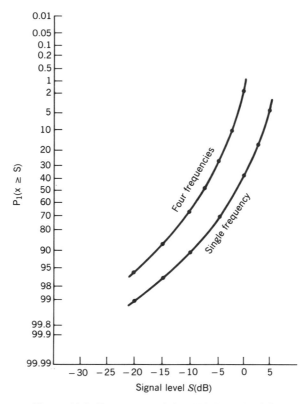

Figure 11.8. Percentage of signal X above level S.

receiver. The diversity scheme is used to reduce signal fading by adding two fading signals received from two antennas. The antenna separation is merely a half wavelength. At 850 MHz it needs only a 6-in. separation. The improvement obtained at a level of -5 dB for a four-frequency system is 72% of location, which is about the same percentage as a single-frequency system without the diversity.

11.2.4 System-Control Considerations

There are several considerations in system control for a combined portable telephone and mobile land system for the following reasons:

First, since the portable telephone system is a three-dimensional system, a portable unit transmitting on the twentieth floor can interfere with mobile units on the street level. To prevent this, the power control at the base station is adjusted to minimize the transmission power of each portable unit in terms of both the system traffic and the interference in the combined system. The portable unit or the mobile unit cannot be allowed to adjust its own power. That would create a "cocktail party" phenomenon, whereby just as at a cocktail party everyone has to talk louder to converse with his or her partner less than two feet away. Such would occur if there were no central control system to control everyone's voice volume. The same phenomenon can happen in the land-mobile/portable system if local controls are used to adjust the transmitted power levels. A well-designed system always has good control regulated by the central office.

Second, the location of each portable unit should be known accurately at the base station or the central office so that it can apply the power control or frequency assignment strategy to avoid interference.

Finally, third, in principle use of low power transceivers for portable units would reduce the possibility of interference and save the batteries for a longer time. How can the low power portable telephone system be made to work within a land-mobile system? The use of repeaters could be a solution. The repeater can be installed in the building just like today's telephone outlets. The repeater receives the radio signal from the portable units and finds ways to get to the mobile switching office by wireline. Since the attenuation due to building penetration will help to isolate interference from the land-mobile system, the repeaters inside the building would receive less interference.

11.3   AIR-TO-GROUND COMMUNICATIONS

Air-to-ground communications is also a mobile communication system, but its system design considerations are quite different from land-mobile communications in many aspects such as propagation path loss, co-channel interference, frequency allocation plans, and hand-off schemes. The different aspects of this system are described in this section.

11.3.1   Propagation Path Loss

The radio path between airplane and ground is usually under line-of-sight conditions following a free-space path-loss rule of 6 dB/oct [see Eq. (2.3.14)] as long as the radio path is clear from the radio horizon. The radio horizon can be calculated as

$$R = \sqrt{2h_a} + \sqrt{2h_b} \qquad (11.3.1)$$

where R is the distance in miles measured along the horizon, h_a is the aircraft altitude in feet above average terrain, and h_b is the ground-station antenna height in feet above the average local terrain as shown in Fig. 11.9. Equation (11.3.1) is plotted in Fig. 11.10, and $R = R_1 + R_2$ can be large if h_a is high. If the aircraft flies away from its ground station and enters into a shadow-diffraction zone beyond the radio horizon, then the loss is about 1.2 dB/mile. There is a transitional region, about 20 miles below the horizon, indicating

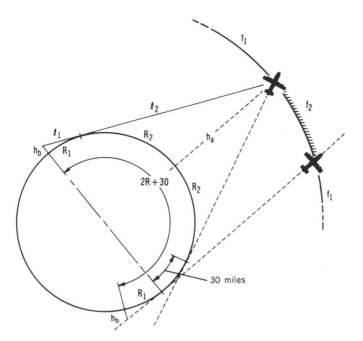

Figure 11.9. The coordinate of air-to-ground system.

a loss of 15 to 20 dB. The typical propagation-loss curve can be found in Fig. 11.11. In the line-of-sight region a Fresnel-zone phenomena occurs, due to the adding and canceling of direct and ground-reflected waves. Reception variations of up to 10 dB can be seen in the line-of-sight region with great regularity. The variation in vertical polarization is usually less when compared with the horizontal polarization.

11.3.2 Co-channel Separation

Since the fading phenomenon due to the Fresnel zone occurs in air-to-ground communication, the requirement of signal-to-interference would be the same as that of the land-mobile S/I, which is 18 to 20 dB at the cell boundary of 210 km (130 miles). If the cell boundary of one co-channel cell is separated from another cell boundary by 32 km (20 miles) beyond the radio horizon, a minimum isolation S/I of 20 dB between two large cells, is shown in Fig. 11.11. Then the co-channel separation is

$$D = 2 \cdot R(h_a, h_b) + 20 \text{ (miles)} \quad \text{(for one existing interferer)} \quad (11.3.2)$$

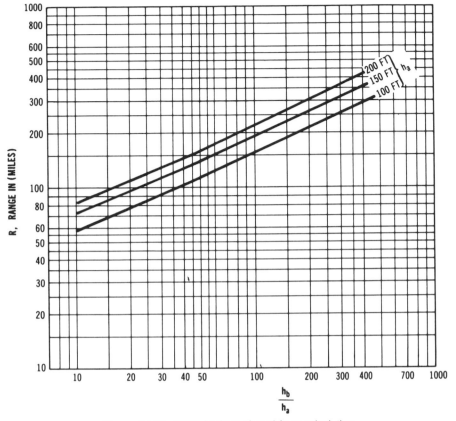

Figure 11.10. Radio horizon plotted from calculation.

One interferer is assumed. Understand that Eq. (11.3.2) is applied only for one co-channel interferer. For the case of six co-channel interferers

$$10 \log_{10} \frac{C}{I} = 10 \log_{10}\left(\frac{C}{\sum_{i=1}^{6} I_i}\right) = 10 \log_{10} \frac{C}{6I_i}$$

$$= 10 \log_{10}\left(\frac{C}{I_i}\right) - 7.78 \text{ dB} \qquad (11.3.3)$$

or

$$10 \log_{10} \frac{C}{I_i} \simeq 28 \text{ dB} \qquad (11.3.4)$$

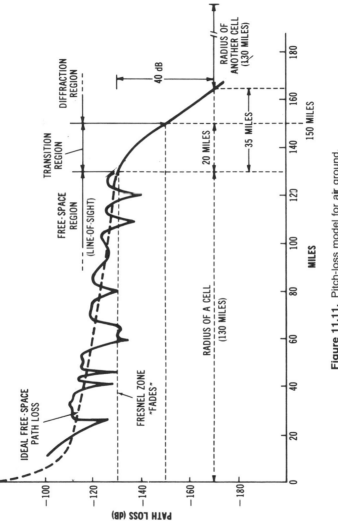

Figure 11.11. Pitch-loss model for air ground.

353

TABLE 11.3 A Suggested Zoning for Six Zones

Zone	Altitude (ft)	Median Altitude (ft)	Cell Radius	Co-channel Ground Station Separation	Transmitted Power at Aircraft (dB)
1st zone	under 2000	2000	$R = 77.4$	$D = 184.78$	0
2nd zone	2000–5000	3500	97.8	225.61	4.86
3rd zone	5000–10,000	7500	136.61	303.23	11.48
4th zone	10,000–20,000	15000	187.35	404.69	17.50
5th zone	20,000–40,000	30000	259.09	548.18	23.52
6th zone	40,000 and above	50000	330.37	690.74	27.96

From Fig. 11.11 the separation beyond the radio horizon is 48 km (30 miles) for a loss of 28 dB. Therefore the co-channel separation D becomes

$$D = 2R(h_a, h_b) + 30 \quad \text{(for six existing interferers)} \quad (11.3.5)$$

In both Eqs. (11.3.2) and (11.3.5), it was found that the co-channel separation for the air-to-ground system is

$$\frac{D}{R} \simeq 2 \quad\quad (11.3.6)$$

Equation (11.3.6) indicates that a group of two frequency-reuse schemes is needed in the air-to-ground system.

11.3.3 Altitude Zoning Considerations

Since the radius R of a cell in an air-to-ground communication system is a function of h_a and h_b, R can be larger if h_a and h_b are higher. Under these conditions different altitude zones must be classified in order to assign frequencies. Six zones are suggested in Table 11.3. When the altitude of the aircraft increases from one zero to another, the radius of the cell for the same ground station becomes larger in that new zone and the propagation path loss is greater because of the higher altitude.

Consider the separation of the ground stations for this system. Although the co-channel reduction factor $a = D/R$ is always close to 2, as shown in Eq. (11.3.6), the cell radius R changes due to the altitude, and the co-channel separation D changes accordingly. In this system the co-channel separation D is not used for co-channel cell separation but for the air-to-ground system. Substituting $h_a = 2000$ ft and $h_b = 100$ ft into Eqs. (11.3.1) and (11.3.5) yields

$$R_1 = \sqrt{2 \times 2000} + \sqrt{2 \times 100} = 77.4 \text{ miles}$$

$$D_1 = 2R_1 + 30 = 184.78 \text{ miles}$$

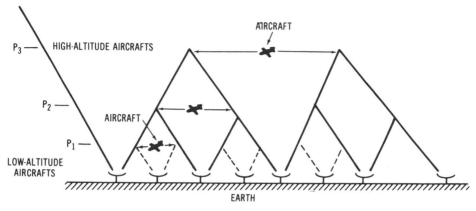

Figure 11.12. The air-to-ground system configuration.

The cell radius and the co-channel ground station are calculated based on the median altitude of each zone, as listed in Table 11.3. From the values listed in Table 11.3, the system illustrated in Fig. 11.12 can be planned using the existing land-mobile sites. As an aircraft flies in low-altitude zones, more ground stations are used than when it flies in high-altitude zones.

11.3.4 Frequency Allocation Plan and Power Control

Since the co-channel reduction factor a is always close to 2, only one set of frequency channels can be assigned in each zone. Proper ground stations must be used to hand off the calls. The number of channels N in each frequency channel set is based on the air traffic. A frequency channel f_{IJ} from a frequency set $\{f_I\}$ can be assigned to the aircraft at the Ith zone—J is the number of the assigned channels in that zone where $1 \le J \le N$. Each frequency channel set consists of N pairs of frequencies. Each pair has a transmitted frequency and a corresponding received frequency assigned to the aircraft. The transmitted frequency and the received frequency at the ground station are the reverse of the frequencies of the aircraft. The allocated frequency bands for ground-station transmission and aircraft transmission have to be separated far enough to avoid any interference between transmission and reception.

Since ground stations cannot be moved and station separations based on different aircraft altitudes cannot be adjusted after they are installed, ground stations must be systematically separated according to the co-channel station separation D of the 1st zone (under 304 m, or 2000 ft). Let the antenna heights of ground stations be 30 m (100 ft), which are the antenna heights in the land-mobile system. In the land-mobile system, the least antenna separation is about 32 km (20 miles); most antenna separations are less than 20 miles. Can these resources be used for the air-to-ground system?

Assignment of Frequencies in Each Zone

Assume that the traffic is given, for example, $M = 300$ users in a given zone, the Ith zone. The average call time per user is $t = 1.76$ min—the same as that of the land-mobile system. The traffic load A is

$$A = M \times \bar{t} = 300 \times \frac{1.76 \text{ min/call}}{60 \text{ min/hr}} = 8.8 \qquad (11.3.7)$$

With a block probability of 2% ($B = 0.02$), and if $A = 8.8$, the number of frequency channels, N_I in the Ith zone can be found from Table 8.1 as $N_I = 15$.

A set of 15 frequency channels in one ground station for the Ith-zone traffic is needed. These 15 frequency channels can be reused at the next cell based on the cell radius, which is dependent on the altitude of the aircraft. The total channels of the system will be

$$N = \sum_{I=1}^{6} N_I \qquad (11.3.8)$$

Where the number 6 indicates six zones. Two ground stations must then be moved closer than $2R$ to keep the continuity of the call by handing off the same frequency channel from one ground station to the next.

Sometimes co-channel interference occurs if the same frequency channel has been assigned to two aircraft in nearby cells, and the aircraft are approaching each other. In this case the ground station should access the frequency channel information from the nearby cells and assign a new frequency channel to one of the aircraft before it comes within co-channel distance.

Power Control

In an air-to-ground system, the near-end-to-far-end interference problem must be solved at a ground station due to the different altitudes of aircraft. The near-end-to-far-end interference problem is much easier to handle in this system than that in the land-mobile system. Since the altitude of the aircraft is always known, the transmitted power can be controlled according to zoning altitude. Table 11.3 lists the transmitted power at the aircraft by different zone. All transmitted powers in six zones are normalized to the transmitted power of the first zone. The power differences are based on the free-space loss rule.

The transmitted power difference between an aircraft in the first zone and in the sixth is 28 dB. If this power is controlled at the aircraft, the ground-station receiver has no near-end-to-far-end interference problems. With ground-station transmission the aircraft's receivers do not have near-end-to-far-end interference problems, and no power control is needed at the ground station.

TABLE 11.4   Satellite Elevation and Azimuth in Each City for Satellite ATC-6 (140°W, 0°N)

Location	Satellite Elevation (degrees)	Satellite Azimuth (degrees)
Chicago, IL	19	242
Des Moines, IA	23	238
Cheyenne, WY	30	227
Denver, CO	32	228
Boulder, CO	32	228
Estes Park, CO	32	228
Salt Lake City, UT	35	219
Reno, NV	40	210
San Francisco, CA	43	207

11.4   LAND-MOBILE/SATELLITE COMMUNICATIONS SYSTEM

Land-mobile-to-satellite communication systems have begun to show their potential. The system can have services allocated to UHF, L band, or Ku band. Implementing this system should not affect the demand for or requirements of the land-mobile system.[6,7] The path loss, noise, and long- and short-term fadings of the land-mobile/satellite system will be discussed here. Other considerations such as traffic load and interference can be found in many textbooks on satellite communications systems.[8]

11.4.1   Propagation Path Loss

In satellite-to-mobile communications the path loss will be based on the relative position of the satellite in space—the elevation angle and the azimuth angle, for example. The satellite elevation and satellite azimuth of ATC-6 (Applications Technology Satellite-6) as measured from different cities are listed in Table 11.4.

The communication link of a mobile-satellite-mobile link is illustrated in Fig. 11.13. The total received signal is obtained by taking the transmitted power P_t, adding the gains at two ground-mobile units and at the satellite, and subtracting the losses from the up-link and down-link:

$$\text{Total received signal} = P_t + (A + B + C - a - b)$$

The path loss calculated from the satellite-to-mobile path (either up-link, or down-link) will be

Path loss (satellite-to-mobile)

= line-of-sight loss + excess path loss (due to the mobile environment)

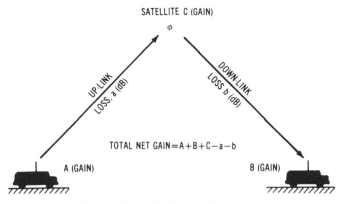

Figure 11.13. Satellite-mobile system.

The Line-of-Sight (LOS) Loss

LOS loss = free-space loss (FSL) + loss due to weather conditions

The free-space loss (FSL) is computed using Eq. (1.2.1), which is converted into dB form

$$\text{FSL} = 36.6 + 20 \log_{10} d_1 + 20 \log_{10} f_1 \quad \text{(for up-link)}$$

$$= 36.6 + 20 \log_{10} d_2 + 20 \log_{10} f_2 \quad \text{(for down-link)} \quad (11.4.1)$$

where d_i is in miles and f_i is in MHz. For $d = 22,200$ miles and $f = 850$ MHz, the FSL is 182 dB. The theoretical prediction of FSL path loss of fixed, isolated ground terminals can be within a fraction of a decibel. The prediction's accuracy is limited by the antenna gain, transmitted power, pointing direction, and so on. For the frequencies above the C band (5 GHz), the signal starts are attenuated by rainfall, fog, clouds, snow, and other weather conditions as the frequency increases. Usually the path loss due to weather conditions is negligible when the operating frequency is under 5 GHz.

Excess Path Loss

Excess path loss occurs when the antenna of the mobile unit is lower than its surroundings, and most of the time, the transmitted or received signal is blocked by human-made structures. This can be seen in Fig. 11.14. Figure 11.14 is formed by taking many pieces of small-scale coverage data, each of them a few hundred wavelengths long with the same attributes, such as the same link, frequency, and environment, and forming a cumulative probability distribution by using only the upper 90% of all small-scale data. Therefore the figure represents large-scale coverage statistics. It shows that human-made structures (in cities, suburbs, etc.) and the street heading relative to the satellite azimuth angle are the two major factors.

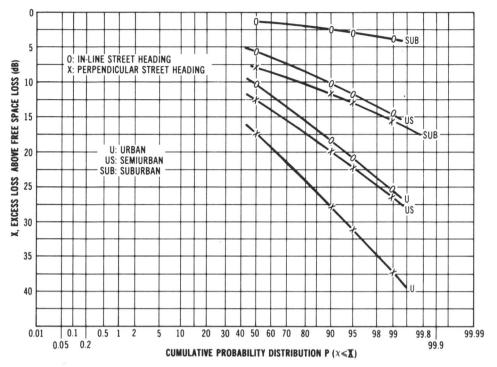

Figure 11.14. The excess path loss in satellite-mobile communication system.

TABLE 11.5   Parameters with Negligible Effects

Environment	Path Loss
Whip antenna mean lobe to the satellite elevation	1 dB loss
Frequencies between 1.5 GHz and 850 MHz	1.3 dB
Street side (toward and away)	1.5 dB
Elevation (19°−43°)	1 dB

The difference in path loss received at the mobile unit between an urban area and a suburban area at the 90% level (i.e., 90% of the received data are above that level) is 8 dB. The difference in path loss at a 90% level in a suburban (or urban) area as compared to the different street headings (in line with the satellite and perpendicular to the satellite) is about 9 dB. The difference in path loss due to different street headings is always smaller at a 50% level and greater at a 99% level, as shown in Fig. 11.14. The differences in path losses among frequencies, elevations, street sides, and mobile antenna main lobes have negligible effects, as shown in Table 11.5. Table 11.6 indicates the excess losses in different environments and street headings at 850 MHz and a satellite elevation of 32°.

TABLE 11.6   Excess Loss (in dB) (90% Data of a Large-Scale Coverage)

Environment		Suburban	Semiurban	Urban
	In-line	2.5	10.5	19
Street orientation	Perpendicular	12	20	28

11.4.2   Noise

Noise consists of thermal noise and external noises. Thermal noise is affected proportional to the absolute temperature $(T$ K$)$ and the bandwidth of frequencies (B).

$$P_n = kTB$$

At the ground external and internal sources of noise are solar noise, galactic noise, atmospheric noise, human-made noise, and equipment noise. At the satellite the noise sources are solar noise (if pointing toward the sun, $T = 10^5$ K), galactic noise, earth noise, and equipment noise.

Figure of Merit (G/T)

Since it is important to avoid losses and reduce the noise at the receiving end, both at the satellite and at the ground station, it is important that the receiving antenna and the electronics should introduce as little noise as possible. The efficiency of these two is usually quoted as the ratio of gain to noise temperature and is called the figure of merit.

$$\text{Figure of merit} = \frac{G}{T} \text{ (dB/K)}$$

where G is antenna and Rf amplifier gain (due to antenna diameter and the frequency) and T is receiver system noise temperature, usually an uncooled receiver with a noise temperature of 316 K (or 43°C or 109°F). At a ground station the antenna size is large, $G/T > 0$, and at a mobile station $G/T < 0$, since the size of the mobile antenna is small.

11.4.3   Fading

In satellite communications the fading phenomenon can be classified as two kinds: long-term and short-term fading. The characteristics of these two kinds of fading are different from those in land-mobile communications (see Chapter 2).

Long-Term Fading

Long-term fading is due to the street heading as related to the satellite azimuth. In the in-line situation the average signal level is higher; in the per-

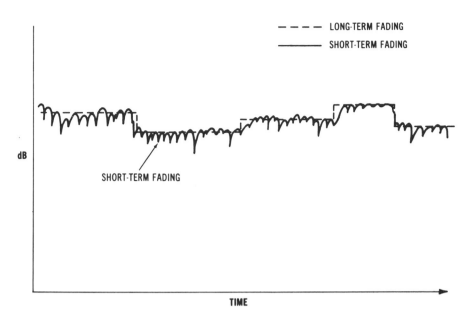

Figure 11.15. Long-term fading of the mobile-satellite communications.

pendicular situation, the average signal level is lower. It looks like a step piecewise function shown as the envelope of the signal in Fig. 11.15. When the street heading is in line with the satellite azimuth, the signal level is high. When the street heading is perpendicular to the satellite azimuth, the signal level is low.

Short-Term Fading

Short-term fading has two different characteristics depending on the street headings. When the street heading is in line with the satellite azimuth, there is a strong direct wave and many weak multipath waves, forming a Rician distribution as shown in the level crossing rate curve of Fig. 11.16A. When the street heading is perpendicular with the satellite azimuth, there are two dominant waves: One is a direct wave at 5 dB above its total reception level, and one is a 180° reflected wave at −10 dB with respect to the total reception level plus other multipath waves, as shown in Fig. 11.16B. With these understandings we can design the signaling format to combat fading.

11.4.4 Applications

In mobile-satellite communications in addition to changes in polarization due to the Faraday rotation effect in the ionosphere, there is crossover from one polarization to another due to the structure of the mobile environment. This precludes the use of two different polarization waves to increase channel capacity. A diversity scheme is needed to combat fading. Space diversity can

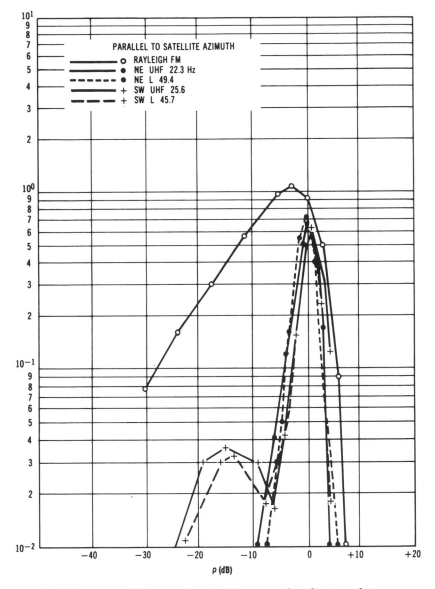

(A) Short-term fading with an in-line street heading condition.

Figure 11.16. Short-term fading with an in-line street heading condition and a perpendicular street heading condition.

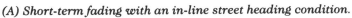

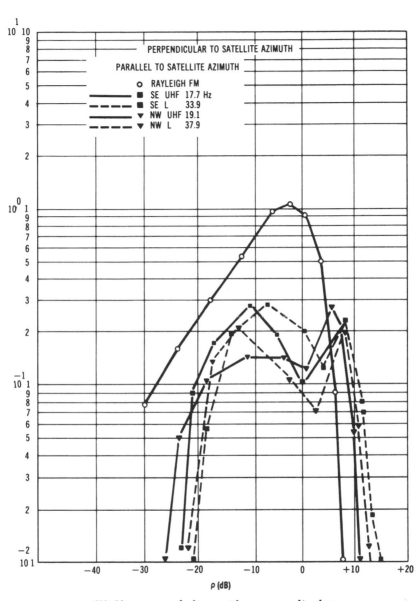

*(B) Short-term fading with a perpendicular
street heading condition.*

Figure 11.16. *(Continued)*

be used to combat short-term fading, and differently located satellites can be used to reduce long-term fading. The major difference between mobile-satellite communications and land-mobile communications are path loss, noise environment, and fading characteristics. Once the differences are understood, the design criteria are straightforward, as provided in books on satellite design.

REFERENCES

1. Sass, P. F., "Propagation Measurements for UHF Spread Spectrum Mobile Communications," *IEEE Trans. Veh. Tech.* 32 (May 1983): 168–176.
2. Kozono, S., and K. Watanabe, "Influence of Environmental Buildings on UHF Land Mobile Radio Propagation," *IEEE Trans. Commun.* Com-25 (Oct. 1977): 1113–1143.
3. Walker, E. H., "Penetration of Radio Signals into Building in the Cellular Radio Environment," *Bell Sys. Tech. J.* 62: 9, Pt. I (Nov. 1983): 2719–2734.
4. Sakamoto, M., S. Kozono, and T. Hattori, "Basic Study on Portable Radio Telephone System Design" (Paper presented at the IEEE Vehicular Technology Conference, San Diego, CA, 1982): 279–284.
5. Lee, W. C. Y., "In Cellular Telephone, Complexity Works," IEEE Circuits & Devices, Vol. 7, No. 1., Jan. 1991, pp. 26–32.
6. Hess, G. C., "Land-Mobile Satellite Excess Path Loss Measurements," *IEEE Trans. Veh. Tech.* VT-29 (1980): 290–297.
7. Reudink, D. O., "Estimates of Path Loss and Radiated Power for UHF Mobile-Satellite Systems," *Bell Sys. Tech. J.* 62: 8, Pt. 1 (1983): 2493–2512.
8. Spilker, J. J., Jr., *Digital Communicating by Satellite* (Prentice Hall, 1977).

PROBLEMS

11.1 To complete a cellular phone call, four frequencies are required. If four frequencies are to be operated in a selective fading environment, why would they not affect the call performance for the mobile units but rather the call performance for the portable unit?

11.2 Assume that all four frequency channels have the same average power when they are received at the portable unit. The percentage of one channel above the threshold of −10 dB (10 dB below the average power) is about 90%. What is the percentage if all four independent channels are above the threshold?

11.3 Can the ground coverage of a portable unit be used to design a portable-unit system? How would the three-dimensional coverage for a portable-unit system be plotted?

11.4 The signal penetration into a building is different in different cities. Why?

11.5 Air-to-ground communications can use single-sideband (SSB) channels, but ground mobile communications cannot. Why?

11.6 If the height of an antenna tower is 200 m and the antenna height of a mobile unit is 3 miles, how far is the radio horizon?

11.7 Show that in a satellite communication system using the geostationary satellite at a high latitude area, the signal received from the in-line heading of the satellite is stronger than the signal received from perpendicular street heading.

11.8 In a geostationary satellite system, what is the satellite orbit as it relates to the earth?

11.9 What is the time delay from a geostationary satellite response after it sends out a signal to the satellite?

11.10 What is the altitude of a satellite when seen in the sky twice a day, and what is the time delay response after sending out a signal to the satellite?

INDEX